# Aromaticity and Antiaromaticity

# Aromaticity and Antiaromaticity

Concepts and Applications

*Miquel Solà*
Universitat de Girona, Spain

*Alexander I. Boldyrev*
Utah State University, UT, USA

*Michał K. Cyrański*
University of Warsaw, Poland

*Tadeusz M. Krygowski*
University of Warsaw, Poland

*Gabriel Merino*
Cinvestav Mérida, Mexico

WILEY

This edition first published 2023


*Registered Office*
John Wiley & Sons Ltd, The Atrium, Southern Gate, Chichester, West Sussex, PO19 8SQ, UK

*Editorial Office*
The Atrium, Southern Gate, Chichester, West Sussex, PO19 8SQ, UK

For details of our global editorial offices, customer services, and more information about Wiley products visit us at www.wiley.com.

Wiley also publishes its books in a variety of electronic formats and by print-on-demand. Some content that appears in standard print versions of this book may not be available in other formats.



*Library of Congress Cataloging-in-Publication Data is Applied for:*

Hardback ISBN: 9781119085898

Cover Design: Wiley
Cover Images: Courtesy of Miquel Solà

Set in 9.5/12.5pt STIXTwoText by Straive, Chennai, India

# Contents

# Foreword

One of the specific features of chemistry compared to physics is the importance of heuristic models, which describe, interpret, and classify the structures and reactivities of molecules. For the evaluation of chemical models, it is important to recognize that they are not right or wrong, but more or less useful for the interpretation and prediction of experiments. To put it in the words of a pioneer of theoretical organic chemistry: "The only criterion for a model is its usefulness, not its 'truth'" (M. J. S. Dewar, *J. Am. Chem. Soc.* **1984**, *106*, 669). Many of the models still valid today were developed before the first physically correct description of the chemical bond was presented by Heitler and London in 1927. This is remarkable because all of the models proposed so far were based on ideas from classical physics, although the chemical bond in molecules is a quantum theoretical phenomenon that is subject to its own postulates, which cannot be derived from classical physics. It seems that the models obtained by experimental observations and filtered by discussions, which served as a neural network, contain quantum theoretical phenomena.

With the advent of quantum theory and its application in chemistry, the models were partially modified and they received a quantum chemical basis that allowed a theoretical understanding and an extension of the rules associated with the model. A prominent example is aromaticity, which is the subject of this book. The concept of aromaticity, which goes back to the introduction of a planar ring structure with three double bonds for benzene by Kekulé in 1865, was given a quantum theoretical basis by Erich Hückel in 1931 that was later translated in 1951 by William Doering into the famous $4n+2$ "Hückel's rule." The emergence of quantum chemical methods combined with numerous theoretical models and the development of sophisticated experimental techniques in recent decades have led to an enormous expansion in the scope of aromaticity far beyond the original chemistry of benzene derived compounds. The concept of aromaticity was extended from unsaturated π-bonded systems in organic chemistry to saturated σ-bonded compounds in inorganic chemistry and the synthesis of metallaaromatic compounds following a theoretical prediction by

Roald Hoffmann among many other discoveries and suggestions greatly expanded its scope. The restriction of aromaticity to cyclic compounds was removed by the proposal of Y-conjugation in tri-substituted molecules by Gund and the relevance for aromatic stabilization in electronically excited states by Baird extended the scope of aromaticity beyond ground state properties. A major impact to the concept of aromaticity is the finding that $\pi$-delocalized aromatic compounds exhibit particular magnetic properties, which were suggested to identify aromatic systems and to estimate the degree of aromaticity.

The scope and limitation of aromaticity and the introduction of new classes of aromatic compounds have been at the center of numerous controversial discussions in recent years. It is appropriate and very timely to summarize the present standing of theoretical and experimental research on the topic of aromaticity. The authors of this book have provided an overview of the current state of the art in 14 chapters. The well-chosen selection summarizes the progress and the most important facets of the current experimental and theoretical research on aromatic compounds and the different points of view on aromaticity, which have been considerably expanded in the last two decades. The question of the occurrence and strength of aromaticity in molecules will continue to be the subject of controversial discussions in the future. It is questionable whether the almost 50 different types of aromaticity proposed in recent decades are a useful addition to the original arsenal. It remains to be seen whether the observation that aromatic compounds exhibit certain magnetic properties also allows the reverse conclusion that the occurrence of such magnetic quantities automatically indicates aromatic stabilization. These are just two of the exciting questions about aromaticity that will be the focus of ongoing research in the field. This book is a timely and important contribution to the current state of the debate, which will be continued.

*Gernot Frenking*
*Philipps-Universität Marburg*

# Preface

The concept of the chemical bond as the link that connects atoms in a molecule is at the core of chemistry. A chemical reaction is nothing but the breaking of some chemical bonds and the formation of others. Despite its importance, the chemical bond is not exactly defined and this is the origin of many past and ongoing controversial debates. Aromatic compounds bear a particular type of chemical bond characterized by extensive electron delocalization in closed circuits. Aromaticity as an idea is more than 150 years old and is still not only alive, but even briskly developing. Like for the chemical bond, aromaticity is a not well-defined property that cannot be measured experimentally. However, the chemical consequences derived from this property are evident and can be quantified. Moreover, this concept is extremely useful to rationalize the structure, stability, and reactivity of many species that otherwise would not be understood.

The initial Kekulé and Erlenmeyer concept of aromaticity was based on structural and reactivity criteria, respectively (cyclic systems resistant to some kind of reactivity). Later it was found that, classically, aromaticity is associated with electron delocalization (at the beginning only by means of $\pi$-electrons), which in many cases was associated with reactivity (and some other properties as magnetic ones). Some root understanding of the aromatic character has in the last decades expanded wider, introducing many aromaticities with various prefixes to underline slightly different understanding of the original meaning. New criteria introduced by means of new "indices" of aromaticity allowed significant broadening of the original term *aromaticity* that now covers the entire Periodic Table and different electronic states. This has been exemplified with the electron localization functions of the ground state of benzene and the first excited state of octahedral $^{5}A_{1g}$ $Be_6$ species depicted in the cover graphics of the book (pictures courtesy of Dr. Ouissam El Bakouri).

Over the last 30 years, there has been a significant expansion in the number and types of aromatic compounds and in our understanding of the concept of aromaticity. This book was written to provide students at the advanced undergraduate level and graduate level with a deep discussion of the recent advances in the field of aromaticity. A good knowledge of standard first-year courses in organic and theoretical chemistry is recommended.

Chapter 1 starts with a brief historical account of the concept of aromaticity. Next, in Chapter 2, we focus on simple electron counting rules that are followed by a large number of organic and inorganic aromatic compounds, such as Hückel's, Baird's, Wade-Mingos', and Hirsch's rules together with other less known aromaticity rules.

In 2001, Boldyrev, Wang, and coworkers detected a series of bimetallic compounds containing $Al_4^{2-}$, the first all-metal aromatic cluster identified, which is σ- and π-aromatic and, therefore, it shows multifold aromaticity. It is now widely accepted that the π-aromaticity of classical organic compounds is not unique and that there are chemical compounds having σ-, δ-, and φ-aromaticity, together with combinations of these different types of aromaticity. Aromaticity and antiaromaticity in metal systems and the new types of aromaticity found in these molecular entities are discussed in Chapters 3 and 11.

The connection among aromaticity, stability, and reactivity is well established. However, among a series of isomers or different electronic states, the most stable and less reactive is not always the most aromatic. The role of aromaticity on stability and reactivity is analyzed in Chapter 4.

The number and variety of molecules that have aromatic character has grown exponentially in the last years. It is important to have methods at hand that allow careful classification of any system as aromatic/nonaromatic/antiaromatic. The last two decades brought a number of powerful tools to diagnose aromaticity. Readers interested in a detailed account of the existing descriptors of aromaticity, are referred to Chapters 5–8. In these chapters, we present the definition of the most used indices of aromaticity based on geometric, energetic, magnetic, and electronic properties of molecules.

The list of the different types of aromaticity is very large. Among them, one of the most important is heteroaromaticity, a particular case of aromaticity found in organic classical molecules in which a —CH= group or C atom of the aromatic ring is replaced by a heteroatom like O, N, or S. Chapter 9 refers to this type of aromaticity.

A conjugated monocyclic molecule of 4n π-electrons with a set of p-orbitals in which there is a single (or an odd number) out-of-phase overlap constitutes a Möbius aromatic species. Chapter 10 provides a detailed description of Möbius aromaticity.

Properties derived from the aromatic character of classical organic molecules are due to the π-electrons. However, it has been shown that the π-electrons of benzene possess a distortive tendency that transforms the $D_{6h}$ molecular symmetry into a $D_{3h}$ symmetry structure. Therefore, $D_{6h}$ symmetry in benzene is not due to the

$\pi$-electrons, but to the $\sigma$-system. Chapter 12 gathers a series of works discussing this paradox.

The concept of aromaticity was initially developed to account for the unusual stability of planar benzene derivatives. However, certain molecules as *closo* boranes and charged fullerenes are examples of species with aromaticity in three dimensions. Chapter 13 analyzes three-dimensional aromaticity with a special emphasis on spherical aromaticity.

In 1972, Baird showed that the lowest lying triplet excited states of conjugated monocyclic compounds with 4n $\pi$-electrons are aromatic, introducing the so-called excited state aromaticity (ESA). Nowadays, it is widely accepted that ESA is important to understand many photochemical reactions. Chapter 14 focuses on the recent developments of ESA, which is presently one of the hot topics in the field of aromaticity.

For all chapters, the references are placed at the end of each to allow the reader to quickly get acquainted with the field.

A number of people have contributed to this book throughout the past years. The authors wish to express their gratitude to all of them, especially, to all PhD and postdoc students that have been working in our groups performing studies that improved our knowledge on aromaticity. Finally, yet importantly, we thank our families for their patience and continuous support.

Finally, the authors wants to dedicate this book to the late Prof. Dr. Paul von Ragué Schleyer for his ground breaking contributions to aromaticity.

*Miquel Solà*
Girona, Catalonia, Spain

*Alexander I. Boldyrev*
Kauai, Hawaii, USA

*Michał K. Cyrański*
Warsaw, Poland

*Tadeusz M. Krygowski*
Warsaw, Poland

*Gabriel Merino*
Mérida, México

# List of Abbreviations

| | |
|---|---|
| 2c-2e | two-center two-electron |
| 2cBO | Orthogonalized two-center Bond Orbitals |
| 2c-ESI | Two-center Electron Sharing Indices |
| 5-MR | Five -Membered-ring |
| 6-MR | Six-Membered ring |
| ACID | Anisotropy of the Induced Current Density |
| AdNDP | Adaptive Natural Density Partitioning |
| AIM | Atoms-in-Molecules |
| AO | Atomic Orbital |
| AOM | Atomic Overlap Matrix |
| ARCS | Aromatic Ring Current Shieldings |
| ARE | Adiabatic Resonance Energy |
| ASE | Aromatic Stabilization Energy |
| ATI | Average Two-Center Index |
| BAC | Bond Alternation Coefficient |
| BE | Bonding Energy |
| BH | Bingel-Hirsch addition |
| BLA | Bond Length Alternation |
| BLE | Bond Length Equalization |
| BLW | Block-Localized Wave function |
| BOIA | Bond Order Index of Aromaticity |
| BV | Bifurcation Value |
| CASSCF | Complete Active Space Self-Consistent Field |
| CCRE | Conjugated Circuits Resonance Energy |
| CMO | Canonical Molecular Orbital |
| COT | Cyclooctatetraene |
| CTED | Corrected Total Electron Density |
| DA | Diels-Alder cycloaddition |
| DBU | 1,8-diazabicycloundec-7-ene |

| | |
|---|---|
| DDQ | 2,3-dichloro-5,6-dicyano-1,4-benzoquinone |
| DFT | Density Functional Theory |
| DI | Delocalization Index |
| DRE | Dewar Resonance Energy |
| EC | Energy Content |
| ECRE | Extra Cyclic Resonance Energy |
| ED | Electron Density |
| EDA | Energy Decomposition Analysis |
| EDLA | Electron Density Localized on Atoms |
| EDDB | Electron Density of Delocalized Bonds |
| EDLB | Electron Density of Localized Bonds |
| ELF | Electron Localization Function |
| EMF | Endohedral Metallofullerene |
| ESI | Electron Sharing Indices |
| ESIPT | Excited State Intramolecular Proton Transfer |
| ETS | Extended Transition State |
| FLU | Aromatic Fluctuation Index |
| HCM | Hollow Cylinder Model |
| HF | Hartree-Fock |
| HMO | Hückel Molecular Orbital |
| HOMA | Harmonic Oscillator Model of Aromaticity |
| HOMED | Harmonic Oscillator Model of Electron Delocalization |
| HOMHED | Harmonic Oscillator Model for Heterocycles of Electron Delocalization |
| HOMO | Highest Occupied Molecular Orbital |
| HOSE | Harmonic Oscillator Stabilization Energy |
| HRE | Hückel Resonance Energy |
| HS | High Spin |
| HSRE | Hess-Schaad Resonance Energy |
| HtFfA | Heat of Formation of a Molecule from Atoms |
| IPR | Isolated Pentagon Rule |
| ISE | Isomerization Stabilization Energy |
| LCAO | Linear Combination of Atomic Orbitals |
| MCI | Multicenter Electron Delocalization Index |
| MHP | Maximum Hardness Principle |
| MO | Molecular Orbital |
| MPP | Minimum Polarizability Principle |
| MSO | Molecular Spin-Orbital |
| NAO | Natural Atomic Orbital |
| NBO | Natural Bond Orbital |
| NCI | Non-Covalent Index |

| | |
|---|---|
| NDDO | Neglect of Diatomic Differential Overlap |
| NICS | Nucleus-Independent Chemical Shift |
| NIR | Near-Infrared |
| NMR | Nuclear Magnetic Resonance |
| LP | Lone Pair |
| LUMO | Lowest Unoccupied Molecular Orbital |
| ON | Occupation Number |
| PAC | Predictive Aromaticity Criteria |
| PAH | Polycyclic Aromatic Hydrocarbon |
| PCH | Polycyclic Conjugated Hydrocarbon |
| PDI | *Para*-Delocalization Index |
| pEDA | π Electron Donor-Acceptor |
| PJT | Pseudo-Jahn-Teller |
| PLR | *Para* Linear Response |
| QC | QuasiClassical |
| QTAIM | Quantum Theory of Atoms in Molecules |
| RAHB | Resonance-Assisted Hydrogen Bond |
| RCBV | Ring-Closure Bifurcation Value |
| RCM | Ring Current Model |
| RE | Resonance Energy |
| REC | Ring Energy Content |
| REPE | Resonance Energy Per π-Electron |
| SA | Salicyldenaniline |
| SCF | Self-Consistent Field |
| sEDA | σ Electron Donor-Acceptor |
| SOJT | Second-Order Jahn-Teller |
| SSAdNP | Solid State Adaptive Natural Density Partitioning |
| TDDA | Dehydrodianthracene |
| TFVC | Topological Fuzzy Voronoi Cell |
| TPA | Two-Photon Absorption |
| TRE | Topological Resonance Energy |
| TS | Transition State |
| VB | Valence Bond |
| VDE | Vertical Detachment Energy |

# 1

# Historical Overview

Source: T. M. Krygowski, H. Szatyłowicz, ChemTexts 2015, 1, 12.

> *"Classification and theory are not ends in themselves. If they generate new experimental work, new compounds, new processes, new methods – they are good; if they are sterile – they are bad."* (E. D. Bergman, in Aromaticity, Pseudo-aromaticity, Anti-aromaticity, Proceedings of an International Symposium held in Jerusalem 31 March – 3 April 1970 [Eds. E. D. Bergmann, B. Pullman], Israel Academy of Science and Humanities, Jerusalem, **1971**, pp. 392–392)

Aromaticity belongs to one of the four most commonly used terms in organic chemistry and related fields, along with conformation, H-bonding, and polymerization [1]. As it can be seen from the histogram in Figure 1.1, the topic of aromaticity appears on average more than 40 times per day in 2020. So great interest

*Aromaticity and Antiaromaticity: Concepts and Applications*, First Edition.
Miquel Solà, Alexander I. Boldyrev, Michał K. Cyrański, Tadeusz M. Krygowski, and Gabriel Merino.

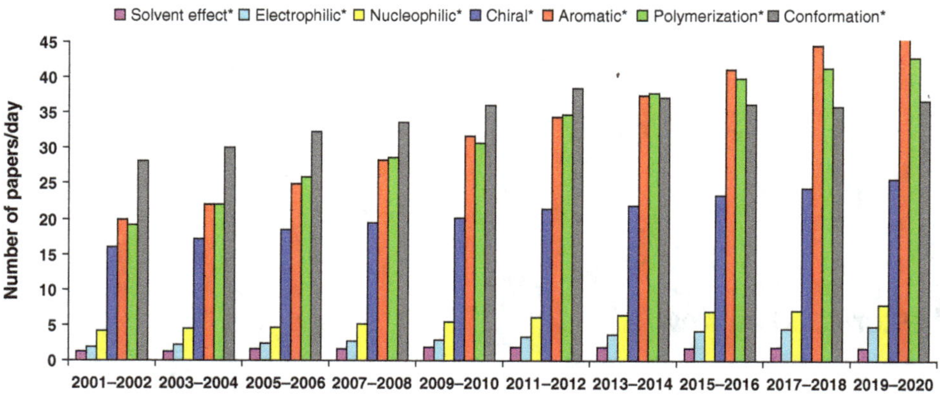


**Figure 1.1** The number of scientific papers, appearing *per* day, in which the indicated term appears in title, abstract or keywords (asterisk denotes any adequate ending, i.e., *s*, *ed*, *ing*, *ity*, or nothing) [1, 2].

in research related to aromaticity is associated with a large number of works on the definition of the term, as well as on experimental and theoretical studies on the physicochemical characteristics attributed to the aromatic character of chemical species in their ground, excited, and transition states [2].

The term *aromatic* applied for the classification of organic species was used for the first time in 1856 by Hoffman in his work on monocarboxylic derivatives of benzene [3]. However, discovery of the archetypic aromatic compound benzene and estimation of its empirical formula as well as its physical properties had been done already 30 years earlier in 1825 by Faraday [4]. Then for almost half of century, most efforts were devoted to the investigation of the structure of benzene and chemical properties of its derivatives. Two aspects were taken into account at that time. Kekulé regarded as aromatic compounds those that are structurally similar to benzene [5, 6], whereas Erlenmeyer [7] considered as aromatic compounds those having alike chemical properties (reactivity) as benzene. Later Kekulé also accepted the importance of a similarity to benzene in chemical properties, and in his *Lehrbuch der organischen Chemie* [8], classified alcohols, aldehydes, and acids into either aromatic or aliphatic ones. From the stoichiometry, it was known that benzene is built up of single and double bonds, but troubles appeared in trying to understand its structure. The first cyclic structure of benzene as a hexagon was suggested by Laurent in 1855 and presented by Loschmidt [9], but much closer to the actual viewpoint is a structural formula by Kekulé [8], as shown in Figure 1.2.

In the Kekulé formula, the carbons are arranged in a hexagon with alternating double and single bonds between them. In 1869, Ladenburg [10] questioned Kekulé's structure, indicating that there should be four isomers for disubstituted benzene derivatives (Figure 1.3), whereas only three of them were known.

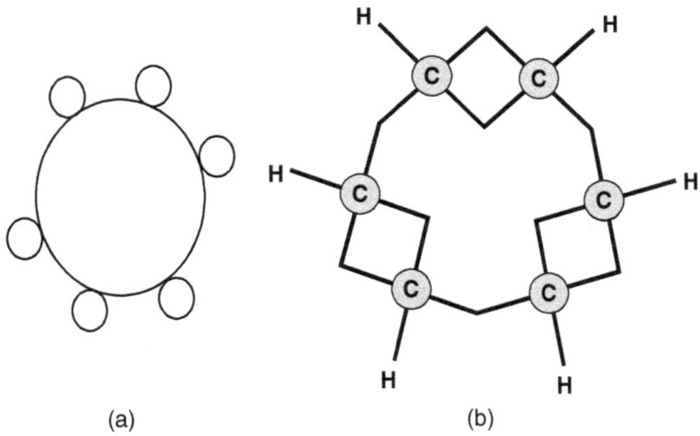


**Figure 1.2** The Loschmidt (a) and the Kekule (b) cyclic formulas of benzene.

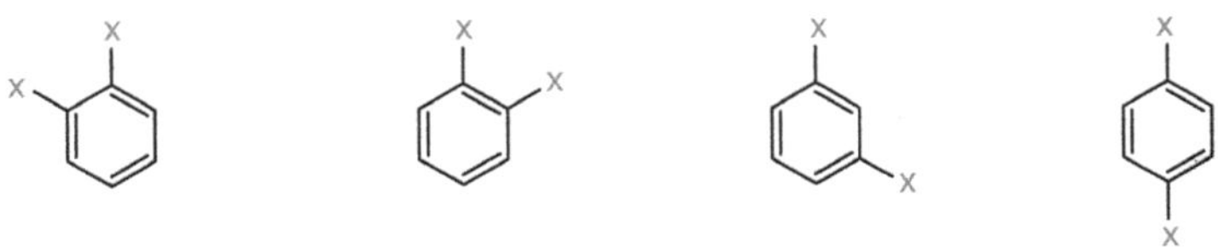


**Figure 1.3** Assumed isomers by Ladenburg for disubstituted benzene derivatives.

Moreover, he suggested another view of its structure, i.e. by presenting it as – actually named – prismane. Figure 1.4 presents relations between isomers of benzene and relates them to isomers of prismane.

In defense of his cyclic structure, Kekulé accepted some assumption that the double and single bonds are in a permanent exchange – again an ingenious intuition? The concept of prismane-like structure of benzene was questioned and rejected by Baeyer [13] since 1,2-disubstituted benzene derivatives did not exhibit optical activity.

At the same time as benzene, another classically aromatic hydrocarbon, naphthalene, was found in 1821 as a result of distillation of tar coal [14]. Its structure as two benzenes fused was proposed in 1866 by Erlenmeyer [7]. Other important aromatic compounds were those containing heteroatoms. They were discovered at the end of the 19th century. Anderson in 1849 [15] obtained the first heteroaromatic molecule named pyridine through his studies on the distillation of bone-oil and other animal matter. In 1870, Limpricht obtained furan [16], whereas in 1890, Hantzsch presented a method of synthesis of pyrrole derivatives [17]. Thiophene accompanying benzene in tar coal was discovered by Meyer in 1883 [18]. It exhibited chemical properties similar to benzene. In 1891, Bamberger [19] proposed a cyclic structure for all these compounds.

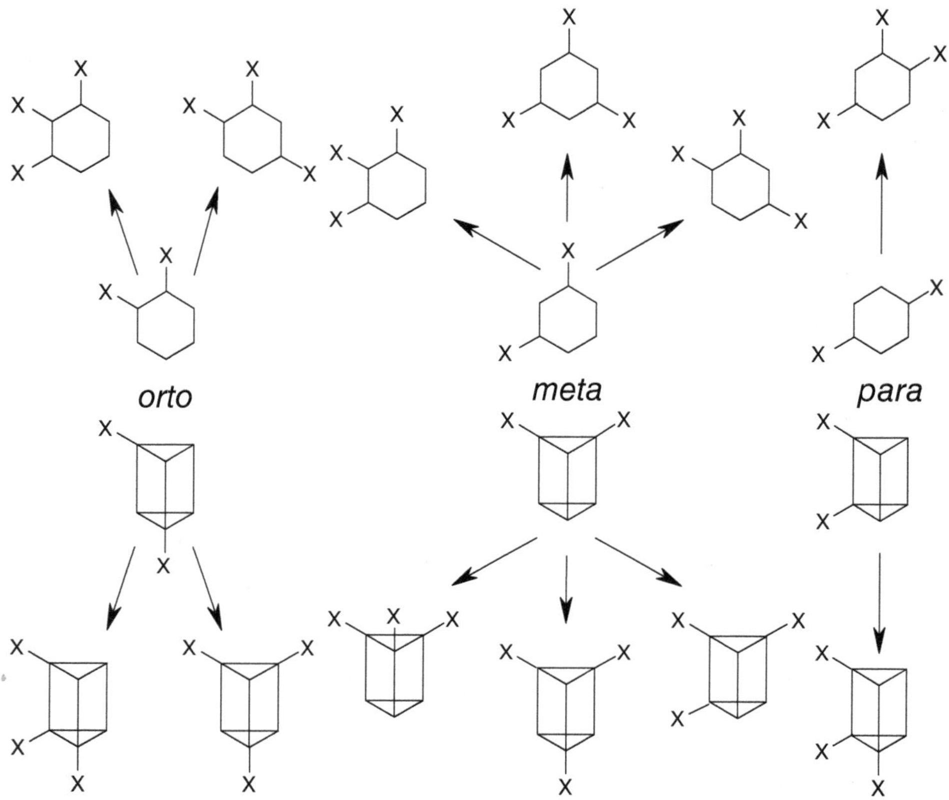


**Figure 1.4** Relationships between substituted benzene and prismane (as a structure of benzene) derivatives [11, 12].

At the end of 19th century, quite a number of aromatic compounds were known, and it was known that these compounds are less reactive than their olefinic analogues. The question that was posed at that time was: Is the resistance of aromatics, thus formally unsaturated compounds, due to their cyclic structure? The question was answered by the work of Willstätter [20], who obtained a derivative of cyclooctatetraene and found that its chemical behavior is similar to a typical olefinic molecule.

A new impetus came from physics. Discovery of the electron and then its role in atoms, and later in molecules, gave a new view on molecular structure. At that time, structural formulae were presented basing on an octet-rule introduced by Lewis [21] and then developed by Langmuir [22]. A modern understanding of the chemical bond was established suggesting that a chemical bond is a pair of electrons shared by two atoms. In 1922, Crocker [23] presented a new version of

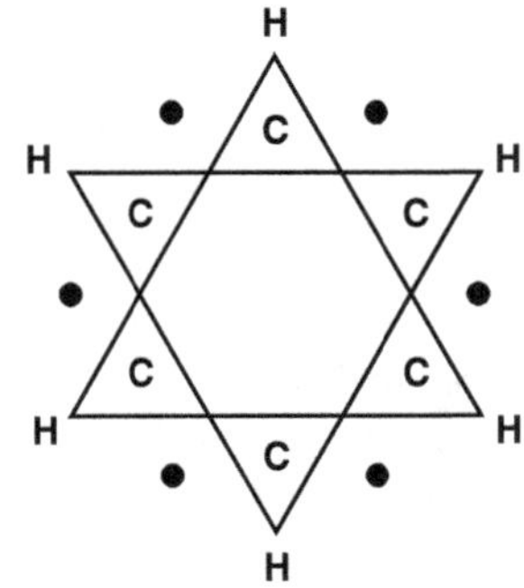


**Figure 1.5** Crocker's electronic structure of benzene.

the benzene structure built up in agreement with the Lewis–Langmuir theory. It is shown in Figure 1.5.

Crocker's structure took into account six equivalent carbon atoms with six electrons in between. This idea was, for a few years, not noticed, until about 1925 when Armit and Robinson [24] made use of it and introduced the concept of electron sextet, and noted that "the group of six electrons...resists disruption," describing for the first time the concept of aromatic sextet. They also introduced an assignment of the aromatic sextet by a circle inside the six membered rings. All these works inspired Clar [25] to formulate the $\pi$-sextet rule as an extension of the Hückel's $4n+2\pi$-electron rule from monocyclic species to benzenoid systems. Moreover, in 1938, Evans and Warhurst [26] already noted the analogy between the $\pi$-electrons of benzene and the six delocalized electrons in the cyclic transition state of the Diels–Alder reaction between butadiene and ethylene, and suggested for the first time the existence of aromatic transition states.

The next important step in description and understanding aromaticity has come as a result of development of quantum mechanics [27] and its applications to chemical problems [28–30]. Then an important contribution by Hückel has to be stressed. He introduced the concept of $\sigma$ and $\pi$ orbitals, describing electrons differing in their properties (symmetry), and then worked out a simplified molecular orbital theory, nowadays known as Hückel Molecular Orbital (HMO) theory [28]. Application of the HMO theory allows interpretation of the nonaromatic properties of cyclobutadiene and cyclooctatetraene, giving a solution for the older question posed in the time of Willstätter. A famous "magic" rule from these works resulted: $4n+2$ saying that cyclic $\pi$-electron systems containing $4n+2$ electrons are stable, whereas those containing $4n$ electrons are less stable. This rule was formulated by Doering et al. [31], who noted that heptatrienyl cation is more stable than cyclopentadienyl cation since the first one possesses $4n+2\pi$-electrons whereas the latter has $4n$. A wide and a strong documentation of this rule has been given by Roberts, Streitwieser, and Regan [32], and the rule may be more generally stated as follows: "those monocyclic coplanar systems of trigonally hybridized atoms which contain $4n+2\pi$ electrons will possess relative electronic stability" [33].

The determination in 1959 of the structure of the *closo* borane $B_{10}H_{10}{}^{2-}$ ion by Lipscomb [34], and the synthesis three years later of the first derivatives of *closo*-dodecaborate and *closo*-decaborate by Muetterties's group [35], released the concept of aromaticity from the two dimensions and moved it to three dimensions. In 1972, Baird [36] predicted the existence of the lowest-lying triplet state aromaticity for monocyclic conjugated rings of $4n$ $\pi$-electrons. This prediction was confirmed by the identification of planar triplet ground states of $C_5H_5{}^+$ and $C_5Cl_5{}^+$ [37, 38]. The concept of $\sigma$-aromaticity was proposed first by Dewar in 1979 [39] to account for the unexpected small strain in cyclopropane. In 1982, Roper et al. [40] synthesized the first metallabenzene, an osmabenzene, and initiated a new group of aromatic species, the so-called metalla-aromatic compounds [41]. Boldyrev, Wang, and coworkers detected in 2001 a series of bimetallic clusters containing $Al_4{}^{2-}$, the first type of all-metal aromatic clusters known, face capped by an $M^+$ cation (M = Li, Na, Cu) [42].

The development of quantum chemical methods as well as computational techniques allowed an effective application of various theoretically based models related to energetic, magnetic properties, and geometry of the $\pi$-electron structure of molecules. There have appeared in various definitions of resonance energy (RE), recalling the most popular and important, the Dewar resonance energy (DRE) [43], and the Hess and Schaad RE's [44, 45] estimated for hundreds of $\pi$-electron hydrocarbons. At that time, appeared also an idea of isodesmic and homodesmotic reactions [46–50], which have given rise to numerous estimations of variously defined aromatic stabilization energy (ASE) values [51]. Parallel to these achievements, theoretical estimations of magnetic properties appeared, as the anisotropy of magnetic susceptibility $\Delta\chi$ [52] and the magnetic susceptibility exaltation $\Lambda$ [53] supplementing the data obtained by experimental measurements. Introduced in 1996 by Schleyer, Nucleus Independent Chemical Shift (NICS) [54] has been recognized as an extremely important tool in estimations of aromaticity [55, 56]. Extremely appealing are the maps of induced ring currents, which illustrate not only the ways of cyclic $\pi$-electron (de)localization, but also quantify its extent [57–64]. At that time, molecular geometry became relatively easily accessible as well as available by theoretical models. Then, Julg et al. [65] defined the variance of the perimeter bond lengths in a molecule as a numerical characteristic of aromaticity. This work was the beginning of a plethora of papers that applied the analysis of molecular geometry for describing the aromaticity of molecules [2, 66]. Development of Quantum Theory of Atoms in Molecules [67] as well as other theories that propose new definitions of atomic charges [68] has resulted in a number of approaches applying atomic charges, electronic delocalization as well as physical properties in bond or ring critical points for the quantification of aromaticity [69, 70]. Presentation of these results as well as the appropriate discussion are, among others, the main aims of this book.

## References


1 Web of Science – Institute for Science Information, Retrieved in October 2021, http://apps.webofknowledge.com/
2 Krygowski, T.M., Szatyłowicz, H., Stasyuk, O.A. et al. (2014). *Chem. Rev.* 114: 6383–6422.
3 Hofmann, A.W. (1856). *Proc. R. Soc. Lond.* 8: 1–3.
4 Faraday, M. (1825). *Philos. Trans. R. Soc. London* 115: 440–446.
5 Kekulé, A. (1865). *Bull. Soc. Chim. Fr. (Paris)* 3: 98–110.
6 Balaban, A.T. (1980). *Pure Appl. Chem.* 52: 1409–1429.
7 Erlenmeyer, E. (1866). *Ann. Chem. Pharm.* 3: 327–359.
8 Kekulé, A. (1866). *Lehrbuch der Organischen Chemie*, vol. 3. Erlangen: Verlag von Ferdinand Enke.
9 Loschmidt, J. (1861). *Chemische Studien – Constitutions – Formulen der organischen Chemie in graphischer Darstellung*. Wien: Carl Gerold's Sohn.
10 Ladenburg, A. (1869). *Ber. Dtsch Chem. Ges.* 19: 140–142.
11 Ciesielski, A. (2014). Energetics and Structure Relations of Carbocyclic Aromatic Systems thesis, University of Warsaw (Warsaw).
12 Ciesielski, A., Krygowski, T.M., and Cyrański, M.K. (2015). *Wiad. Chem.* 69: 789–808.
13 Baeyer, A. (1886). *Ber. Dtsch. Chem. Ges.* 19: 1797–1810.
14 Kidd, J. (1821). *Philos. Trans. R. Soc. London* 111: 209–221.
15 Anderson, T. (1849). *Justus Liebigs Ann. Chem.* 70: 32–38.
16 Limpricht, H. (1870). *Ber. Dtsch. Chem. Ges.* 3: 90–91.
17 Hantzsch, A. (1890). *Ber. Dtsch. Chem. Ges.* 23: 1474–1476.
18 Meyer, V. (1883). *Ber. Dtsch. Chem. Ges.* 16: 1465–1478.
19 Bamberger, E. (1891). *Ber. Dtsch. Chem. Ges.* 24: 1758–1764.
20 Willstätter, R. and Waser, E. (1911). *Ber. Dtsch. Chem. Ges.* 44: 3423–3445.
21 Lewis, G.N. (1916). *J. Am. Chem. Soc.* 38: 762–785.
22 Langmuir, I. (1919). *J. Am. Chem. Soc.* 41: 868–934.
23 Crocker, E.C. (1922). *J. Am. Chem. Soc.* 44: 1618–1630.
24 Armit, T.W. and Robinson, R. (1925). *J. Chem. Soc. Trans.* 127: 1604–1618.
25 Clar, E. (1972). *The Aromatic Sextet*. New York: Wiley.
26 Evans, M.G. and Warhust, E. (1938). *Trans. Faraday Soc.* 34: 614–624.
27 Pauling, L. and Wilson, E.B. (1935). *Introduction to Quantum Mechanics*. New York: McGraw-Hill.

**28** Hückel, E. (1938). *Grundzüge der Theorie ungessättiger and aromatuscher Vewrbindungen*. Berlin: Verlag Chemie.

**29** Syrkin, Y.K. and Dyatkina, M.E. (1950). *Structure of Molecules and the Chemical Bond*. London: Butterworths Scientific Publications.

**30** Pullman, B. and Pullman, A. (1952). *Les theories electroniques de la chimie organique*. Paris: Masson and Cie.

**31** Doering, W.v.E. and Detert, F.L. (1951). *J. Am. Chem. Soc.* 73: 876–877.

**32** Roberts, J.D., Streitwieser, A., and Regan, C.M. (1952). *J. Am. Chem. Soc.* 74: 4579–4582.

**33** Streitwieser, A. Jr., (1961). *A Molecular Orbital Theory for Organic Chemists*. New York: Wiley.

**34** Lipscomb, W.N., Pitochelli, A.R., and Hawthorne, M.F. (1959). *J. Am. Chem. Soc.* 81: 5833–5834.

**35** Knoth, W.H., Miller, H.C., England, D.C. et al. (1962). *J. Am. Chem. Soc.* 84: 1056–1057.

**36** Baird, N.C. (1972). *J. Am. Chem. Soc.* 94: 4941–4948.

**37** Breslow, R., Chang, H.W., Hill, R., and Wasserman, E. (1967). *J. Am. Chem. Soc.* 89: 1112–1119.

**38** Saunders, M., Berger, R., Jaffe, A. et al. (1973). *J. Am. Chem. Soc.* 95: 3017–3018.

**39** Dewar, M.J.S. (1979). *Bull. Soc. Chim. Belg.* 88: 957–967.

**40** Elliott, G.P., Roper, W.R., and Waters, J.M. (1982). *J. Chem. Soc., Chem. Commun.* 811–813.

**41** Feixas, F., Matito, E., Poater, J., and Solà, M. (2013). *WIREs Comput. Mol. Sci.* 3: 105–122.

**42** Li, X., Kuznetsov, A.E., Zhang, H.-F. et al. (2001). *Science* 291: 859–861.

**43** Dewar, M.J.S. and de Llano, C. (1969). *J. Am. Chem. Soc.* 91: 789–795.

**44** Hess, B.A. and Schaad, L.J. (1971). *J. Am. Chem. Soc.* 93: 305–310.

**45** Hess, B.A. and Schaad, L.J. (1971). *J. Org. Chem.* 36: 3418–3423.

**46** Hehre, W.J., McIver, R.T., Pople, J.A., and Schleyer, P.v.R. (1974). *J. Am. Chem. Soc.* 96: 7162–7163.

**47** Radom, L. (1974). *J. Chem. Soc., Chem. Commun.* 403–404.

**48** George, P., Trachtman, M., Bock, C.W., and Brett, A.M. (1976). *J. Chem. Soc., Perkin Trans. 2* 1222–1227.

**49** George, P., Trachtman, M., Brett, A.M., and Bock, C.W. (1977). *J. Chem. Soc., Perkin Trans. 2* 1036–1047.

**50** Hehre, W.J., Radom, L., Schleyer, P.v.R., and Pople, J.A. (1986). *Ab Initio Molecular Orbital Theory*. New York: Wiley.

**51** Cyrański, M.K. (2005). *Chem. Rev.* 105: 3773–3811.

**52** Flygare, W.H. (1974). *Chem. Rev.* 74: 653–687.

**53** Dauben, H.J., Wilson, J.D., and Laity, J.L. (1968). *J. Am. Chem. Soc.* 90: 811–813.

**54** Schleyer, P.v.R., Maerker, C., Dransfeld, A. et al. (1996). *J. Am. Chem. Soc.* 118: 6317–6318.

**55** Chen, Z., Wannere, C.S., Corminboeuf, C. et al. (2005). *Chem. Rev.* 105: 3842–3888.

**56** Stanger, A. (2010). *J. Org. Chem.* 75: 2281–2288.

**57** Steiner, E., Fowler, P.W., Soncini, A., and Jenneskens, L.W. (2007). *Faraday Discuss.* 135: 309–323.

**58** Fowler, P.W., Steiner, E., Havenith, R.W.A., and Jenneskens, L.W. (2004). *Magn. Reson. Chem.* 42: S68–S78.

**59** Fias, S., Fowler, P.W., and Delgado, J.L. (2008). *Chem. - A Eur. J.* 14: 3093–3099.

**60** Fliegl, H., Sundholm, D., Taubert, S. et al. (2009). *J. Phys. Chem. A* 113: 8668–8676.

**61** Lazzeretti, P. (2004). *Phys. Chem. Chem. Phys.* 6: 217–223.

**62** Ciesielski, A., Krygowski, T.M., Cyrański, M.K. et al. (2009). *Phys. Chem. Chem. Phys.* 11: 11447–11455.

**63** Gomes, J.A.N.F. and Mallion, R.B. (2001). *Chem. Rev.* 101: 1349–1383.

**64** Geuenich, D., Hess, K., Köhler, F., and Herges, R. (2005). *Chem. Rev.* 105: 3758–3772.

**65** Julg, A. and François, P. (1967). *Theor. Chim. Acta* 8: 249–259.

**66** Krygowski, T.M. and Cyrański, M.K. (2001). *Chem. Rev.* 101: 1385–1419.

**67** Bader, R.F.W. (1990). *Atoms in Molecules: A Quantum Theory*. Oxford: Clarendon.

**68** Fonseca-Guerra, C., Handgraaf, J.-W., Baerends, E.J., and Bickelhaupt, F.M. (2004). *J. Comput. Chem.* 25: 189–210.

**69** Poater, J., Duran, M., Solà, M., and Silvi, B. (2005). *Chem. Rev.* 105: 3911–3947.

**70** Feixas, F., Matito, E., Poater, J., and Solà, M. (2015). *Chem. Soc. Rev.* 44: 6434–6451.

# 2

# Simple Electron Counting Rules

*"Rules are rules but no laws. They can be broken."*

G. Frenking (the 11th Congress on Electronic Structure: Principles and Applications 2018 held in Toledo)

## 2.1 Introduction

Archetypal aromatic compounds such as $C_6H_6$, $C_5H_5^+$ in its lowest lying triplet state, $Al_4^{2-}$, $B_{12}H_{12}^{2-}$, or $C_{60}^{10+}$ present high symmetry. Symmetry is a common characteristic of quintessential aromatic compounds. In many cases, these symmetric molecular structures bring about shells of degenerate highest-occupied molecular orbitals (MOs) that can be fully occupied, resulting in a closed-shell structure, or can can have the highest-energy occupied shell only half-filled with same-spin electrons. These two types of electronic structure provide an extra energetic stabilization to aromatic molecules that resembles the stabilization gained by the main-group elements having eight valence electrons in its valence shell (octet rule), transition metals following the eighteen valence electron rule or having $d^5$ parallel spin occupation, and f-elements that obey the 32-electron principle. This closed-shell or half-filled shell electronic structure is the source of several rules of aromaticity, the most relevant being Hückel's $4n+2$ rule. Given the imprecise character of the aromaticity concept, the existence of a series of very simple mathematical rules based solely on the counting of valence or π-electrons that can account for the aromaticity of a large number of organic and inorganic molecules is remarkable. Substituents, counterions, external fields, and so on may reduce the symmetry of these representative aromatic compounds and remove the degeneracy of highest-occupied MOs. Still, in many cases, the electronic structure of these less symmetric species resembles that of typical aromatic

*Aromaticity and Antiaromaticity: Concepts and Applications*, First Edition.
Miquel Solà, Alexander I. Boldyrev, Michał K. Cyrański, Tadeusz M. Krygowski, and Gabriel Merino.

compounds and, consequently, they retain a significant aromatic character. This means that the rules that are discussed in the next sections of this chapter can be applied not only to compounds with high symmetry but also to these compounds with less or nonsymmetry at all.

## 2.2 Hückel's $4n + 2$ Rule

In 1931, Hückel [1–4] applied his simplified MO theory to conjugated molecules. The results obtained were used by Doering and Detert to derive the $4n+2$ $\pi$-electron rule [5]. This was the first and most important of a series of rules for aromaticity. The Hückel's $4n+2$ $\pi$-electron rule [1–4] states that monocyclic conjugated hydrocarbons (annulenes, $C_nH_n$) of $D_{nh}$ symmetry with $4n+2$ $\pi$-electrons are aromatic, whereas those with $4n$ $\pi$-electrons are antiaromatic. The origin of this rule is the MO distribution of $\pi$-orbitals in $D_{nh}$ annulenes that can be determined with the Hückel molecular orbital (HMO) method or with more sophisticated methodologies. In the HMO method, the following assumptions are made: (i) only $\pi$ valence electrons are considered, $\sigma$ electrons are ignored; (ii) the diagonal matrix elements are set equal to $\alpha$ (Coulomb integral), the energy of an electron in a carbon *2p* orbital; and (iii) the off-diagonal elements are set equal to $\beta$ (resonance integral) if the two atoms are next to each other and zero otherwise. The secular equation to be solved for a cyclic conjugated $\pi$-system of $D_{nh}$ symmetry with N atoms is the following N × N secular determinant:

$$\begin{vmatrix} \alpha-\varepsilon & \beta & \cdots & \cdots & \beta \\ \beta & \alpha-\varepsilon & \beta & \vdots & \vdots \\ \vdots & \vdots & \vdots & \vdots & \vdots \\ \beta & \cdots & \cdots & \beta & \alpha-\varepsilon \end{vmatrix} \tag{2.1}$$

The solution of this Hückel determinant is:

$$\varepsilon_j = \alpha + 2\beta\cos\left(\frac{2j\pi}{N}\right) \qquad (j = 0, 1, 2, \cdots N-1) \tag{2.2}$$

The most stable MO (j = 0) is formed by the in-phase interactions of all p-atomic orbitals. The rest of the orbitals, except the last one for even N values that has all p-orbitals out of phase, come by pairs (j = i and j = N − i have the same energy). As can be seen in Figure 2.1, closed-shell electronic structures are then obtained for the following number of $\pi$-electrons: 2, 6, 10, 14,..., that is, $4n+2$ $\pi$-electrons ($n$ = 0, 1, 2,...). This closed-shell electronic structure provides aromatic stabilization. On the contrary, the singlet ground state of $4n$ $\pi$-electrons annulenes does not lead to a closed-shell electronic structure. As a result, these systems (e.g., butadiene, cyclooctatetraene,...) suffer Jahn–Teller distortions that break the $D_{nh}$ symmetry and localize the $\pi$-electrons. These annulenes with $4n$ $\pi$-electrons are

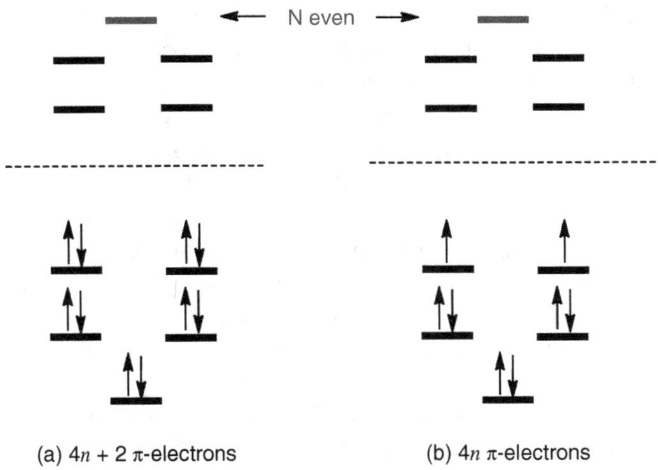


**Figure 2.1** The distribution of π-molecular orbitals in $D_{nh}$ annulenes and the origin of the (a) Hückel's $4n + 2$ π-electron rule and (b) Baird's $4n$ π-electron rule.

antiaromatic. The preparation of the tropylium cation, $C_7H_7^+$, by Doering and Knox [6] in 1954 is considered the first experimental verification of Hückel's $4n + 2$ π-electron rule. From a theoretical point of view, the first well-documented proof of the $4n + 2$ π-electron rule has to be attributed to Roberts, Streitwieser, and Regan who computed the resonance energy for a series of cyclic conjugated hydrocarbons [7]. The Hückel's rule can also be applied to species different from classical aromatic annulenes. For instance, the inorganic cluster $Li_3^+$ is considered a σ-aromatic species [8] with two delocalized σ-electrons that follows Hückel's rule.

## 2.3 Baird's 4*n* π-Electron Rule for the Lowest-Lying Triplet Excited State

The Baird's $4n$ π-electron rule [9] for the lowest-lying triplet excited state, $T_1$, asserts that annulenes of $D_{nh}$ symmetry that are antiaromatic in their singlet ground state are aromatic in their lowest-lying triplet state ($T_1$), and vice versa for those that are aromatic in the ground state but antiaromatic in the $T_1$ state. The basis of this rule is the open-shell half-filled electronic structure that is obtained in the lowest-lying triplet state of monocyclic conjugated hydrocarbons of $D_{nh}$ symmetry for the following number of π-electrons (see Figure 2.1b): 4, 8, 12,…, that is, $4n$ π-electrons ($n = 1, 2, 3,\ldots$). This open-shell half-filled electronic structure provides aromatic stabilization in a similar way transition metals are stabilized by having a $d^5$ parallel spin occupation. According to this rule, the lowest-lying triplet excited states of cyclobutadiene or cyclooctatetraene are aromatic. This rule was confirmed by the identification of the planar triplet ground states of some cyclopentadienyl cations [10]. In fact, the relative stability of photochemically obtained cyclobutadiene [11] and spectroscopic observation of the triplet diradical state of cyclobutadiene [12] is in line with the aromaticity of the $T_1$ state of this species. Moreover, isolation of a triplet benzene dianion [13] provided additional support for this rule. It is generally accepted that excited state aromaticity is not only found in the $T_1$ state of $4n$ monocycles but also in their lowest-lying singlet excited states ($S_1$) corresponding to the occupation of the same orbitals but with unparalleled spins. For benzene, the $T_1$ and $S_1$ states are antiaromatic. Indeed, while the ground state of benzene is chemically quite inert, the $S_1$ state is characterized by a rich reactivity. For this reason, Ottosson et al. say that benzene could be viewed as a "Dr. Jekyll and Mr. Hyde" molecule with the reactive antiaromatic excited state representing its "Mr. Hyde" character [14].

The existence of the lowest-lying triplet aromaticity for monocyclic conjugated rings of $4n$ $\pi$-electrons was predicted by Baird using perturbational MO theory in 1972 [9] and confirmed by the identification of planar triplet ground states of $C_5H_5^+$ and $C_5Cl_5^+$ [10, 15]. Triplet state aromaticity was applied to rationalize the singlet-triplet energy gaps of substituted fulvenes [16] or the dipole moments of fulvenes, fulvalenes, and azulene [17]. As can be seen in Figure 2.2, electron-donating groups X stabilize the ground state of pentafulvenes by increasing the contribution of the resonance structure *b* and destabilize its $T_1$ state. The result is an increase of the singlet-triplet energy gap. The reverse is true when X is an electron withdrawing group. Heptafulvenes have the opposite

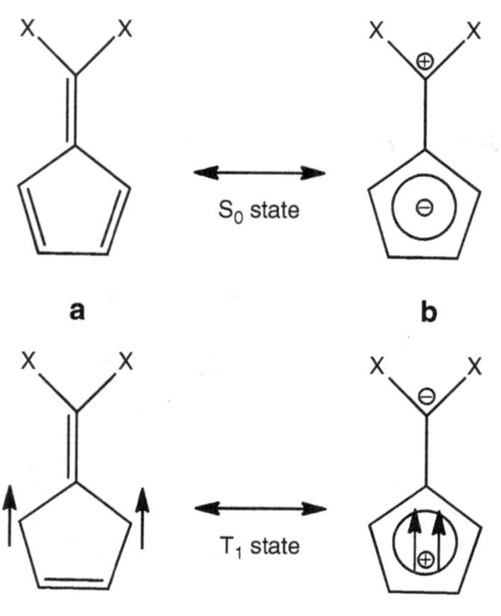


**Figure 2.2** The two most important resonant structures for the ground state and $T_1$ state of pentafulvenes. Resonances structures b are particularly important due to aromaticity. Source: Adapted from Ref. [18].

behavior that can also be explained by the effect of substituents on the Hückel and Baird aromaticities of the $S_0$ and $T_1$ states, respectively, so the singlet-triplet energy gap for heptafulvenes increases with electron withdrawing groups. The change of the sign of the dipole moment in fulvenes when going from $S_0$ to $T_1$ is also explained by the nonnegligible contribution of the ionic resonance structures *b* in the description of these two states. Recent spectroscopic data of the reversal of (anti)aromaticity in the $T_1$ state as compared to the ground state in bis-rhodium hexaphyrins has been taken as an additional support of Baird's rule [19].

Moreover, Baird's rule has been corroborated by a number of computational studies. Already in 1982, Fratev et al. showed that the equilibrium structure of the $T_1$ state of cyclobutadiene presents $D_{4h}$ symmetry with all C—C bonds of the same length [20] as expected from the Baird's rule. Not only cyclobutadiene, but also the optimized geometry of the $T_1$ state of $C_5H_5^+$, $C_7H_7^-$, $C_8H_8$, and $C_9H_9^+$ was found to be of $D_{nh}$ symmetry with C—C bond lengths close to those of benzene [21]. The Baird's rule was also confirmed through nucleus-independent chemical shifts (NICS), magnetic susceptibility, and aromatic stabilization energy calculations by Schleyer et al. [21, 22] as well as from the study of ring currents in $4n\pi$-electron monocycles [23]. Further support came from a work [24] based on the analysis of the bifurcation of the π-contribution to the electron localization function (ELF) for the lowest-lying triplet state of $4n\pi$-electron monocycles. Complete active space self-consistent field (CASSCF) calculations of different variants of NICS [25, 26] and several electron delocalization indices of aromaticity [27] on the ground and low-lying excited states of cyclobutadiene, benzene, and cyclooctatetraene reinforced the validity of the Baird's rule not only for the $T_1$ state, but also for the analogous $S_1$ state as can be seen in the values of Table 2.1. In this

**Table 2.1** Values of NICS(1), $NICS(1)_{zz}$, and MCI for low-lying singlet and triplet states of cyclobutadiene, benzene, and cycloctatetraene at the CASSCF level of theory. NICS in ppm and MCI in au. Values from Refs. [25–27].

| Symmetry/Species | State | NICS(1) | $NICS(1)_{zz}$ | MCI |
|---|---|---|---|---|
| $D_{4h}/C_4H_4$ | $S_0$ | 28.23 | 88.14 | 0.0092 |
| | $T_1$ | −6.54 | −16.47 | 0.0361 |
| | $S_1$ | −4.28 | −16.38 | 0.0491 |
| $D_{6h}/C_6H_6$ | $S_0$ | −9.53 | −27.83 | 0.0435 |
| | $T_1$ | 30.10 | 90.61 | 0.0023 |
| | $S_1$ | 34.67 | 102.76 | 0.0041 |
| $D_{8h}/C_8H_8$ | $S_0$ | 32.33 | 98.22 | 0.0005[a)] |
| | $T_1$ | −8.98 | −25.60 | 0.0047[a)] |
| | $S_1$ | −9.02 | −26.34 | 0.0061[a)] |

a) Obtained with a $D_{4h}$ symmetry.

table, the more negative the NICS values and the larger the multicenter index (MCI) value, the higher the aromaticity of the ring. It is important to note that MCI values decrease significantly when the size of the ring increases.

## 2.4 Soncini and Fowler's Rule

A generalization of the Hückel's and Baird's rules was proposed by Soncini and Fowler [28]. Their rule states that the monocyclic structures of $4n + 2$ $\pi$-electrons at the lowest-lying electronic states with even spin (singlet, quintet,...) and the monocycles of $4n$ $\pi$-electrons at the lowest-lying states with odd spin (triplet, septet,...) are aromatic. Thus, following this rule, the lowest-lying singlet and quintet state of benzene are aromatic whereas its lowest-lying triplet and septet states are antiaromatic. The reverse is true for cyclooctatetraene. The MCI values obtained with the Becke's three-parameter hybrid exchange functional with non-local Lee-Yang-Parr correlation functional (B3LYP) method collected in Table 2.2 provide evidence on the validity of this rule. MCI results indicate that benzene is aromatic in its $S_0$ and $Q_1$ states and cyclooctatetraene in its $T_1$ and Septet1 states. It is worth noting that CASSCF results by the same authors [27] do not support the aromaticity of the $Q_1$ and $Septet_1$ states of benzene and cyclooctatetraene, respectively. So, the study was inconclusive about the validity of the Soncini and Fowler's rule.

## 2.5 Möbius' $4n$ $\pi$-Electron Rule

Möbius aromaticity is the particular electron delocalization found in cyclic conjugated species having a molecular topology that resembles that of a Möbius

**Table 2.2** MCI values for low-lying singlet, triplet, quintet, and septet states of $C_6H_6$ and $C_8H_8$ obtained at the B3LYP/6−311++g(d,p) level of theory. All units are a.u. Values from Ref. [27].

| Symmetry/Species | State | MCI |
|---|---|---|
| $D_{6h}/C_6H_6$ | $S_0$ | 0.0721 |
| | $T_1$ | −0.0015 |
| | $Q_1$ | 0.0451 |
| $D_{4h}/C_8H_8$ | $S_0$ | −0.0005 |
| | $T_1$ | 0.0271 |
| | $Q_1$ | 0.0013 |
| | $Septet_1$ | 0.0178 |

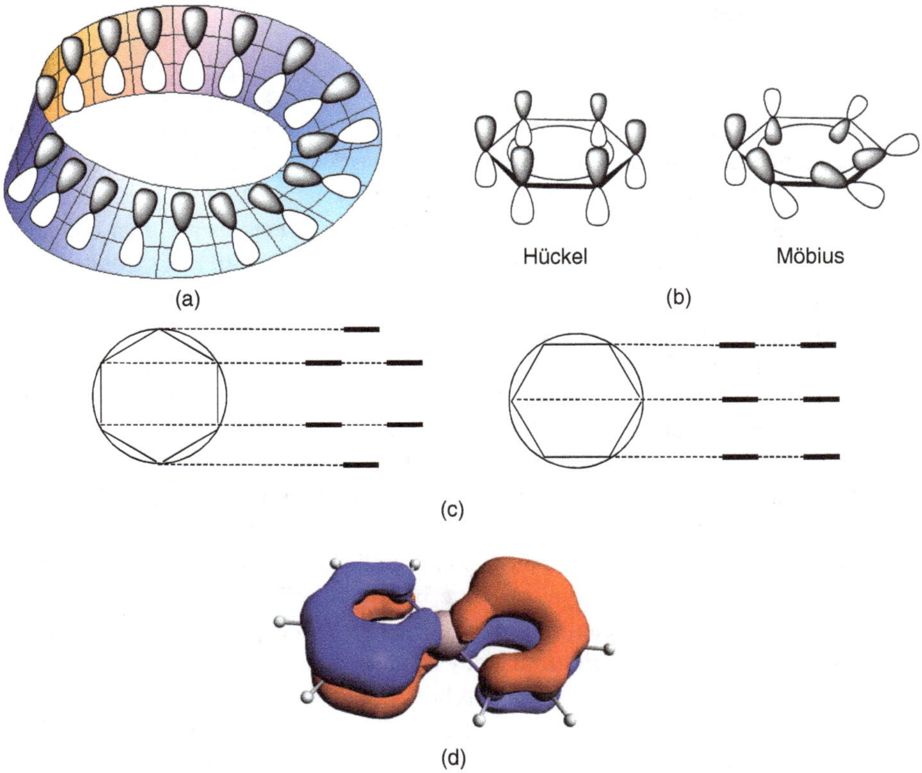


**Figure 2.3** (a) Schematic representation of the distribution of the p-orbitals on a Möbius strip. (b) Comparison between Hückel and Möbius HOMO π-orbital in a six-membered ring. (c) Distribution of π-orbitals in a six-membered ring with Hückel (left) and Möbius (right) topologies. (d) HOMO of $Cu^{II}[HN(CH)_3O]_2$ with Möbius aromatic character [29].

strip (see Figure 2.3a) [30–32]. These species present a phase inversion in the most stable π-MO. In the HMO method, the secular equation to be solved for a cyclic π-system with N atoms in a Möbius topology is the following N × N secular determinant:

$$\begin{vmatrix} \alpha-\varepsilon & \beta & \cdots & \cdots & -\beta \\ \beta & \alpha-\varepsilon & \beta & \vdots & \vdots \\ \vdots & \vdots & \vdots & \vdots & \vdots \\ -\beta & \cdots & \cdots & \beta & \alpha-\varepsilon \end{vmatrix} \tag{2.3}$$

Because of the phase inversion, the resonance integral corresponding to the interaction of the 1st and Nth atom changes its sign (compare Eqs. [2.1] and [2.3]).

The solution of this Hückel determinant is:

$$\varepsilon_j = \alpha + 2\beta \cos\left(\frac{\pi(2j+1)}{N}\right) \qquad (j = 0, 1, 2, \cdots N - 1) \tag{2.4}$$

The most stable MO is obtained for j = 0 and j = N-1 and therefore it is double degenerate. The rest of the orbitals, except the last one for odd N values also come by pairs (j = i and j = N-1-i have the same energy) as shown in Figure 2.3c for a six-membered ring (6-MR). Closed-shell electronic structures are then obtained for the following number of π-electrons: 4, 8, 12,..., that is, $4n$ π-electrons ($n$ = 1, 2, 3,...). This closed-shell electronic structure provides aromatic stabilization. The electron counting is opposed to that of the Hückel's rule and, consequently, Möbius-type annulenes with $4n + 2$ π-electrons are antiaromatic. This type of aromaticity was first described theoretically by Craig and Paddock [33] in 1958, and later was more elaborated by Heilbronner [30] in 1964. However, the chemistry of Möbius aromatic species has not been developed until the last decades. Real Möbius aromatic systems proved elusive until the work Herges et al. [31, 34, 35] in 2003. Several porphyrinoids that show a dynamic Hückel-Möbius topological interconversion in solution through conformational changes, metalation, redox processes, or acid/base chemistry have been synthesized recently. These new compounds have controllable aromatic behaviors and, depending on the external conditions, can switch their aromaticity from Hückel to Möbius or the other way round [36–40]. Möbius aromaticity has been also detected theoretically in several planar metallacycles [29, 41]. These metallacycles are $4n\pi$ aromatic species that shows one phase inversion in their most stable π-orbital (see Figure 2.3d).

## 2.6 The Linking Number Rule

The linking number ($L_k$), which is used to classify the varied winding of DNA strands [42], is an integer number that describes the linking of two closed curves in three-dimensional space [43]. The linking number represents the number of times that each curve winds (twist, $T_w$, and writhe, $W_r$, $L_k = T_w + W_r$) around the other (see Figure 2.4). When applied to cyclic conjugated annulenes, the linking number rule asserts that annulenes with an even $L_k$ value obey Hückel's $4n + 2$ π-electron rule of aromaticity for the closed-shell ground state, whereas those having an odd $L_k$ follow the Möbius' $4n$ π-electron rule. This rule represents a generalization of the Hückel's and Möbius' rules for species having several writhes or twists. In 2014, the group of Herges [44, 45] designed and synthesized the first high-order Möbius molecule, a triply twisted Möbius dehydroannulene with $T_w = 3$, $W_r = 0$, and $L_k = 3$.

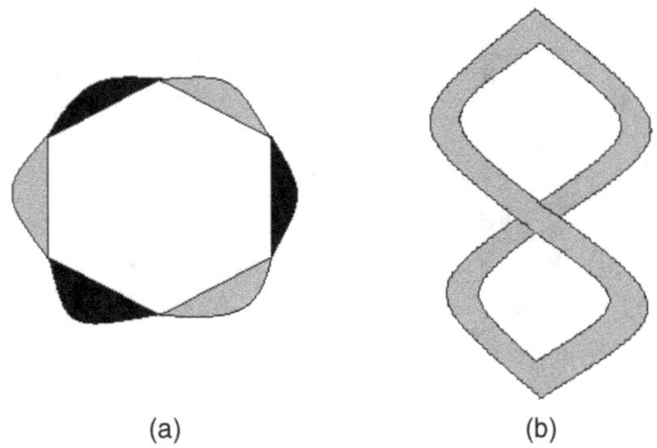


**Figure 2.4** Schematic illustration of a strip with (a) three twists ($T_w = 3$; $W_r = 0$; $L_k = 3$) and (b) one writhe ($T_w = 0$; $W_r = 1$; $L_k = 1$).

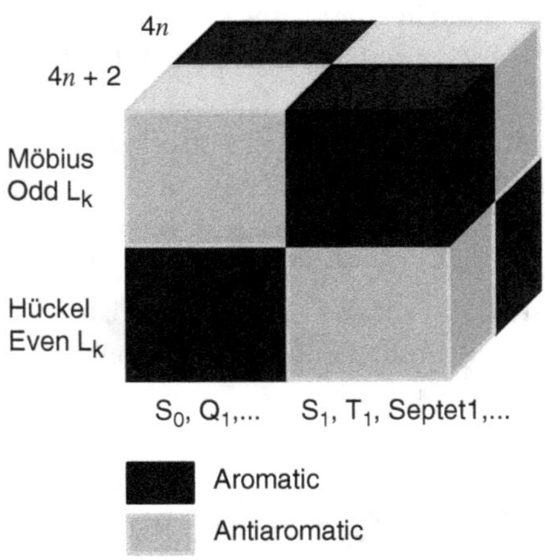


**Figure 2.5** Ottosson's cube summarizing several combinations of (anti)aromatic annulenes depending on the number of π-electrons, twists, and writhes, and electronic state. Source: Adapted from reference [18].

The Hückel, Möbius, Baird, Soncini and Fowler, and linking number rules discussed in the previous sections can be summarized in the so-called Ottosson's cube shown in Figure 2.5. For instance, an annulene in the triplet state with an odd linking number and with $4n$ π-electrons is antiaromatic, whereas in the ground state is aromatic.

## 2.7 Platt's Ring Perimeter Model

The $4n + 2$ Hückel's rule strictly holds for monocyclic systems like benzene and cyclooctatetraene. Polycyclic aromatic hydrocarbons (PAHs) such as pyrene and coronene in Figure 2.6 do not obey this rule. They have 16 and 24 π-electrons, respectively, and, therefore, they should be considered antiaromatic according

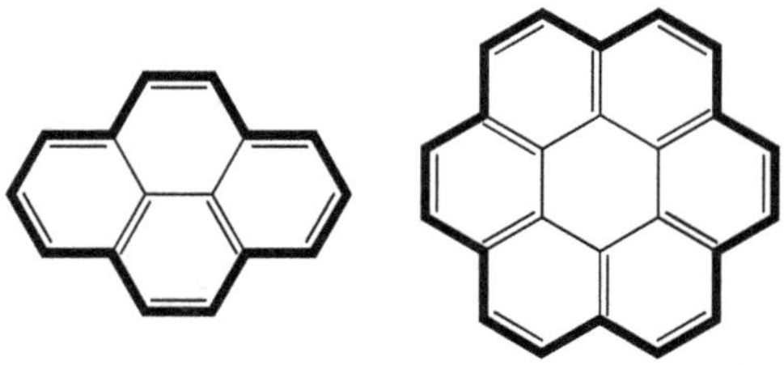

(a) (b)

**Figure 2.6** A resonance structure of (a) pyrene and (b) coronene with the perimeter underlined in bold.

Hückel's rule. However, these molecules behave as aromatic species. The violation of $4n + 2$ Hückel's rule in PAHs was already well recognized in the beginning of the fifties [7]. A first attempt to extent the Hückel's $4n + 2$ $\pi$-electron rule from monocyclic annulenes to PAHs corresponded to the Platt's ring perimeter model [46]. According to this model, PAHs can be divided into two parts: a perimeter and an inner core. The perimeter is considered as an annulene and the inner core represents only a perturbation of the perimeter. The aromatic character of the PAH is that of the annulene of the perimeter as derived from the Hückel's rule. Then, in the case of pyrene and coronene, they have a perimeter with 14 and 18 $\pi$-electrons, respectively, and, consequently, they follow Hückel's rule for aromatic species (Figure 2.6). Unfortunately, this simple rule fails to account for the aromaticity of many nonbenzenoid polycyclic conjugated hydrocarbons (PCHs) [47].

## 2.8 Clar's $\pi$-Sextet Rule

A very useful extension of the Hückel's $4n + 2$ $\pi$-electron rule from monocyclic species to benzenoid systems was proposed in 1972 by Clar [48, 49], who formulated the $\pi$-sextet rule in the book *The Aromatic Sextet*. This model can be traced back through Crocker [50] in 1922, and Armit and Robinson in 1925 [51] to Armstrong in 1890 [52, 53], who was the first person to depict several benzenoids with aromatic $\pi$-sextets in the same way Clar did eighty years later [54]. In this model, aromaticity is regarded as a local property of the six-membered rings (6-MRs). Clar's $\pi$-sextet rule states that the Kekulé resonance structure with the largest number of disjoint aromatic $\pi$-sextets is the most important resonance structure for characterization of PAHs properties. The aromatic $\pi$-sextet is defined as six $\pi$-electrons localized in a single benzene-like ring separated from adjacent rings by formal C—C single bonds. For instance, in the resonance structure *a* of Figure 2.7, only the central ring has a $\pi$-sextet. Then, application of this rule to phenanthrene indicates that the resonance structure *a* in Figure 2.7 is less important than resonance structure *b* with two $\pi$-sextets to describe the properties and aromaticity

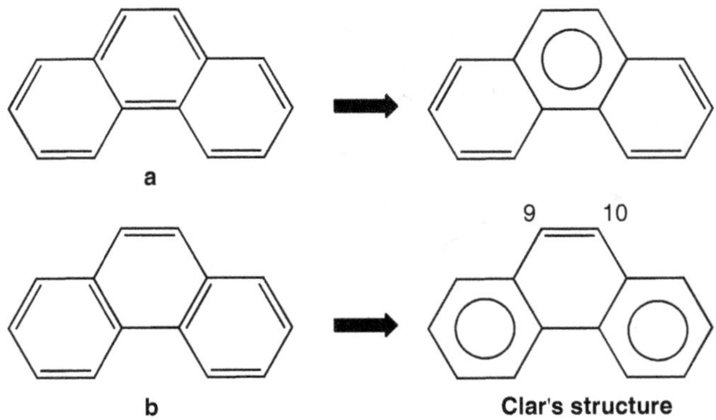


**Figure 2.7** Two out of the five Kekulé resonance structures of phenanthrene and their corresponding Clar aromatic π-sextets indicated with a circle. The structure with the largest number of aromatic π-sextets is the so-called Clar's structure.

of phenanthrene. Therefore, outer rings in phenanthrene are expected to have a larger fraction of local aromaticity than the central ring, which in fact is what has been confirmed using different descriptors of local aromaticity [55].

When we consider anthracene, the situation is a little bit different. As can be seen in Figure 2.8, all possible resonance structures for anthracene have a unique π-sextet. In this case, the Clar structure is better described as a combination of these three structures and this situation is usually represented by Clar's structure of Figure 2.8 in which the arrow indicates the presence of a migrating π-sextet. The Clar structure of anthracene indicates that the inner and outer rings of anthracene should have a similar aromaticity. This prediction has been confirmed by several indicators of local aromaticity [55, 56].

At variance with Platt's ring perimeter model, Clar's π-sextet rule can be used to make predictions and explain the structure and reactivity of benzenoid species. For instance, there is a good correspondence between the single, double, or intermediate character of a given C—C bond according to Clar's structure and the experimental observed bond lengths. The X-ray structure of phenanthrene confirms that the 9,10 bond (see Figure 2.7) is the shortest (1.341 Å) of all its C—C bond lengths [57]. Another interesting example concerning the reactivity of PAHs is the explanation offered by this model of the different reactivity of isomers tribenzo[*fg,ij,rst*]pentaphene and benzo[*qr*]naphto[2,1,8,7-*fghi*]pentacene (see Figure 2.9). The latter reacts readily with maleic anhydride whereas the former is unreactive [58]. The simple picture of the Clar structures of the two isomers shows that benzo[*qr*]naphto[2,1,8,7-*fghi*]pentacene has double bonds available to act as a diene in a Diels–Alder reaction whereas tribenzo[*fg,ij,rst*]pentaphene

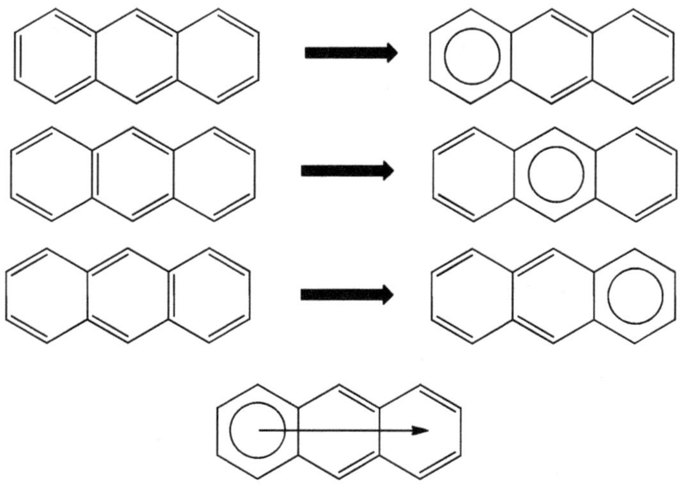


**Figure 2.8** Three out of the four Kekulé resonance structures of anthracene and their corresponding Clar aromatic $\pi$-sextets indicated with a circle. The Clar structure is a combination of these three structures.

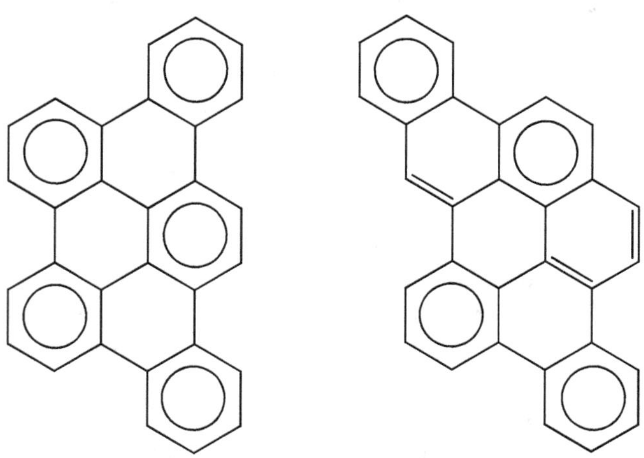


**Figure 2.9** Clar's structures of isomers tribenzo[*fg,ij,rst*]pentaphene and benzo[*qr*]naphto[2,1,8,7-*fghi*]pentacene.

has only π-sextets and empty rings and, therefore, is chemically inert in front of a dienophile.

Theory and computation have also provided extensive support to the predictions made by the Clar's π-sextet rule [49, 59]. Moreover, this model developed to explain the properties of benzenoid compounds has been applied successfully to describe the local aromaticity of carbon nanotubes, graphene nanoribbons, carbon nanocones, and carbon nanotori [60–65].

### 2.8.1 Glidewell and Lloyd's Rule

An important limitation of the Clar's π-sextet rule is that it can be applied only to benzenoid species. To overcome this limitation and to extent Clar's rule to non-benzenoid PCHs, Glidewell and Lloyd formulated a new, more general rule [66] stating that the total population of π-electrons in PCHs tends to form the smallest $4n + 2$ groups and to avoid the formation of the smallest $4n$ groups. For benzenoid systems (i.e., PCHs having only 6-MRs), Glidewell and Lloyd's rule reduces to Clar's π-sextet rule. Figure 2.10 shows four examples of nonbenzenoid PCHs in which application of the Glidewell and Lloyd's rule leads to conclusion that one of the resonance structures (the one shown in bold) is more relevant than the other(s) to explain their electronic and molecular properties. A systematic study [47] in an extensive series of PCHs confirmed that the Glidewell and Lloyd's rule is obeyed for most of the PCHs.

### 2.8.2 The Y-Rule

There are some benzenoids that have a unique Clar structure (e.g., phenanthrene), whereas several Clar structures are possible for other PAHs (e.g., anthracene). For these latter, Clar's rule cannot differentiate which of the corresponding resonance structures has more influence in determining the local aromaticity of the benzenoid compound. The Y-rule [67, 68] by Ruiz-Morales is an attempt to solve this limitation. The Y-rule is a modification of Clar's rule for benzenoid

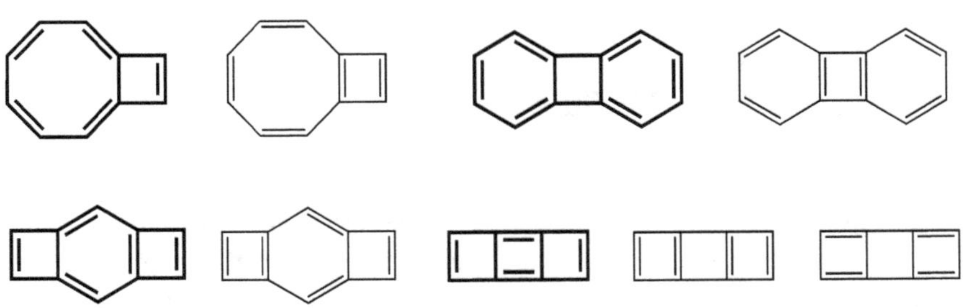

**Figure 2.10** Application of the Glidewell–Lloyd's rule to a series of nonbenzenoid PAHs leads to the conclusion that the most representative resonance structures are those depicted in bold.

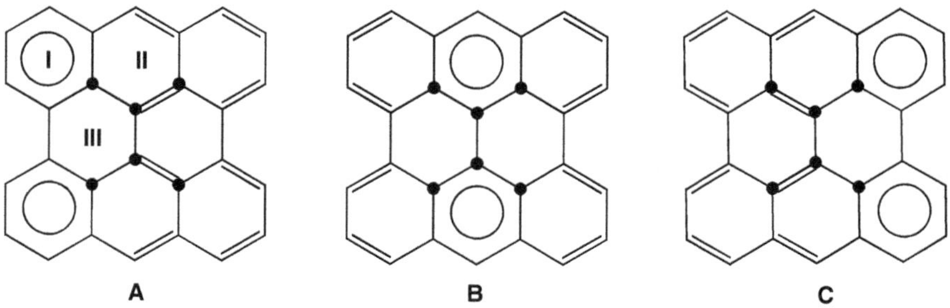


**Figure 2.11** Among the three possible Clar structures for one of the isomers of $C_{28}H_{14}$, the Y-rule considers structure B to be the most important.

hydrocarbons possessing internal carbon atoms that simultaneously belong to three hexagons. To apply this rule one has to follow five steps: (i) the aromatic sextets has to be located so as to cover all internal carbon atoms; (ii) among different Clar-type structures satisfying condition (i), one has to choose the structure with the large number of aromatic sextets; (iii) among different Clar-type structures obeying (i) and (ii), the one with the highest symmetry, provided that it exists, is considered the most important structure; (iv) if the Clar-type structure that obeys criteria (i)–(iii) is unique, this structure is expected to be the best one to represent the electronic structure and aromaticity of the benzenoid species; and (v) if there are several Clar-type structures that follow requirements (i)–(iii), then one has to consider their superposition as the best approximation to the electronic structure and aromaticity of the benzenoid compound. The selected Clar-type structure based on the Y-rule can be different from the one chosen with the Clar's $\pi$-sextet rule. In Figure 2.11, we depict the different possible Clar structures for a $C_{28}H_{14}$ isomer. The only structure that covers all internal C atoms marked with a black dot in Figure 2.11 is structure *B*. Therefore, according to the Y-rule this structure is the one that better describes the local aromaticity of the system. As expected from Clar's rule, the aromaticity of ring *III* diagnosed with NICS is much lower than that of rings *I* and *II*. NICS(1) values however indicate the same aromaticity for rings *I* and *II* in $C_{28}H_{14}$ [67].

## 2.9 Hirsch's $2(n+1)^2$ Rule

Hirsch's $2(n+1)^2$ rule [69] is a rule for spherical aromaticity. It is based on the fact that the Schrödinger equation of an uniform electron gas surrounding the surface of a sphere is the same as that of the rigid rotor and, therefore, the wave functions of this model are characterized by the angular momentum quantum number $l$ ($l = 0, 1, 2, \ldots$) similar to the situation found for atomic orbitals. Each energy level is $2l + 1$ times degenerated, and, therefore, $\pi$-shells are completely filled when we have 2, 8, 18, 32, 50, 72…electrons, that is, $2(n+1)^2$ electrons (see

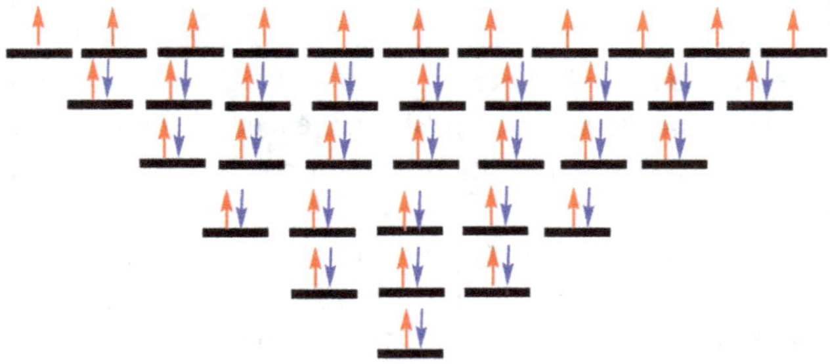


**Figure 2.12** Hirsch's $2(n+1)^2$ and $2n^2+2n+1$ ($S = n + ½$) rules in spherical aromatic species with s, p, d, f, g…distribution of molecular orbitals.

Figure 2.12). Consequently, spherical species with $2(n+1)^2$ π-electrons are aromatic in a way analogous to annulenes of $(4n+2)$ π-electrons. If we consider that the behavior of the π-electron system of an icosahedral fullerene can be approximated to that of a spherical electron gas, then icosahedral $C_{20}{}^{2+}$, $C_{60}{}^{10+}$, and $C_{80}{}^{8+}$ among many others are aromatic fullerenes, as they have 18, 50, and 72 electrons, respectively. Table 2.3 includes a series of $C_{60}$ and $C_{80}$ species with different

**Table 2.3** $NICS(1)_{zz}$ (in ppm), MCI (in electrons) and BLA (in Å) values for $C_{60}$ and $C_{80}$ with different charges and multiplicities.

| System | Symmetry | Ring | $NICS(1)_{zz}$ | MCI | BLA | Spin |
|---|---|---|---|---|---|---|
| $C_{60}$ | $I_h$ | 6-MR | 0.8 | 0.018 | 0.058 | S = 0 |
| | | 5-MR | 21.5 | 0.011 | | |
| $C_{60}{}^{1-}$ | $I_h$ | 6-MR | −1.4 | 0.017 | 0.002 | S = 11/2 |
| | | 5-MR | −19.9 | 0.049 | | |
| $C_{60}{}^{19+}$ | $I_h$ | 6-MR | −14.9 | 0.019 | 0.013 | S = 9/2 |
| | | 5-MR | −25.3 | 0.041 | | |
| $C_{60}{}^{10+}$ | $I_h$ | 6-MR | −18.6 | 0.011 | 0.030 | S = 0 |
| | | 5-MR | −29.5 | 0.017 | | |
| $C_{80}$ | $S_6$ | 5-MR | 10.7 | 0.019 | | S = 0 |
| | | 6-MR | −5.2 | 0.012 | 0.025 | |
| | | 5-MR | 26.3 | 0.018 | | |
| | | 6-MR | 11.3 | 0.014 | 0.001 | |
| | | 6-MR | −5.1 | 0.012 | 0.025 | |
| $C_{80}{}^{8+}$ | $I_h$ | 6-MR | −7.2 | 0.011 | 0.015 | S = 0 |
| | | 5-MR | −4.0 | 0.017 | | |
| $C_{80}{}^{5-}$ | $I_h$ | 6-MR | −20.8 | 0.019 | 0.012 | S = 13/2 |
| | | 5-MR | −5.5 | 0.034 | | |

charges and multiplicities from Ref. [70]. The aromaticity analysis is performed by means of the magnetic NICS(1)$_{zz}$, the electronic MCI, and the bond length alternation (BLA) indicators. The BLA is the difference between the longest and shortest C—C bond length in the molecule. Low values of BLA are expected for aromatic compounds. For $C_{60}$, all descriptors point out that it has a nonaromatic or only slightly aromatic character.

$C_{60}^{10+}$ appears to be more aromatic than $C_{60}$ with more negative NICS and smaller BLA, in line with Hirsch's rule for a system with 50 electrons. Similarly, $C_{80}^{+8}$ with 72 electrons and, therefore, obeying the Hirsch's rule is found to have more negative NICS values and smaller BLA than $C_{80}$. The fact that MCI does not show an increase when going from $C_{60}$ to $C_{60}^{10+}$ or from $C_{80}$ to $C_{80}^{+8}$ is probably due to the decrease in the number of electrons that translates into a reduction in the number of delocalized electrons.

## 2.10 The $2n^2 + 2n + 1$ ($S = n + ½$) Rule

The $2n^2 + 2n + 1$ ($S = n + ½$) rule [70] applies to open-shell spherical aromatic compounds. Extension of Baird's rule to spherical compounds leads to the conclusion that spherical systems having the degenerate highest-occupied MOs (HOMOs) occupied with half-filled electrons in parallel spin with the rest of lower levels orbitals being full filled are aromatic. Taking the distribution of orbitals corresponding to a uniform electron gas surrounding the surface of a sphere, this situation is reached for a number of electrons equal to 1, 5, 13, 25, 41, 61, 85,..., that is, for spherical species with $2n^2 + 2n + 1$ π-electrons and with an electronic spin of $S = n + ½$ (see Figure 2.12). For instance, according to this rule, $C_{60}^{19+}$ with $S = 9/2$, $C_{60}^{1-}$ with $S = 11/2$ and $C_{80}^{5-}$ with $S = 13/2$ having 41, 61, and 85 electrons, respectively, are open-shell spherical aromatic fullerenes. NICS(1)$_{zz}$, MCI, and BLA results in Table 2.3 show that this is indeed the case and that $C_{60}^{19+}$, $C_{60}^{1-}$ and $C_{80}^{5-}$ following the $2n^2 + 2n + 1$ ($S = n + ½$) are at least as aromatic as $C_{60}^{10+}$ and $C_{80}^{5-}$ that follow the Hirsch's $2(n + 1)^2$ rule.

## 2.11 Wade–Mingos' $2n + 2$ Rule

The determination in 1959 of the structure of the *closo* borane $B_{10}H_{10}^{2-}$ by Lipscomb [71] and the synthesis in 1962 of the first derivatives of *closo*-dodecaborate and *closo*-decaborate by Muetterties's group [72] moved the concept of aromaticity from two to three dimensions (see Chapter 13). *Closo* borane clusters are stable and aromatic boron clusters with a general formula of $[B_nH_n]^{2-}$. This stability is at odds with most borohydrides that are electron-deficient compounds, highly reactive, and unstable. *Closo* borane clusters have a structure of polyhedron with

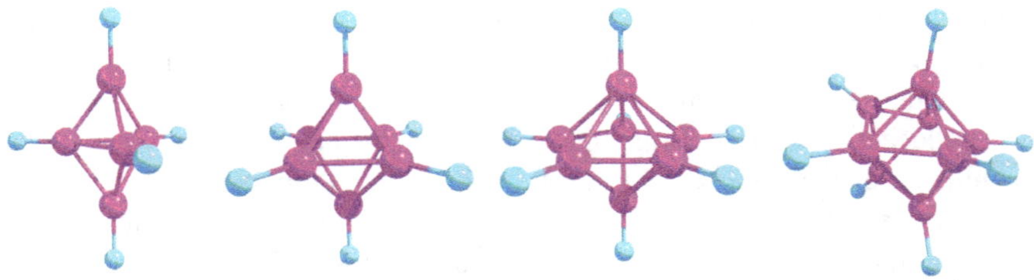


**Figure 2.13** Closo borane anions, $B_nH_n^{2-}$, $n = 5-8$.

triangular faces (e.g., Figure 2.13 shows the structure of $B_nH_n^{2-}$ with n = 5–8). They obey Wade's $2n + 2$ electron rule, where $n$ is the number of vertexes of the polyhedron, or Mingos' rule that is $4n + 2$ [73, 74]. Wade's rule refers to the skeletal electrons (all valence electrons except those of the B–H bonds), whereas Mingos' rule also incorporates the exo electrons of the B—H bonds, thus the $4n + 2$ in Mingos' rule designates the total number of valence electrons. Therefore, Wade and Mingos' rules are equivalent. For $B_8H_8^{2-}$, the total number of valence electrons is $8 * 3 + 8 * 1 + 2 = 34 = 4 * 8 + 2$ ($n = 8$ in Mingos' rule), while the number of valence electrons without the exo electrons of the B—H bonds is $34 - 8 * 2 = 18 = 2 * 8 + 2$ (n = 8 in Wade's rule). When referring to *electron pairs*, the Wade's rule is $n + 1$ and the Mingos' rule is $2n + 1$. Although Wade–Mingos' rule was initially developed for borane clusters, it has been shown that it is applicable to other main group clusters as Zintl phases and ions [75]. Again the justification of the Wade–Mingos' rule comes from the MO theory. BH units are assumed to be *sp* hybridized. Combination of *sp* outward orbitals with 1*s* orbitals of H forms the exo electron pairs of the B–H bonds. The rest of the *n sp* inward orbitals form a radial in-phase orbital that is the most stable skeletal bonding MO and $n - 1$ unoccupied orbitals higher in energy. The $2n$ p unhybridized orbitals yield $n$ bonding and $n$ antibonding tangential orbitals. $[B_nH_n]^{2-}$ clusters with $n + 1$ electron pairs have closed-shell structures that result from filling the lowest-lying *sp* radial orbital and the $n$ bonding tangential orbitals. It is worth noting that the Wade–Mingos' rule refers to the *inner* aromaticity of skeletal electrons of the cage, whereas the Hirsch's rule describes the *outer* aromaticity of the electrons delocalized on the surface of the spherical species.

### 2.11.1 Jemmis' *mno* Rule

In the same way the $4n + 2$ Hückel's rule fails for PAHs, the Wade–Mingos' rule does not work for condensed polyhedral boranes in which monomeric borane cluster units share one or more boron vertices. Jemmis' *mno* rule [76] was formulated as an extension of the Wade–Mingos' rule to condensed polyhedral boranes. According to this rule, the number of skeletal electron pairs required for a condensed polyhedral borane, carborane, heteroborane, metallaborane, or metallocene cluster to be aromatic is given by $m+n+o+p-q$, where $m$ = number of subclusters, $n$ = number of vertices, $o$ = number of single-vertex shared condensations, $p$ = the number of missing vertices, and $q$ = number of caps. As a rule of aromaticity, it is known as the Jemmis' *mno* rule because $p$ and $q$ are not necessary for *closo* borane aromatic clusters and are only required for *nido* and *arachno* clusters. For instance, $B_{20}H_{16}$ in Figure 2.14 is a condensed polyhedral borane formed by the sharing of 4 vertices between two icosahedra ($m = 2$) and with 20 vertices ($n = 20$). This compound was the first synthesized condensed polyhedral borane [77]. For this species, numbers $o$, $p$, and $q$ are zero. Then 22 skeletal electron pairs are necessary to generate an aromatic borane as required by Jemmis' rule $(2+20+0+0+0)$. Sixteen BH units provide 16 pairs and four shared boron atoms provide 6 pairs. Then $B_{20}H_{16}$ formed by two condensed polyhedra and with 22 skeletal electron pairs is aromatic and stable according to Jemmis' *mno* rule. In $Si(C_2B_9H_{11})_2$ shown in Figure 2.13, we have $m = 2$; $n = 23$; and $o = 1$, and therefore, Jemmis' rule requires 26 skeletal electron pairs. Four CH units provide 6 pairs, 18 BH units deliver 18 pairs, and Si provide 2 pairs giving the 26 pairs that satisfy the Jemmis' *mno* rule [78].

### 2.11.2 Equivalence between Hückel's and Wade–Mingos' Rules

A direct connection between the Hückel's $4n+2$ $\pi$-electron rule for two-dimensional aromaticity and the Wade–Mingos' $2n+2$ rule for three-dimensional

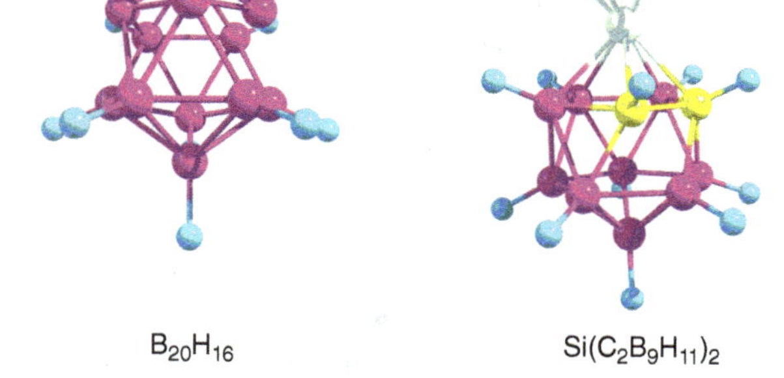


**Figure 2.14** The molecular structure of $B_{20}H_{16}$ and $Si(C_2B_9H_{11})_2$.

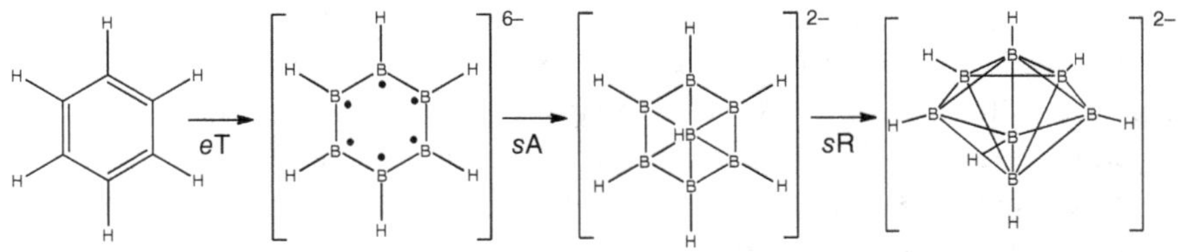


**Figure 2.15** Link established between benzene and $B_7H_7^{2-}$.

aromaticity has been established [79] by showing that there is an equivalence between *closo* borane cluster and classical organic annulenes. To find the equivalent aromatic *closo* borane cluster of an annulene, the following procedure has to be applied: (i) state the model organic compound; (ii) transmute each C atom into a B atom and one electron (eT) [80]; (iii) replace extra electrons by sacrificial atoms or agents (sA); (iv) and finally (if necessary) generate the new borohydride compound by structural relaxation (sR). During the whole process, the number of valence electrons remains unaltered. $H^+$, $B^{3+}$ or $[BH]^{4+}$ among others can be used as sacrificial atoms or agents. These cations have empty valence orbitals perfectly suited to form multicenter bonds. As an example, let us find the borohydride equivalent to benzene that is the archetype of the aromatic molecules (see Figure 2.15). In this case the total number of valence electrons that remains constant through the process is 30 electrons. The electronic transmutation of C with $B^-$ drives us to $[B_6H_6]^{6-}$ with chair-like structure. The sacrificial agent we add is $[BH]^{4+}$ leading to hexagonal pyramid $[B_7H_7]^{2-}$ that is not a minimum, but relaxes to a pentagonal bipyramid $[B_7H_7]^{2-}$, as experimentally found [81]. The NICS(0) of −22.8 ppm points out that $[B_7H_7]^{2-}$ is at least as aromatic as benzene [79]. The relatively low BLA value of 0.17 Å seems to confirm the aromatic character of this *closo* borane. Similar links are found for $C_4H_4^{2-}$, $C_5H_5^-$, and $C_7H_7^+$ with $[B_5H_5]^{2-}$, $[B_6H_6]^{2-}$, and $[B_8H_8]^{2-}$, respectively [79]. Therefore, there is a connection between highly stable monocyclic hydrocarbon reference compounds fulfilling $(4n+2)\pi$ Hückel's electron rule with the corresponding monodeltahedral *closo* borane clusters that obey the $4n+2$ Wade–Mingos' rule. Notice that the value of $n$ has not the same meaning in one equation and the other. In $(4n+2)\pi$ Hückel's rule $n$ can be any integer and has no direct relationship with the structure. On the contrary, in $4n+2$ Wade–Mingos' rule $n$ is structure dependent and refers to the number of vertexes occupied usually by boron atoms.

## 2.12 Other Rules

The observed experimental abundances found in experimental mass spectra of alkali, alkaline earth metals, and gold clusters of 2, 8, 18, 20, 34, 40,…atoms can be successfully explained with the spherical jellium model [82]. The energy levels of valence electrons for such a model are $1S^2 1P^6 1D^{10} 2S^2 1F^{14} 2P^6 1G^{18} 2D^{10}$...and, therefore, with 2, 8, 18, 20, 34, 40,...electrons one gets a closed-shell electronic structure. These species are particularly stable and they are considered superatoms, although they can also be regarded as jellium aromatic in the same way *closo* boranes are considered aromatic. An open-shell jellium aromaticity rule has been also formulated [83].

In some large cylindrical molecules such as nanotubes, cycloparaphenylenes, or certain boron and boron nitride clusters, among others, the aromaticity can be described by the hollow cylinder model (HCM) [84]. Solution of the Schrödinger equation for a particle in this HCM gives MOs that can be separated into radial (located in the different layers of the cylinder and perpendicular to the cylinder axis) and tangential (located in the outer part of the cylinder and parallel to the cylinder axis). This distribution gives rise to the $4N + 2M$ rule of radial and tangential aromaticity, N and M being the number of degenerated and nondegenerated MOs, respectively. Finally, certain orbital occupations have been associated with disk [85], cubic [86], or tetrahedral [87] aromaticities.

## References


1 Hückel, E. (1931). *Z. Physik* 70: 104–186.
2 Hückel, E. (1931). *Z. Physik* 72: 310–337.
3 Hückel, E. (1932). *Z. Physik* 76: 628–648.
4 Hückel, E. (1937). *Z. Elektrochemie* 43: 752–788. 827-849.
5 Doering, W.v.E. and Detert, F.L. (1951). *J. Am. Chem. Soc.* 73: 876–877.
6 Doering, W.v.E. and Knox, L.H. (1954). *J. Am. Chem. Soc.* 76: 3203–3206.
7 Roberts, J.D., Streitwieser, A., and Regan, C.M. (1952). *J. Am. Chem. Soc.* 74: 4579–4582.
8 Alexandrova, A.N. and Boldyrev, A.I. (2003). *J. Phys. Chem. A* 107: 554–560.
9 Baird, N.C. (1972). *J. Am. Chem. Soc.* 94: 4941–4948.
10 Breslow, R., Chang, H.W., Hill, R., and Wasserman, E. (1967). *J. Am. Chem. Soc.* 89: 1112–1119.
11 Bally, T. and Masamune, S. (1980). *Tetrahedron* 36: 343–370.
12 Kostenko, A., Tumanskii, B., Kobayashi, Y. et al. (2017). *Angew. Chem. Int. Ed.* 56: 10183–10187.
13 Gould, C.A., Marbey, J., Vieru, V. et al. (2021). *Nature Chem.* 13: 1001–1005.
14 Papadakis, R. and Ottosson, H. (2015). *Chem. Soc. Rev.* 44: 6472–6493.
15 Saunders, M., Berger, R., Jaffe, A. et al. (1973). *J. Am. Chem. Soc.* 95: 3017–3018.
16 Ottosson, H., Kilså, K., Chajara, K. et al. (2007). *Chem. Eur. J.* 13: 6998–7005.

**17** Möllerstedt, H., Piqueras, M.C., Crespo, R., and Ottosson, H. (2004). *J. Am. Chem. Soc.* 126: 13938–13939.

**18** Rosenberg, M., Dahlstrand, C., Kilså, K., and Ottosson, H. (2014). *Chem. Rev.* 114: 5379–5425.

**19** Sung, Y.M., Yoon, M.-C., Lim, J.M. et al. (2015). *Nature Chem.* 7: 418–422.

**20** Fratev, F., Monev, V., and Janoschek, R. (1982). *Tetrahedron* 38: 2929–2932.

**21** Gogonea, V., Schleyer, P.v.R., and Schreiner, P.R. (1998). *Angew. Chem. Int. Ed.* 37: 1945–1948.

**22** Jiao, H.J., Schleyer, P.v.R., Mo, Y. et al. (1997). *J. Am. Chem. Soc.* 119: 7075–7083.

**23** Fowler, P.W., Steiner, E., and Jenneskens, L.W. (2003). *Chem. Phys. Lett.* 371: 719–723.

**24** Villaume, S., Fogarty, H.A., and Ottosson, H. (2008). *ChemPhysChem* 9: 257–264.

**25** Karadakov, P.B. (2008). *J. Phys. Chem. A* 112: 7303–7309.

**26** Karadakov, P.B. (2008). *J. Phys. Chem. A* 112: 12707–12713.

**27** Feixas, F., Vandenbussche, J., Bultinck, P. et al. (2011). *Phys. Chem. Chem. Phys.* 13: 20690–20703.

**28** Soncini, A. and Fowler, P.W. (2008). *Chem. Phys. Lett.* 450: 431–436.

**29** Rimola, A., Alí-Torres, J., Rodríguez-Rodríguez, C. et al. (2011). *J. Phys. Chem. A* 115: 12659–12666.

**30** Heilbronner, E. (1964). *Tetrahedron Lett.* 5: 1923–1928.

**31** Ajami, D., Oeckler, O., Simon, A., and Herges, R. (2003). *Nature* 426: 819–821.

**32** Rzepa, H.S. (2005). *Chem. Rev.* 105: 3697–3715.

**33** Craig, D.P. and Paddock, N.L. (1958). *Nature* 181: 1052–1053.

**34** Ajami, D., Hess, K., Köhler, F. et al. (2006). *Chem. Eur. J.* 12: 5434–5445.

**35** Herges, R. (2006). *Chem. Rev.* 106: 4820–4842.

**36** Pawlicki, M. and Latos-Grażyński, L. (2015). *Chem. Asian J.* 10: 1438–1451.

**37** Stępień, M., Latos-Grażyński, L., Sprutta, N. et al. (2007). *Angew. Chem. Int. Ed.* 46: 7869–7873.

**38** Tanaka, Y., Saito, S., Mori, S. et al. (2008). *Angew. Chem. Int. Ed.* 47: 681–684.

**39** Higashino, T., Soya, T., Kim, W. et al. (2015). *Angew. Chem. Int. Ed.* 54: 5456–5459.

**40** Naoda, K., Mori, H., Oh, J. et al. (2015). *J. Org. Chem.* 11726–11733.

**41** Mauksch, M. and Tsogoeva, S.B. (2010). *Chem. Eur. J.* 16: 7843–7851.

**42** Mirkin, S.M. (2001). DNA Topology: fundamentals. In: *Encyclopedia of Life Sciences*, 1–11. Wiley.

**43** Rappaport, S.M. and Rzepa, H.S. (2008). *J. Am. Chem. Soc.* 130: 7613–7619.

**44** Schaller, G.R., Topić, F., Rissanen, K. et al. (2014). *Nature Chem.* 6: 608–613.

**45** Wang, E., He, Z., Zhao, E. et al. (2015). *Chem. Eur. J.* 21: 11707–11711.

**46** Platt, J.R. (1949). *J. Chem. Phys.* 17: 484–495.

**47** El Bakouri, O., Poater, J., Feixas, F., and Solà, M. (2016). *Theor. Chem. Acc.* 135: 205.

**48** Clar, E. (1972). *The Aromatic Sextet*. New York: Wiley.

**49** Solà, M. (2013). *Front. Chem.* 1: 22.

**50** Crocker, E.C. (1922). *J. Am. Chem. Soc.* 44: 1618–1630.

**51** Armit, T.W. and Robinson, R. (1925). *J. Chem. Soc., Trans.* 127: 1604–1618.

**52** Armstrong, H.E. (1887). *J. Chem. Soc., Trans.* 51: 258–268.

**53** Armstrong, H.E. (1890). *Proc. Chem. Soc. Lond.* 6: 95–106.

**54** Balaban, A.T., Schleyer, P.v.R., and Rzepa, H.S. (2005). *Chem. Rev.* 105: 3436–3447.

**55** Portella, G., Poater, J., Bofill, J.M. et al. (2005). *J. Org. Chem.* 70: 2509–2521.

**56** Schleyer, P.v.R., Manoharan, M., Jiao, H.J., and Stahl, F. (2001). *Org. Lett.* 3: 3643–3646.

**57** Kay, M.I., Okaya, Y., and Cox, D.E. (1971). *Acta Cryst. B* 27: 26–33.

**58** Clar, E. and Zander, M. (1958). *J. Chem. Soc.* 1861–1865.

**59** Randić, M. (2014). *Chem. Phys. Lett.* 601: 1–5.

**60** Martín-Martínez, F.J., Fias, S., Van Lier, G. et al. (2013). *Phys. Chem. Chem. Phys.* 15: 12637–12647.

**61** Martín-Martínez, F.J., Fias, S., Van Lier, G. et al. (2012). *Chem. Eur. J.* 18: 6183–6194.

**62** Martín-Martínez, F.J., Melchor, S., and Dobado, J.A. (2008). *Org. Lett.* 10: 1991–1994.

**63** Balaban, A.T. and Klein, D.J. (2009). *J. Phys. Chem. C* 113: 19123–19133.

**64** Wassmann, T., Seitsonen, A.P., Saitta, A.M. et al. (2010). *J. Am. Chem. Soc.* 132: 3440–3451.

**65** Song, X., Wang, D., Gao, X.J., and Gao, X. (2020). *Chem. Phys. Lett.* 750: 137487.

**66** Glidewell, C. and Lloyd, D. (1984). *Tetrahedron* 40: 4455–4472.

**67** Ruiz-Morales, Y. (2004). *J. Phys. Chem. A* 108: 10873–10896.

**68** Gutman, I. and Ruiz-Morales, Y. (2007). *Polycycl. Aromat. Comp.* 27: 41–49.

**69** Hirsch, A., Chen, Z., and Jiao, H. (2000). *Angew. Chem. Int. Ed.* 39: 3915–3917.

**70** Poater, J. and Solà, M. (2011). *Chem. Commun.* 47: 11647–11649.

**71** Lipscomb, W.N., Pitochelli, A.R., and Hawthorne, M.F. (1959). *J. Am. Chem. Soc.* 81: 5833–5834.

**72** Knoth, W.H., Miller, H.C., England, D.C. et al. (1962). *J. Am. Chem. Soc.* 84: 1056–1057.

**73** Wade, K. (1971). *J. Chem. Soc. D: Chem. Commun.* 792–793.

**74** Mingos, D.M.P. (1972). *Nature Phys. Sci.* 236: 99–102.

**75** Liu, C., Popov, I.A., Chen, Z. et al. (2018). *Chem. Eur. J.* 24: 14583–14597.

**76** Jemmis, E.D., Balakrishnarajan, M.M., and Pancharatna, P.D. (2002). *Chem. Rev.* 102: 93–144.

**77** Friedman, L.B., Dobrott, R.D., and Lipscomb, W.N. (1963). *J. Am. Chem. Soc.* 85: 3505–3506.

**78** Priyakumari, C.P. and Jemmis, E.D. (2014). Electron-counting rules in cluster bonding – polyhedral boranes, elemental boron, and boron-rich solids. In: *The Chemical Bond: Chemical Bonding Across the Periodic Table* (ed. G. Frenking and S. Shaik), 113–147. Weinheim: Wiley-VCH.

**79** Poater, J., Solà, M., Viñas, C., and Teixidor, F. (2014). *Angew. Chem. Int. Ed.* 53: 12191–12195.

**80** Olson, J.K. and Boldyrev, A.I. (2012). *Chem. Phys. Lett.* 523: 83–86.

**81** Klanberg, F., Eaton, D.R., Guggenberger, L.J., and Muetterties, E.L. (1967). *Inorg. Chem.* 6: 1271–1281.

**82** Cohen, M.L., Chou, M.Y., Knight, W.D., and De Heer, W.A. (1987). *J. Phys. Chem.* 91: 3141–3149.

**83** Poater, J. and Solà, M. (2019). *Chem. Commun.* 55: 5559–5562.

**84** Duong, L.V., Pham, H.T., Tam, N.M., and Nguyen, M.T. (2014). *Phys. Chem. Chem. Phys.* 16: 19470–19478.

**85** Tai, T.B., Havenith, R.W.A., Teunissen, J.L. et al. (2013). *Inorg. Chem.* 52: 10595–10600.

**86** Cui, P., Hu, H.-S., Zhao, B. et al. (2015). *Nature Commun.* 6: 6331.

**87** Glukhovtsev, M.N., Schleyer, P.v.R., and Stein, A. (1993). *J. Phys. Chem.* 97: 5541–5546.

# 3

# Aromaticity from Organic to Inorganic Compounds

*"Human science fragments everything in order to understand it, kills everything in order to examine it."*

Leo Tolstoy

## 3.1 Introduction

Initially the concept of aromaticity was introduced in organic chemistry. Thus, after Michael Faraday reported the isolation of benzene by distillation in 1825 [1] and noted that it was much less reactive than other unsaturated hydrocarbons, Kekulé introduced the term *aromatic* for a general classification of benzene derivatives [2]. Kekulé also observed a characteristic odor or fragrance of these substances and related them to relatively low chemical reactivity. Subsequent research in the area of aromaticity revealed that the low reactivity of "aromatic" compounds with unsaturated carbon–carbon bonds is not associated with the aroma. Instead, the exceptional stability and low reactivity was recognized to originate from peculiarities of the chemical bonding and electronic structure. Hückel [3] demonstrated that a closed-shell monocyclic system should have $4n + 2$ valence π-electrons in order to be aromatic. The concept of antiaromaticity, relating to the corresponding destabilization seen in cyclic systems with $4n$ π-electrons, was introduced by Breslow [4, 5]. Dewar introduced σ-aromaticity in order to explain the conjugation pattern in cyclopropane [6]. However, Schleyer and coworkers proved that cyclopropene is not a σ-aromatic molecule [7]. The first doubly aromatic system, the 3,5-dehydrophenyl cation, was identified by Chandrasekhar, Jemmis, and Schleyer as being doubly (σ- and π-) aromatic [8]. All these are milestone works in advancing aromaticity, antiaromaticity, and double aromaticity concepts in organic chemistry. The concept of aromaticity

*Aromaticity and Antiaromaticity: Concepts and Applications*, First Edition.
Miquel Solà, Alexander I. Boldyrev, Michał K. Cyrański, Tadeusz M. Krygowski, and Gabriel Merino.

has been extended into inorganic chemistry. It turns out that aromaticity in inorganic chemistry is more complex than in organic chemistry. Participation of *d*-atomic orbitals (AOs) and *f*-AOs in chemical bonding metal systems allows to introduce δ-aromaticity/antiaromaticity and ϕ-aromaticity/antiaromaticity [9] (see chapter 12 of Ref. [10] for further details). These new types of aromaticity lead to complicated combination of aromaticities and antiaromaticities. Finally, a term conflicting aromaticity introduced by Boldyrev and Wang [8] deals with the simultaneous presence of one of more types of aromaticity and one or more types of antiaromaticity simultaneously. There are a few reviews [9–23] on aromaticity in inorganic chemistry, where all these complicated issues have been discussed with the many examples of inorganic aromatic/antiaromatic molecules. In this chapter we discuss a set of typical aromatic inorganic compounds in order to make this discussion manageable.

## 3.2 π-Aromatic Inorganic Species

While aromaticity in inorganic molecules is a rather new phenomenon, one group of inorganic compounds containing tetraatomic aromatic building blocks composed out of group 15 and 16 elements: $M_4^{2-}$ (M = P, As, Sb, and Bi) and $M_4^{2+}$ (M = S, Se, and Te) has been known since the 1970s. Cisar and Corbett reported the synthesis of a remarkable compound, namely, 2,2,2-crypt-potassium tetra-bismuthide (2-), $(C_{18}H_{36}N_2O_6K^+)_2Bi_4^{2-}$, containing an aromatic $Bi_4^{2-}$ unit [24], which is valence isoelectronic to the classical aromatic $C_4H_4^{2-}$ dianion with 6 π-electrons [25]. Seven years later, Critchlow and Corbett [26] synthesized another potassium-crypt salt of tetraantimonide (2-), namely, $(2,2,2\text{-crypt-K+})_2Sb_4^{2-}$ containing the aromatic $Sb_4^{2-}$ dianion [26]. Korber and coworkers [27] later synthesized the $Cs_2P_4{\cdot}2NH_3$ compound, which contains another representative of the aromatic tetraatomic dianion $P_4^{2-}$. All these $M_4^{2-}$ (M = P, As, Sb, and Bi) dianions have a planar square structure in the binary $NaM_4^-$ pyramidal gas phase compounds as it was shown by joint photoelectron and theoretical studies [28]. While isolated $M_4^{2-}$ dianions are not electronically stable, they are stabilized in the binary $NaM_4^-$ compounds due to the presence of the $Na^+$ cation. This research also showed that the planar square structure is a global minimum for $M_4^{2-}$ dianions since photoelectron spectroscopy usually produces global minimum structures. Chemical bonding analysis of the π-bonding in the prototypical dianion $P_4^{2-}$ recovered by the Adaptive Natural Density Partitioning (AdNDP) method [21] is shown in Figure 3.1.

The AdNDP method leads to the partitioning of the charge density into the components with the highest possible degree of localization of electron pairs, started with lone pairs (LPs) and up to nc-2e bonds (see detailed discussion

**Figure 3.1** 4c-2e π-bonds recovered by the AdNDP analysis in $P_4^{2-}$.

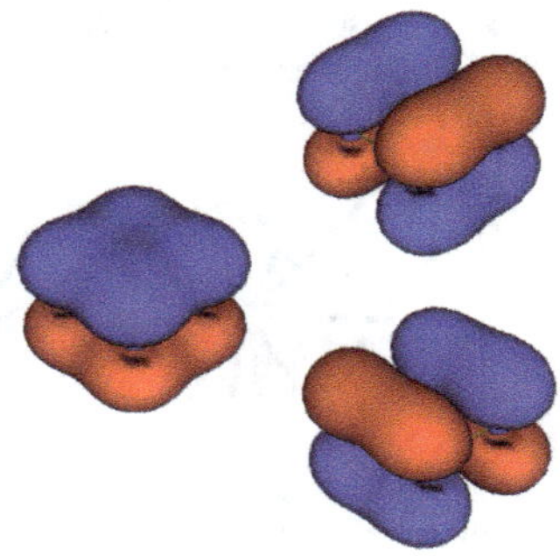

of the AdNDP method in Ref. [21]). In the $P_4^{2-}$ dianion, the AdNDP method recovers four lone pairs (one lone pair on each P atom), four two center – two electron (2c-2e) σ-bonds, and three completely delocalized π-bonds, which renders π-aromaticity in this species.

Gillespie and coworkers reported in 1968 the isoelectronic group 16 chalcogen $Se_4^{2+}$ and $Te_4^{2+}$ dications in solution [29, 30]. Three years later, Gillespie and coworkers also reported $S_4(SO_3F)_2$ containing the $S_4^{2+}$ dication [31]. In the same year, Corbett and coworkers [32] reported synthesis and crystal structure of $Te_4^{2+}(AlCl_4^-)_2$ and showed that the $Te_4^{2+}$ unit has a nearly perfect square structure. Theoretical analyses [33, 34] of the chemical bonding in $S_4^{2+}$, $Se_4^{2+}$, and $Te_4^{2+}$ have confirmed that all of these species are aromatic with 6 π electrons. Theoretical studies on the lightest member of this family [35], namely, the $O_4^{2+}$ dication, have shown it to be the first all-oxygen aromatic species.

The first set inorganic antiaromatic compounds $[K([2.2.2]crypt)]_3[Ln(\eta^4\text{-}Sb_4)_3]\cdot$ 4py (py = pyridine, Ln = La, 1; Y, 2; Ho, 3; Er, 4; Lu, 5) were reported by Sun and coworkers [36]. The $[Ln(\eta^4\text{-}Sb_4)_3]^{3-}$ anion feature three all-metal π antiaromatic $Sb_4$ rhombus units as coordination ligands and have quasi-$D_{3h}$ symmetry (Figure 3.2).

A simple electron count reveals that the Ln atom possesses a positive charge of 3+ in the ionic limit in all the $[Ln(\eta^4\text{-}Sb_4)_3]^{3-}$ complexes, which would lead to a negative charge of 2- on each $Sb_4$ block, as also confirmed by an effective oxidation state analysis. It was previously discussed that an isolated $Sb_4^{2-}$ species had 6 π electrons and it was aromatic with a square-planar structure. However, the bonding situation in the $Sb_4$ units in the $[Ln(\eta^4\text{-}Sb_4)_3]^{3-}$ cluster is different from that in an isolated $Sb_4^{2-}$ dianion due to the interactions between the $Sb_4$ units and the central Ln atom. To be more precise, there exists appreciable equatorial Sb-Sb bonding between the neighboring $Sb_4$ units, as well as between the Sb units and the central Ln atom, via the 3c-2e σ bonds in $[Ln(\eta^4\text{-}Sb_4)_3]^{3-}$ (Figure 3.3c).

In fact, approximately two electrons on each $Sb_4$ unit participate in the formation of the 3c-2e intracluster equatorial bonds, thus leaving only four electrons for the π framework on the $Sb_4$ unit. The rhombus distortion of the $Sb_4$ units can

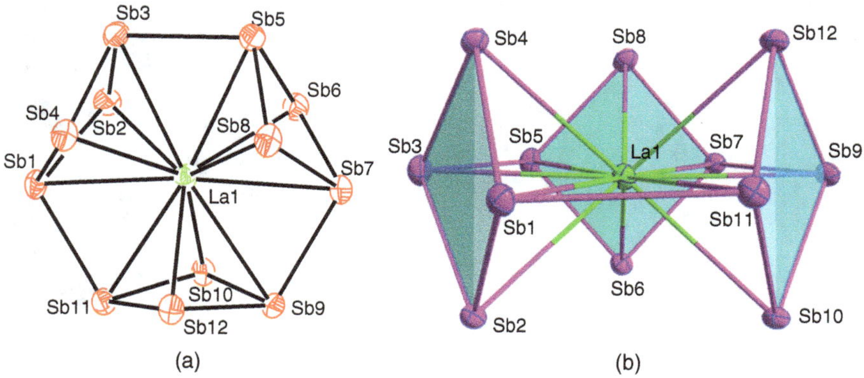


**Figure 3.2** The structure of $[La(\eta_4\text{-}Sb_4)_3]^{3-}$ (1) (thermal ellipsoid plot was drawn at 50 percent probability). (a) Top view from central projection. (b) Side view.

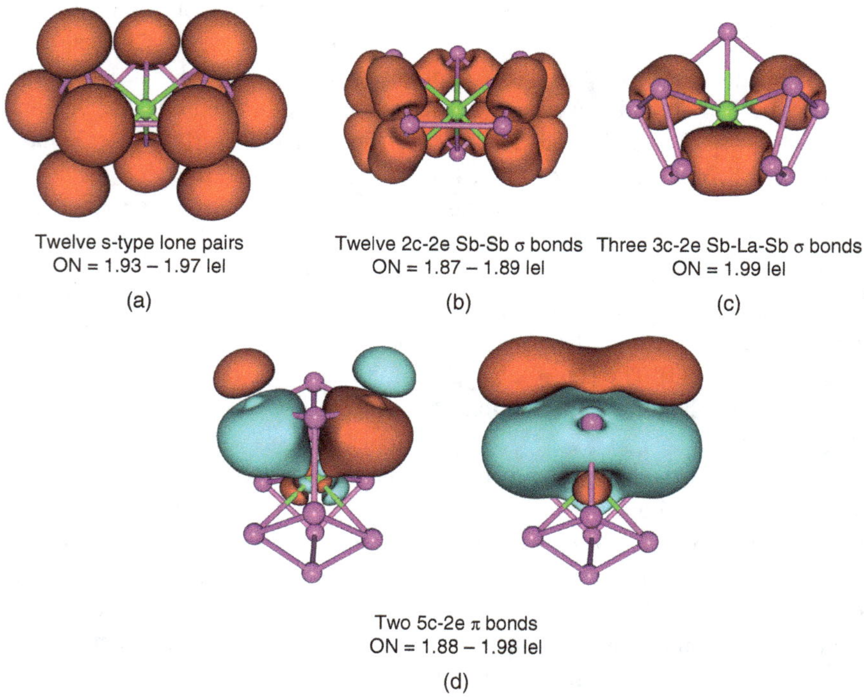


**Figure 3.3** Chemical bonding analysis obtained via the AdNDP method. (a) The 12 *s*-type lone-pairs (1c-2e bonds) on12 Sb atoms (one per Sb atom), shown superimposed onto the molecular framework. (b) The 12 2c-2e Sb-Sb σ bonds, shown superimposed onto the molecular framework. (c) The three 3c-2e σ bonds, shown superimposed onto the molecular framework. (d) Two 5c-2e π bonds in one *cyclo*-$Sb_4$ unit. There are a total of six 5c-2e π bonds in the whole $[La(\eta^4\text{-}Sb_4)_3]^{3-}$ cluster.

be considered as a direct consequence of its antiaromaticity. We think the local π antiaromaticity of the $Sb_4$ units is also consistent with the very high air-sensitivity and reactivity of all the $[K([2.2.2]crypt)]_3[Ln(\eta^4\text{-}Sb_4)_3]\cdot 4py$ compounds. In addition to that, the multicenter indices calculations reveal strong resemblance with cyclobutadiene, thus confirming antiaromatic character of the $Sb_4$ fragment.

Crystalline inorganic compounds containing pentaatomic aromatic building blocks composed of group 14 elements, namely, $M_5^{6-}$ (M = Si, Ge, Sn, and Pb), and group 15 elements, namely, $M_5^-$ (M = N, P, As, Sb, and Bi), which are valence isoelectronic to the prototypical $C_5H_5^-$ cyclopentadienyl anion are also well known. Schnering, and coworkers in 1986 reported synthesis and structure of a remarkable compound, $Li_8MgSi_6$ [37]. Instead of the hexaatomic $Si_6$ building blocks as the molecular formula might suggest, detailed analyses revealed that this compound has a truly unexpected structure with columns of pentagonal aromatic rings $Si_5^{6-}$ stacked exactly on top of each other. The lithium cations are found exactly halfway between the ring planes in a ferrocene-like geometry. The isolated $Si^{4-}$ monoatomic anions are located within the plane of the pentagonal rings. Thus, instead of using all six Si atoms to form cluster building blocks, this compound contains aromatic pentagonal $Si_5^{6-}$ building blocks and individual Si atoms.

Todorov and Sevov [38] recently reported the syntheses and structural characterization of three additional compounds, namely, $Na_8BaPb_6$, $Na_8BaSn_6$, and $Na_8EuSn_6$, which are similar to the previously discussed $Li_8MgSi_6$. These three compounds are constructed with columns of pentagonal aromatic $Sn_5^{6-}$ or $Pb_5^{6-}$ rings stacked exactly on top of each other. These compounds are unequivocal proof of the importance of aromaticity in the formation of the $Si_5^{6-}$, $Sn_5^{6-}$, or $Pb_5^{6-}$ building blocks and suggest that many more compounds with similar inorganic aromatic building blocks are likely to be synthesized in the future. In order to rationalize chemical bonding in those compounds we performed the Solid State Adaptive Natural Density Partitioning (SSAdNDP) analysis of the $Na_8BaSn_6$ solid-state compound. The SSAdNDP method [39] is an extension of the AdNDP method to the periodically boundary conditions and thus it is well suitable for deciphering chemical bonding in solid-state materials. Results of the SSAdNDP analysis are shown in Figure 3.4.

The general SSAdNDP search reveals one lone pair on each of the Sn atoms, and a 2c-2e σ-bond between every pair of Sn atoms around the pentagonal rings (Figure 3.4b). Three 5c-2e π bonds on each of the $Sn_5$ moieties highlight each ring's aromaticity. The remaining six electron pairs could not be localized on the two nonring Sn atoms, which would satisfy the $Sn^{4-}$ octet configuration of a purely ionic system. However, only the single *s*-type lone pair was found on each of the

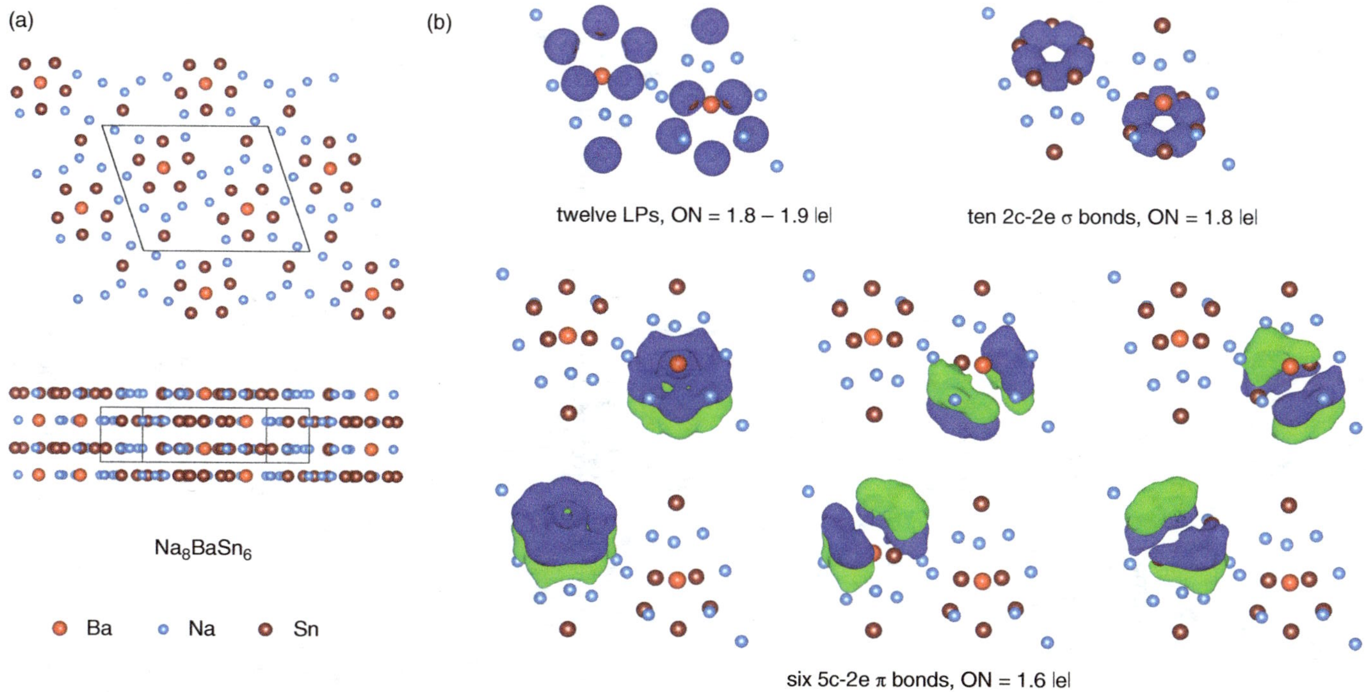


**Figure 3.4** (a) Structure, (b) SSAdNDP chemical bonding pattern the $Na_8BaSn_6$ Zintl phase. The unit cell is shown in black.

two atoms; no *p*-type lone pairs could be localized. We believe this is due to the metallic nature of these remaining electrons, making them intrinsically nonlocalizable. This interpretation is reinforced by lack of a +1 charge on each Na atom (the natural atomic orbital [NAO] population analysis shows +0.6–0.8 e charges on Na atoms), which would be required for Sn to assume a − 4 state.

The valence isoelectronic aromatic pnictogen $M_5^-$ (M = N, P, As, Sb, and Bi) anions have been extensively studied theoretically and experimentally [40–55]. The cyclic aromatic $P_5^-$ anion has been found in alkali metal salts [44, 45] and incorporated in several mixed sandwich complexes such as $[(\eta^5\text{-}C_5Me_5)Fe(\eta^5\text{-}P_5)]$ and $[(\eta^5\text{-}C_5Me_5)Cr_2(\mu,\eta^5\text{-}P_5)]$ [53–55]. A carbon-free sandwich complex $[Ti(\eta^5\text{-}P_5)_2]^-$ has been prepared and characterized by Brennessel and coworkers [47]. Scherer synthesized a sandwich compound $[CpFe(As_5)Fe(C_5Me_5)]PF_6$ containing a planar aromatic $As_5^-$ anion [51, 52]. The $M_5^-$ (M = N, P, As, Sb, and Bi) anions have been prepared in molecular beams and their photoelectron spectra have been recorded [40–43, 46]. Their perfect pentagon structures have been confirmed through excellent agreement between theoretical and experimental vertical detachment energies (VDEs) [42]. Aromaticity in these anions was established through molecular orbital (MO) analysis [40–50], nucleus independent chemical shifts (NICS) and ring current calculations [50]. The lightest member of group 15 pentagonal aromatic clusters is $N_5^-$, which was identified mass spectrometrically by Christe and coworkers [56].

Wang et al. [57] reported syntheses of $La_{10}Si_8O_3$ and $Ce_{10}Si_8O_3$ featuring the concurrence of strong Ln—Ln bonds and a planar $Si_6$ Zintl anion analogous to benzene reported. The formal electron count on the $Si_6$ unit in these substances is $[Si_6]^{6-}$, which indeed makes it analogous to benzene. We performed model computations for the isolated $Si_6^{6-}$ species in order to characterize chemical bonding in this structural unit. The results of the AdNDP analysis are shown in Figure 3.5.

The bonding in the $Si_6^{6-}$ unit can be described as a combination of the localized (six lone pairs and six 2c-2e Si-Si σ-bonds forming a ring) and delocalized (three 6c-2e π-bonds) units. The presence of three 6c-2e π-bonds is a sign of π-aromaticity according to the $4n+2$ Hückel's rule. Again, aromaticity proves to be an important factor for the explanation of the stability of $Si_6^{6-}$ as a building block in these solid-state structures. Hexapnictogen $M_6$ (M = N, P, As, Sb, and Bi) molecules being a valence isoelectronic analogue of benzene is anticipated to have a planar aromatic structure. Indeed, in 1985 Scherer et al. [58] synthesized and characterized the $\{(\eta_5\text{-}Me_5C_5)Mo\}_2(\mu,\eta^6\text{-}P_6)$ triple-decker sandwich complex containing a planar $P_6$ ring with almost equal P—P bond lengths, which is considered to be an all-phosphorus analogue of benzene. However, theoretical calculations revealed [59–69] that there are at least seven nonplanar $P_6$ isomers lower in energy than the planar benzene-like $D_{6h}$ structure. Moreover,

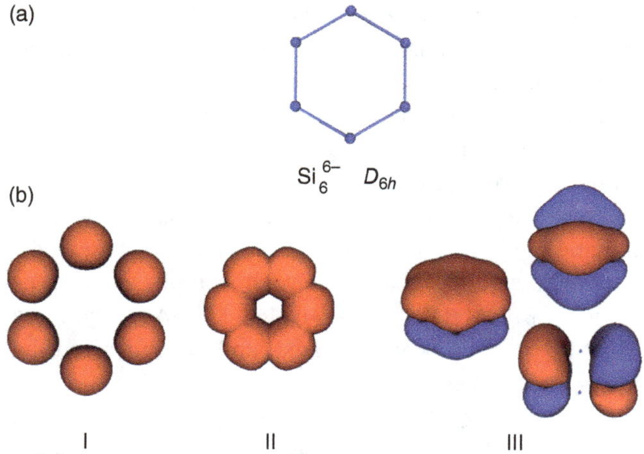


**Figure 3.5** (a) structure of $Si_6^{6-}$ ($D_{6h}$, $^1A_{1g}$); (b) results of the AdNDP localization: I – six lone pairs (1c-2e ON = 1.81 |e|) superimposed on the single molecular frame, II – six 2c-2e σ-bonds (ON = 1.96 |e|) superimposed on the single molecular frame, III – three 6c-2e π-bonds (ON = 2.00 |e|).

it was shown that the $D_{6h}$ benzene-like structure is a second order saddle point. Geometry optimization along the imaginary modes leads to the $D_2$ distorted structure, which is 2.7 kcal/mol lower in energy. Galeev and Boldyrev explained [69] that the out-of-plane distortion in the $D_{6h}$ $P_6$ is due to a second-order Jahn–Teller (SOJT) effect resulting from vibronic coupling of the MO before the highest occupied MO (HOMO-1, $e_{2g}$) and the lowest-unoccupied MO (LUMO, $e_{2u}$). Analysis of molecular orbitals involved in vibrational coupling demonstrated that filling the unoccupied molecular orbitals involved in vibronic coupling with electron pairs of Mo atoms suppresses the pseudo-Jahn-Teller (PJT) effect in the $CpMoP_6MoCp$ sandwich, with the $P_6$ ring becoming planar ($D_{6h}$) upon complex formation [70]. The planar regular hexagon $As_6$ is the central deck of the $[\{(\eta^5\text{-}EtC_5H_4)Mo\}_2(\mu,\eta^6\text{-}As_6)]$ triple-decker sandwich with bond lengths between those for As-As single and double bonds [71].

As one can see from this short overview, inorganic analogs of prototypical aromatic $C_4H_4^{2-}$, $C_5H_5^-$, and $C_6H_6$ aromatic molecules are well documented in chemistry. However, inorganic chemistry provides many more opportunities for aromaticity and aromaticity.

## 3.3 Aromaticity in Main Group Metal Compounds

Metal bonding in chemistry assumes complete electron delocalization in metals. One would think that metal compounds would be considered for presence of aromaticity. However, only recently aromaticity was extended to initially main group metal clusters and more recently to transition metal clusters. Though, as we discussed in Section 3.2, metal clusters such as $Sb_4^{2-}$ and $Bi_4^{2-}$ were known in chemistry for four decades; aromaticity in such compounds have not been discussed until recently.

Robinson and coworkers [72] reported in 1995 the synthesis of the first organometallic compound $Na_2[(Mes_2C_6H_3)Ga]_3$ (Mes = 2,4,6-$Me_3C_6H_2$) containing an aromatic triangular ring composed out of three Ga atoms embedded in a large organometallic compound. The authors [72] concluded that their compound is π-aromatic system on the basis of the electron count and the 60° Ga-Ga-Ga angle. In subsequent works, Robinson, Schaefer, and coworkers [73–76] showed computationally using model compounds that this compound indeed contains two π-electrons and thus fulfills the $4n + 2$ requirement for π-aromatic molecules. Recently, Schleyer and coworkers [77] confirmed aromaticity in this compound by high negative calculated NICS values. Power and coworkers [78] recently reported synthesis of the first $Na_2[(AlAr'')_3]$ (Ar'' = $C_6H_3$–2,6-($C_6H_2$–2,4,6-$Me_3)_2$) compound containing an aromatic triangular ring composed out of three Al atoms.

In 2000, Twamley and Power [79] reported the synthesis of the compound $K_2[Ga_4(C_6H_3$–2,6-$Trip_2)_2]$ (Trip = $C_6H_2$–2,4,6-$iPr_3$) containing an almost perfect square made of four Ga atoms. The follow-up chemical bonding analysis [80, 81] revealed the presence of two completely delocalized π-electrons, indicating a π-aromatic system.

In 2001, Li et al. [82] showed that the experimentally observed bimetallic clusters $LiAl_4^-$, $NaAl_4^-$, and $CuAl_4^-$ in their most stable isomeric forms contain a square-planar $Al_4^{2-}$ dianion. In order to rationalize chemical bonding in this cluster, the authors looked at chemical bonding in the $Al_4^{2-}$ dianion, assuming that bonding between counter-cations ($Li^+$, $Na^+$, and $Cu^+$) and $Al_4^{2-}$ is primarily ionic. The AdNDP analysis of the chemical bonding in $Al_4^{2-}$ is presented in Figure 3.6.

The bonding is clearly a combination of a set of four lone pairs at Al atoms and three completely delocalized bonds. The delocalized bonding framework of $Al_4^{2-}$ includes one 4c-2e σ-radial bond, one 4c-2e σ-tangential bond, and one 4c-2e π-bond. Application of the $4n + 2$ electron Hückel's rule separately to the π-bonding set and the $4n + 4$ rule the σ-bonding set shows that the system is doubly (σ- and π) aromatic (see Chapter 11 for more details on this counting rule). The double aromaticity (simultaneous presence of σ- and π-aromaticity) of the $Al_4^{2-}$ dianion has been established using a variety of theoretical techniques, namely, ring current maps [83, 84], aromatic ring-current shielding (ARCS) [85, 86], NICS [87], valence bond (VB) assessment of σ- and π-aromaticity [88], bifurcation analysis of the electron localization function (ELF) [89], and resonance energy (RE) estimations [90, 91]. Two years later, the first antiaromatic metal cluster

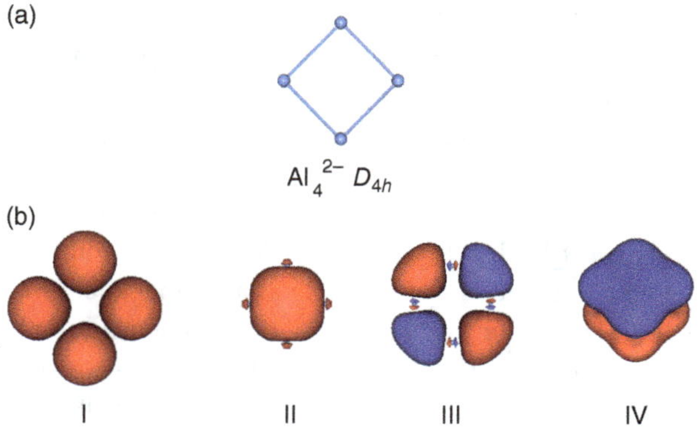


**Figure 3.6** (a) structure of $Al_4^{2-}$ ($D_{4h}$, $^1A_{1g}$); (b) results of the AdNDP localization: I – four lone pairs (1c-2e ON = 1.67 |e|) superimposed on the single molecular frame, II – 4c-2e σ-radial bond (ON = 2.00 |e|), III – 4c-2e σ-tangential bond (ON = 2.00 |e|), IV – 4c-2e π-bond (ON = 2.00 |e|).

$Li_3Al_4^-$ was discovered in a molecular beam by Wang and coworkers [92] (see more discussion of this cluster in Chapter 11). Clusters $LiAl_4^-$ and $Li_3Al_4^-$ helped to advance the concepts of aromaticity and antiaromaticity into all-metal systems (clusters composed out of metal atoms only).

## 3.4 Aromaticity in Transition Metal Compounds

Aromaticity and antiaromaticity concepts were further extended to transition metal compounds. In 2001, Boldyrev and coworkers [93] suggested that the remarkable stability of the $Hg_4^{6-}$ building block in the $Na_3Hg_2$ sodium-mercury amalgam is related to its multiple aromaticity. The square-planar $Hg_4^{6-}$ is valence isoelectronic to the all-metal aromatic cluster $Al_4^{2-}$, which was shown to be a doubly (σ- and π-) aromatic system. The AdNDP analysis of the isolated $Hg_4^{6-}$ anion and this anion embedded into a crystalline lattice of the $Na_3Hg_2$ amalgam is shown in Figure 3.7.

The AdNDP analysis of the isolated $Hg_4^{6-}$ cluster revealed that neither 5*d*- nor 6*s*-AOs participate in chemical bonding. The 6*p*-AO-based MOs are responsible for bonding in this cluster. There are four *s*-AO type lone pairs in the cluster (one on each atom). There are also three completely bonding orbitals formed by the $\sigma_t$-MO (σ-tangential MO), $\sigma_r$-MO (σ-radial MO) and π-MO, which are responsible for σ- and π-aromaticity in the cluster (Figure 3.7a). The results of the AdNDP analysis for the embedded $Hg_4^{6-}$ clusters are shown in Figure 3.7b. One can see that the isolated and embedded clusters have essentially the same bonding patterns

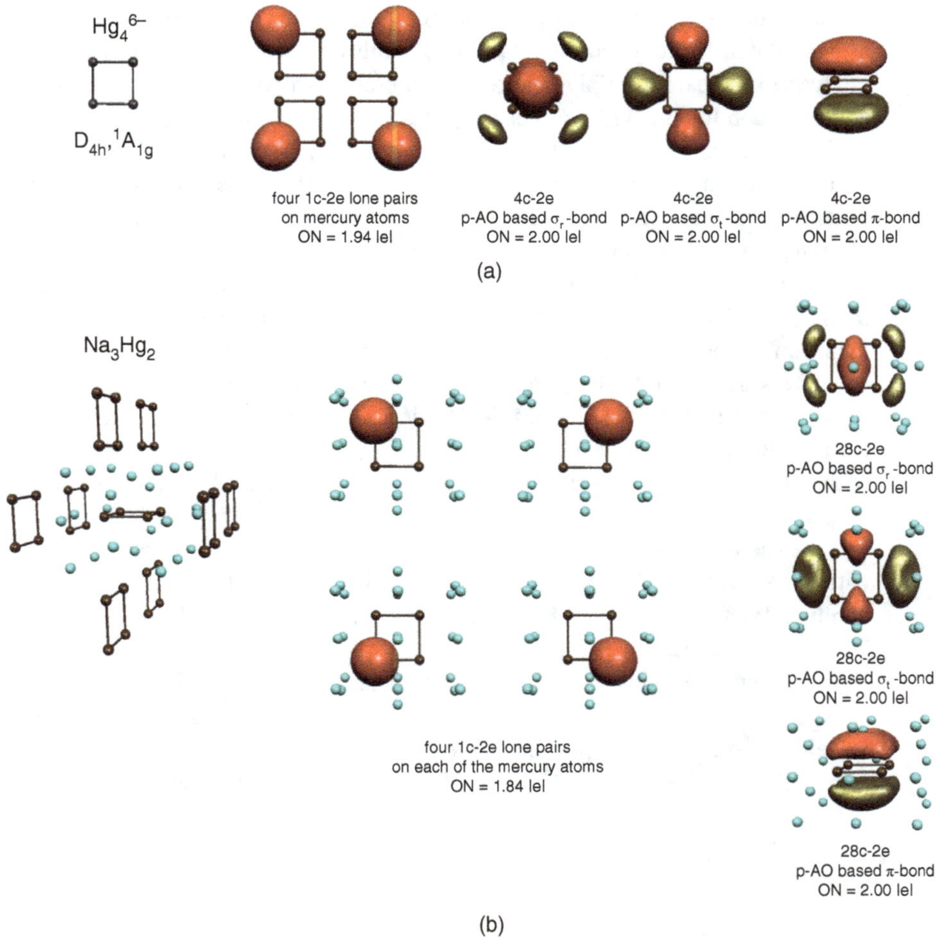


**Figure 3.7** (a) The bonding elements recovered by the AdNDP analysis for the bare $Hg_4^{6-}$ cluster; (b) The bonding elements recovered by the AdNDP analysis for the $Hg_4^{6-}$ cluster embedded in a part of the crystalline lattice of the $Na_3Hg_2$ amalgam comprised of 24 sodium cations placed at their position in the real crystalline structure.

with some distortion imposed by the influence of sodium counterions. The delocalized bonds in the embedded cluster can no longer be presented as 4c-2e bonds. They are 28c-2e bonds, but the nodal structure of those 28c-2e bonds is the same as in 4c-2e bonds of the bare $Hg_4^{6-}$ cluster. The AdNDP method allows restricting the search to 4c-2e bonds. The occupation numbers (ONs) of the recovered 4c-2e $\sigma$- and $\pi$-bonds dropped significantly (2.00 |e| to 1.18 and 0.88 |e| for 4c-2e $\sigma$-delocalized bonds, and to 0.49 |e| for 4c-2e $\pi$-delocalized bond) showing covalent bonding between sodium and mercury atoms. The embedded $Hg_4^{6-}$ cluster is still a doubly σ- (both $\sigma_r$- and $\sigma_t$-) and π-aromatic system. Thus, though, the evaluation of the influence of counter-cations is highly desirable, the chemical bonding analysis of the bare multiply-charged anion provides a good interpretation of the chemical bonding in $Hg_4^{6-}$ as a part of the amalgam.

Since the recognition of the aromaticity in this first transition metal compound, aromaticity was found in transition metal clusters [94–103] and transition metal oxide clusters [104–107]. In more recent years, the number of transition metal compounds containing an aromatic transition metal core is steadily growing [108–114]. The sandwich-like compounds containing aromatic planar square $Pd_4$ core [108–110] are discussed in Chapter 11. Here, we focus on other transition metal aromatic compounds. In 2012, Sadighi and coworkers reported [111] synthesis of the compound containing $[(LAu)_3]^+$ (L = 1,3-bis(2,6-iPr$_2$C$_6$H$_4$)imidazole-2-ylidene) with the $Au_3^+$ σ-aromatic gold core. In 2014, Maestri and coworkers reported [112] synthesis of the compound with the $[C_{72}H_{60}P_3S_3Pd_3]^+$ cation with the $Pd_3^+$ σ-aromatic $Pd_3^+$ core. We further analyzed chemical bonding in this cluster using the AdNDP method. We simplified the calculations of the real complex by substituting bulky groups at P atoms with three H atoms because we believe it will not affect the overall chemical bonding. We found (Figure 3.8) the 3c-2e delocalized σ-bond with somewhat low ON 1.4 |e| instead of 2.0 |e|.

If we extend this bond to 3 S atoms (to form 6c-2e bond) the ON increases up to 1.7 |e| and if we further include P atoms in bonding (to form 9c-2e bond) its ON is 2.0 |e|. However, about 70 percent of the density of this pair of electrons is indeed responsible for direct Pd-Pd 3c-2e bonding, which makes this system aromatic. We also found four lone pairs (*d*-AOs) on every Pd atom (lower line in Figure 3.8).

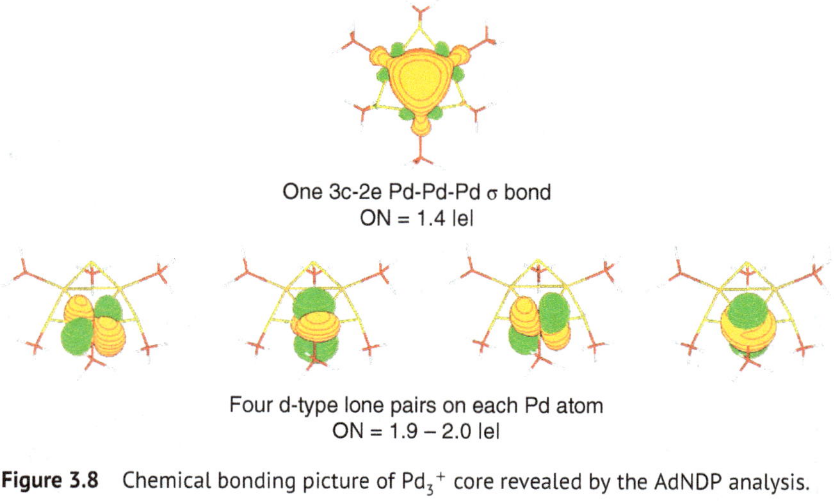


**Figure 3.8** Chemical bonding picture of $Pd_3^+$ core revealed by the AdNDP analysis.

Additionally, we found that every Pd atom forms two classical 2c-2e sigma bonds with two S atoms and one classical sigma Pd-P bond (not shown in the figure).

Though gas-phase clusters are deemed to be σ-aromatic when they satisfy the $4n+2$ rule of aromaticity for delocalized σ electrons with the range of $n$, there were no solid state compounds containing aromatic transition metal cores with $n>0$. Sun and coworkers recently showed [113] that a bottleable compound, which was isolated as a stable $[K([2.2.2]crypt)]^+$ salt, featuring a $[Au_2Sb_{16}]^{4-}$ cluster core possessing two all-metal aromatic $AuSb_4$ fragments with six delocalized σ electrons each ($n = 1$). The structure of the $[Au_2Sb_{16}]^{4-}$ cluster core is shown in Figure 3.9. Each Au atom in $[Au_2Sb_{16}]^{4-}$ cluster core is coordinated by four Sb atoms, and the quadrangles formed by the coordinating Sb atoms are slightly deviated from the plane. In the $AuSb_4$ units, the Sb–Au–Sb bond angles are in the range of 82.35°–105.69° for Au1, and 82.67°–107.96° for Au2, respectively. The two $AuSb_4$ faces are almost parallel to each other. Hence, the core of the cluster core can be described as a distorted tetragonal prism made up of two $AuSb_4$ species, which are further stabilized by $Sb_4$ fragments on the sides (Figure 3.9). The Au–Sb bond lengths (2.69 ± 0.02 Å) span a very narrow range, thus underlining the importance of Au-Sb interactions within the two $AuSb_4$ units. The chemical bonding analysis performed for the $[Au_2Sb_{16}]^{4-}$ cluster core by the AdNDP method revealed that we have 10 $d$–type electron lone pairs on two Au atoms (five on each) with ONs in the range of 1.83–1.99 |e|; 16$s$–type LPs on 16 Sb atoms with ON = 1.93–1.97 |e|; 2 $p$–type lone pairs on two peripheral Sb atoms (Sb1 and Sb16) with ON = 1.71 |e|; 19 classical two-center two-electron (2c–2e) Sb–Sb σ bonds (ON = 1.93–1.98 |e|) responsible for the direct bonding between Sb atoms constituting the framework of the cluster. Slight involvement of the $d_{x^2-y^2}$ orbital of Au (~9%) in the direct bonding with four Sb atoms at two quasi-planar $AuSb_4$ units enhances flattening of the square-like $Sb_4$ fragments featuring equal Au–Sb distances. Somewhat lower ON values of the two $p$-type LPs on

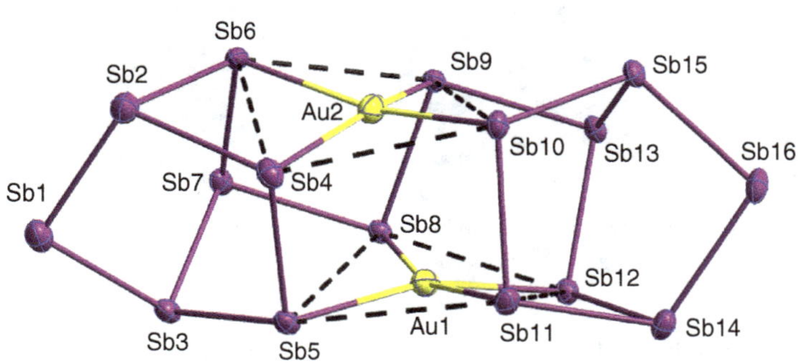


**Figure 3.9** Thermal ellipsoid plot of the molecular cluster anion $[Au_2Sb_{16}]^{4-}$ (drawn at 50 percent probability).

the peripheral Sb1 and Sb16 atoms is a direct sign of π-conjugation with the two neighboring atoms Sb2, Sb3 and Sb14, Sb15, respectively, that perfectly explains the slightly decreased Sb–Sb (2.72–2.75 Å) bond lengths at these fragments. As a matter of fact, these two LPs can be found as two 3c-2e π-bonds with larger ON values (1.85 |e|), thus confirming the π-delocalization over the Sb1-Sb2-Sb3 and Sb14-Sb15-Sb16 fragments (see Ref. [113] for detailed discussion). The remaining 12 electrons were found to participate only in a delocalized bonding: there are three five-center two-electron (5c–2e) σ-bonds with ON = 1.86–1.99 |e| found within each quasi-planar $AuSb_4$ fragment (Figure 3.10).

Noteworthy, contribution of the 6*s* orbital of Au in the formation of the completely bonding orbital is assessed to be 39 percent. Essentially, these bonds look

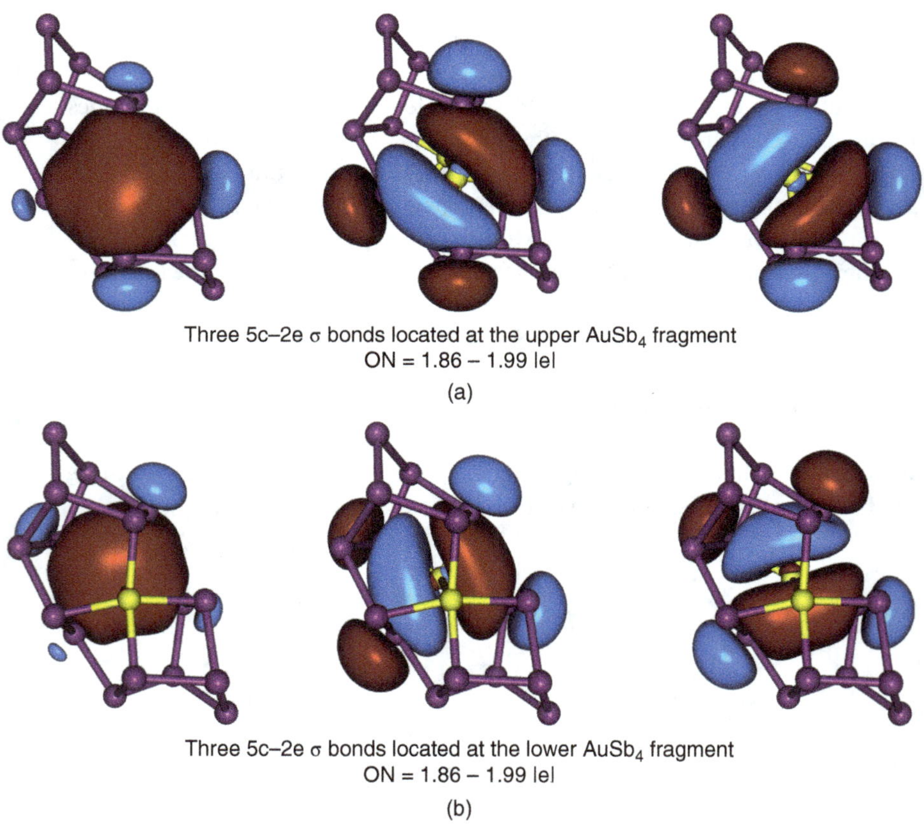


**Figure 3.10** Delocalized chemical bonding elements deciphered for the complex $[Au_2Sb_{16}]^{4-}$. Six 5c-2e delocalized σ bonds with ON = 1.86–1.99 |e| found at: (a) upper $AuSb_4$ (Au1-Sb5-Sb8-Sb11-Sb12) and (b) lower $AuSb_4$ (Au2-Sb4-Sb6-Sb9-Sb10) fragments.

like classical σ-aromatic bonds: one completely bonding and two bonds possessing one nodal plane (mutually perpendicular) with $n = 1$ according to $4n+2$ rule. Thus, it was demonstrated that the $[Au_2Sb_{16}]^{4-}$ anion represents the first all-metal σ-aromatic cluster in a solid state that possesses two peculiar σ-aromatic $AuSb_4$ fragments featuring six delocalized σ electrons each, which separately satisfy the Hückel's rule of aromaticity. That pushes the boundaries of the original idea of August Kekulé and proves the beauty and power of the σ-aromaticity concept in its generality: from short-lived gas-phase or matrix-isolated clusters to bottleable solid-state compounds. This discovery is especially important since the sextet of delocalized electrons was initially set at the heart of the aromaticity concept.

So far, aromaticity in transition metal compounds had been a curiosity, but Maestri, Malacria and coworkers later on reported [114] that the aromatic $[Pd_3]^+$ clusters can catalyze the selective semi-reduction of internal alkynes to *cis*-alkenes. Studies on factors governing the formation of all-metal aromatics enabled the design of an optimized catalytic system that delivers *cis*-alkenes with almost complete selectivity on a gram scale with very low catalyst loadings (0.03 mol%). We believe that aromatic transition metal compounds have a great future in chemistry.

## 3.5 Conclusions

We have shown that the concepts of aromaticity and antiaromaticity can be valuable tools for explaining structure, stability, chemical bonding, and other molecular properties of inorganic compounds and their building blocks. Extending the concepts of aromaticity and antiaromaticity to include metal systems provides a new characteristic feature of their chemical bonding. In particular, the building blocks based on metal clusters frequently possess multiple aromaticity, multiple antiaromaticity, and conflicting aromaticity. We still believe that the major advances of these concepts in inorganic chemistry and material science are lying ahead of us.

## References


1 Faraday, M. (1825). *Philos. Trans. R. Soc. London* 115: 440–446.
2 Kekulé, A. (1865). *Bull. Soc. Chim. Fr. (Paris)* 3: 98–110.
3 Hückel, E. (1931). *Z. Phys.* 70: 104–186.
4 Breslow, R. (1965). *Chem. Eng. News* 43: 90–100.
5 Breslow, R. (1973). *Acc. Chem. Res.* 6: 393–398.
6 Dewar, M.J.S. (1979). *Bull. Soc. Chim. Belg.* 88: 957–967.

7 Wu, W., Ma, B., Wu, J.I. et al. (2009). *Chem. Eur. J.* 15: 9730–9736.

8 Chandrasekhar, J., Jemmis, E.D., and Schleyer, P.v.R. (1979). *Tetrahedron Lett.* 3707–3710.

9 Boldyrev, A.I. and Wang, L.S. (2005). *Chem. Rev.* 105: 3716–3757.

10 Sergeeva, A.P., Averkiev, B.B., and Boldyrev, A.I. (2010). All-transition metal aromaticity and antiaromaticity. In: *Metal-Metal Bonding. Structure and Bonding* (ed. G. Parkin), 275–306. Berlin/Heidelberg: Springer.

11 Alexandrova, A.N., Boldyrev, A.I., Zhai, H.-J., and Wang, L.S. (2006). *Coord. Chem. Rev.* 250: 2811–2866.

12 Tsipis, C.A. (2005). *Coord. Chem. Rev.* 249: 2740–2762.

13 Tsipis, C.A. (2010). *Aromaticity/Antiaromaticity in "Bare" and "Ligand-Stabilized" Rings of Metal Atoms*, Metal-Metal Bonding. Structure and Bonding, vol. 136 (ed. G. Parkin), 217–274. Berlin/Heidelberg: Springer.

14 Zubarev, D.Y., Sergeeva, A.P., and Boldyrev, A.I. (2009). *Chemical Reactivity Theory. A Density Functional View* (ed. P.K. Chattaraj), 439–452. New York: CRC Press. Taylor & Francis Group.

15 Zubarev, D.Y. and Boldyrev, A.I. (2009). *Computational Inorganic and Bioinorganic Chemistry* (ed. E.I. Solomon, R.A. Scott and R.B. King), 551–562. Chistester, UK: John Wiley & Sons, Ltd.

16 Sergeeva, A.P. and Boldyrev, A.I. (2010). *Aromaticity and Metal Clusters* (ed. P.K. Chattaraj), 55–68. Boca Raton: CRC Press.

17 Galeev, T.R. and Boldyrev, A.I. (2011). *Annu. Rep. Prog. Chem., Sect. C.* 107: 124–147.

18 Galeev, T.R. and Boldyrev, A.I. (2013). *Reference Module in Chemistry, Molecular Sciences and Chemical Engineering, from Comprehensive Inorganic Chemistry II*, 2e (ed. J. Reedijk and K. Poeppelmeier), 245–275. Elsevier, Amsterdam.

19 Romanescu, C., Galeev, T.R., Li, W.-L. et al. (2013). *Acc. Chem. Res.* 46: 350–358.

20 Popov, I.A. and Boldyrev, A.I. (2014). *The Chemical Bond. Chemical Bonding Across the Periodic Table* (ed. G. Frenking and S. Shaik), 421–444. Weinheim: Wiley-VCH.

21 Sergeeva, A.P., Popov, I.A., Piazza, Z.A. et al. (2014). *Acc. Chem. Res.* 47: 1349–1358.

22 Mercero, M., Boldyrev, A.I., Merino, G., and Ugalde, J.M. (2015). *Chem. Soc. Rev.* 44: 6519–6534.

23 Feixas, F., Matito, E., Poater, J., and Solà, M. (2013). *WIREs Comput. Mol. Sci.* 3: 105–122.

24 Cisar, A. and Corbett, J.D. (1977). *Inorg. Chem.* 16: 2482–2487.

25 Sekiguchi, A., Matsuo, T., and Watanabe, H. (2000). *J. Am. Chem. Soc.* 122: 5652–5563.

**26** Critchlow, S.C. and Corbett, J.D. (1984). *Inorg. Chem.* 23: 770–774.

**27** Kraus, F., Aschenbrenner, J.C., and Korber, N. (2003). *Angew. Chem. Int. Ed.* 42: 4030–4033.

**28** Kuznetsov, A.E., Zhai, H.-J., Wang, L.S., and Boldyrev, A.I. (2002). *Inorg. Chem.* 41: 6062–6070.

**29** Gillespie, R.J., Barr, J., Crump, D.B. et al. (1968). *Can. J. Chem.* 46: 3607.

**30** Zubarev, D.Y. and Boldyrev, A.I. (2008). *Phys. Chem. Chem. Phys.* 10: 5207–5217.

**31** Gillespie, R.J., Passmore, J., Ummat, P.K., and Vaidya, O.C. (1971). *Inorg. Chem.* 10: 1327–1332.

**32** Couch, T.W., Lokken, D.A., and Corbett, J.D. (1972). *Inorg. Chem.* 11: 357–361.

**33** Tanaka, K., Yamabe, T., Terama-E, H., and Fukui, K. (1979). *Inorg. Chem.* 18: 3591–3595.

**34** For a review of $M_4^{2+}$ VI chalcogen dications see Passmore, J. (1992; Chapter 19). In The Chemistry of Inorganic Ring Systems. In: *Studies in Inorganic Chemistry*, vol. 14 (ed. R. Steudel), 373–406. Amsterdam: Elsevier.

**35** Elliott, B.M. and Boldyrev, A.I. (2005). *J. Phys. Chem. A* 109: 236–239.

**36** Min, X., Popov, I.A., Pan, F.-X. et al. (2016). *Angew. Chem. Int. Ed.* 55: 5531–5535.

**37** Nesper, R., Curda, J., and Schnering, H.G.v. (1986). *J. Solid State Chem.* 62: 191–199.

**38** Todorov, I. and Sevov, S.C. (2004). *Inorg. Chem.* 43: 6490–6494.

**39** Galeev, T.R., Dunnington, B.D., Schmidt, J.R., and Boldyrev, A.I. (2013). *Phys. Chem. Chem. Phys.* 15: 5022–5029.

**40** Gausa, M., Kascher, R., Lutz, H.O. et al. (1994). *Chem. Phys. Lett.* 230: 99–102.

**41** Gausa, M., Kascher, R., Seifert, G. et al. (1996). *J. Chem. Phys.* 104: 9719–9728.

**42** Zhai, H.J., Wang, L.S., Kuznetsov, A.E., and Boldyrev, A.I. (2002). *J. Phys. Chem. A* 106: 5600–5606.

**43** Lippa, T.P., Xu, S.-J., Lyapustina, S.A. et al. (1998). *J. Chem. Phys.* 109: 10727–10731.

**44** Baudler, M., Akpapoglou, S., Ouzounis, D. et al. (1988). *Angew. Chem. Int. Ed.* 27: 280–281.

**45** Baudler, M., Duster, D., and Ouzounis, D. (1987). *Z. Anorg. Allg. Chem.* 544: 87–94.

**46** Jones, R.O., Ganteför, G., Hunsicker, S., and Pieperhoff, P. (1995). *J. Chem. Phys.* 103: 9549–5962.

**47** Urnezius, E., Brennessel, W.W., Cramer, C.J., and Schleyer, P.v.R. (2002). *Science* 295: 832–834.

**48** Beck, V., Cowley, A.R., and O'Hare, D. (2004). *Organometallics* 23: 4265–4270.
**49** Dransfeld, A., Nyulaszi, L., and Schleyer, P.v.R. (1998). *Inorg. Chem.* 37: 4413–4420.
**50** De Proft, F., Fowler, P.W., Havenith, R.W.A. et al. (2004). *Chem. Eur. J.* 10: 940–950.
**51** Scherer, O.J. (1990). *Angew. Chem. Int. Ed.* 29: 1104–1122.
**52** Scherer, O.J. (1999). *Acc. Chem. Res.* 32: 751–762.
**53** Scherer, O.J. and Bruck, T. (1987). *Angew. Chem. Int. Ed.* 26: 59–59.
**54** Scherer, O.J., Schwalb, J., Wolmershauser, G. et al. (1986). *Angew. Chem. Int. Ed.* 25: 363–364.
**55** Chamizo, J.A., Ruizmazon, M., Salcedo, R., and Toscano, R.A. (1990). *Inorg. Chem.* 29: 879–880.
**56** Vij, A., Pavlovich, J.G., Wilson, W.W. et al. (2002). *Angew. Chem. Int. Ed.* 41: 3051–3054.
**57** Wang, L., Tang, Z., Lorenz, B., and Guloy, A.M. (2008). *J. Am. Chem. Soc.* 130: 11258–11259.
**58** Scherer, O.J., Sitzmann, H., and Wolmershäuser, G. (1985). *Angew. Chem. Int. Ed.* 24: 351–352.
**59** Hiberty, P.C. and Volatron, F. (2007). *Heteroatom Chem.* 18: 129–134.
**60** Nguyen, M.T. and Hegarty, A.F. (1986). *J. Chem. Soc. Chem. Comm.* 383–385.
**61** Jones, R.O. and Hohl, D.J. (1990). *J. Chem. Phys.* 92: 6710–6721.
**62** Jones, R.O. and Seiferd, G. (1992). *J. Chem. Phys.* 96: 7564–7572.
**63** Warren, D.S. and Gimarc, B.M. (1992). *J. Am. Chem. Soc.* 114: 5378–5385.
**64** Kobayashi, K., Miura, H., and Nagase, S. (1994). *J. Mol. Struct.: THEOCHEM* 311: 69–77.
**65** Gimarc, B.M. and Zhao, M. (1997). *Coord. Chem. Rev.* 158: 385–412.
**66** Haeser, M., Schneider, U., and Alhrichs, R. (1992). *J. Am. Chem. Soc.* 114: 9551–9559.
**67** Ballone, P. and Jones, R.O. (1994). *J. Chem. Phys.* 100: 4941–4946.
**68** Häser, M. and Treutler, O. (1995). *J. Chem. Phys.* 102: 3703–3711.
**69** Galeev, T.R. and Boldyrev, A.I. (2011). *Phys. Chem. Chem. Phys.* 13: 20549–20556.
**70** Ivanov, A.S., Bozhenko, K.V., and Boldyrev, A.I. (2012). *Inorg. Chem.* 51: 8868–8872.
**71** Scherrer, O.J., Sitzmann, H., and Wolmershäuser, G. (1989). *Angew. Chem. Int. Ed.* 28: 212–213.
**72** Li, X.W., Pennington, W.T., and Robinson, G.H. (1995). *J. Am. Chem. Soc.* 117: 7578–7579.
**73** Li, X.-W., Xie, Y., Schreiner, P.R. et al. (1996). *Organometallics* 15: 3798–3803.
**74** Xie, Y., Schreiner, P.R., Schaefer, H.F. III, et al. (1996). *J. Am. Chem. Soc.* 118: 10635–10639.

75 Xie, Y., Schreiner, P.R., Schaefer, H.F. III, et al. (1998). *Organometallics* 17: 114–122.

76 Robinson, G.H. (1999). *Acc. Chem. Res.* 32: 773–782.

77 Chen, Z., Wannere, C.S., Corminboeuf, C. et al. (2005). *Chem. Rev.* 105: 3842–3888.

78 Wright, R.J., Brynda, M., and Power, P.P. (2006). *Angew. Chem. Int. Ed.* 45: 5953–5956.

79 Twamley, B. and Power, P.P. (2000). *Angew. Chem., Int. Ed.* 39: 3500–3503.

80 Kuznetsov, A.E., Boldyrev, A.I., Li, X., and Wang, L.-S. (2001). *J. Am. Chem. Soc.* 123: 8825–8831.

81 Phillips, A.D. and Power, P.P. (2002). *J. Cluster Sci.* 13: 569–586.

82 Li, X., Kuznetsov, A.E., Zhang, H.F. et al. (2001). *Science* 291: 859–861.

83 Fowler, P.W., Havenith, R.W.A., and Steiner, E. (2001). *Chem. Phys. Lett.* 342: 85–90.

84 Fowler, P.W., Havenith, R.W.A., and Steiner, E. (2002). *Chem. Phys. Lett.* 359: 530–536.

85 Juselius, J., Straka, M., and Sundholm, D. (2001). *J. Phys. Chem. A* 105: 9939–9944.

86 Lin, Y.-C., Juselius, J., Sundholm, D., and Gauss, J. (2005). *J. Chem. Phys.* 122: 214308.

87 Chen, Z., Cominboeuf, C., Heine, T. et al. (2003). *J. Am. Chem. Soc.* 125: 13930–13931.

88 Havenith, R.W.A. and van Lenthe, J.H. (2004). *Chem. Phys. Lett.* 385: 198–201.

89 Santos, J.C., Tiznado, W., Contreras, R., and Fuentealba, P. (2004). *J. Chem. Phys.* 120: 1670–1673.

90 Zhan, C.-G., Zheng, F., and Dixon, D.A. (2002). *J. Am. Chem. Soc.* 124: 14795–14803.

91 Boldyrev, A.I. and Kuznetsov, A.E. (2002). *Inorg. Chem.* 41: 532–537.

92 Kuznetsov, A.E., Birch, K.A., Boldyrev, A.I. et al. (2003). *Science* 300: 622–625.

93 Kuznetsov, A.E., Corbett, J.D., Wang, L.-S., and Boldyrev, A.I. (2001). *Angew. Chem. Int. Ed.* 40: 3369–3372.

94 Averkiev, B.B. and Boldyrev, A.I. (2007). *J. Phys. Chem. A* 111: 12864–12866.

95 Tsipis, A.C., Kefalidis, C.E., and Tsipis, C.A. (2008). *J. Am. Chem. Soc.* 130: 9144–9155.

96 Chi, X.X. and Liu, Y. (2007). *Int. J. Quant. Chem.* 107: 1886–1896.

97 Wang, B., Zhai, H.-J., Huang, X., and Wang, L.-S. (2008). *J. Phys. Chem. A* 112: 10962–10967.

98 Ivanov, A.S., Zhang, X., Wang, H. et al. (2015). *J. Phys. Chem. A* 119: 11293–11303.

**99** Tanaka, H., Neukemans, S., Janssens, E. et al. (2003). *J. Am. Chem. Soc.* 125: 2862–2863.
**100** Höltzl, T., Janssens, E., Veldeman, N. et al. (2008). *Chem. Phys. Chem.* 9: 833–838.
**101** Höltzl, T., Veldeman, N., Veszpremi, T. et al. (2009). *Chem. Phys. Lett.* 469: 304–307.
**102** Lin, L., Lievens, P., and Nguyen, M.T. (2010). *J. Mol. Struct. THEOCHEM* 943: 23–31.
**103** Lin, L., Höltzl, T., Gruene, P. et al. (2008). *Chem. Phys. Chem.* 9: 2471–2474.
**104** Huang, X., Zhai, H.-J., Kiran, B., and Wang, L.-S. (2005). *Angew. Chem. Int. Ed.* 44: 7251–7254.
**105** Zhai, H.-J., Averkiev, B.B., Zubarev, D.Y. et al. (2007). *Angew. Chem. Int. Ed.* 46: 4277–4280.
**106** Zhai, H.-J., Wang, B., Huang, X., and Wang, L.-S. (2009). *J. Phys. Chem. A* 113: 3866–3875.
**107** Chen, W.J., Zhai, H.J., Zhang, Y.F. et al. (2010). *J. Phys. Chem. A* 114: 5958–5966.
**108** Muñoz-Castro, A., Mac-Leod Carey, D., and Arratia-Perez, R. (2010). *J. Chem. Phys.* 132: 164308.
**109** Sergeeva, A.P. and Boldyrev, A.I. (2010). *Phys. Chem. Chem. Phys.* 12: 12050–12054.
**110** Guo, J.C. and Li, S.-D. (2010). *Eur. J. Inorg. Chem.* 5156–5160.
**111** Robilotto, T.J., Bacsa, J., Gray, T.B., and Sadighi, J.P. (2012). *Angew. Chem. Int. Ed.* 51: 12077–12080.
**112** Blanchard, S., Fensterbank, L., Gontard, G. et al. (2014). *Angew. Chem. Int. Ed.* 53: 1987–1991.
**113** Popov, I.A., Pan, F.-X., You, X.-R. et al. (2016). *Angew. Chem. Int. Ed.* 55: 15344–15346.
**114** Monfredini, A., Santacroce, V., Deyris, P.-A. et al. (2016). *Dalton Trans.* 45: 15786–15790.

# 4

# Stability and Reactivity in Aromatic Compounds

*"Aromaticity is like the icing on a cake. Icing contributes only a little to holding the cake together, but it is the most delicious part"*

Zhongfang Chen, Paul von Ragué Schleyer et al. (Chem. Rev. **2005**, 105, 3842)

## 4.1 Introduction

One important aspect of aromatic compounds is their enhanced energetic/thermodynamic and chemical/kinetic stabilities. In classical aromatic planar organic annulenes, thermodynamic and kinetic stabilities had been demonstrated by hydrogenation experimental studies, calculation of aromatic stabilization and resonance energies, and preference for substitution reactions rather than addition. Aromatic *closo* boron hydrides, $[B_nH_n]^{2-}$, $n = 5$ to 12, which geometrically have the shape of deltahedra, are also very stable (e.g., $Li_2[B_{12}H_{12}]$ [1] and $Na_2[B_{12}H_{12}]$ [2] do not decompose below 600°C) and favor substitution versus addition reactions. Indeed, Hoffmann [3] argued against considering aromatic certain new forms of aromaticity, like novel metalla-aromatic species, because these new aromatic compounds have minimal kinetic persistence and they cannot be put it into a bottle and keep it there for a day. Although this view is not supported by all chemists [4], the connection between stability and aromaticity is widely accepted. There are many examples showing that the most stable isomers (or electronic states) are also the most aromatic. However, there are also examples of the opposite situation. Indeed, aromaticity is just one of the many factors that affect the relative energy of isomers. Steric repulsions, conjugation, hyperconjugation, electrostatic interactions, orbital interactions, weak or long-range interactions, and other effects may compensate the stabilization due to aromaticity. In these cases, the least aromatic species can be the most

*Aromaticity and Antiaromaticity: Concepts and Applications*, First Edition.
Miquel Solà, Alexander I. Boldyrev, Michał K. Cyrański, Tadeusz M. Krygowski, and Gabriel Merino.

thermodynamically stable one among several isomers. Therefore, the relationship between aromaticity and stability sometimes is complex.

Not only aromaticity and stability, but also aromaticity and reactivity are two deeply connected concepts. If one considers that approximately two-thirds of all compounds known are fully or partially aromatic [5], it is clear that many chemical reactions will be influenced by the increase or decrease in aromaticity all over the reaction. Chemical structures (reactants, intermediates, and products) and transition states (TSs) are often influenced by aromatic stabilization or antiaromatic destabilization. The most important group of this kind of reactions are the pericyclic ones [6, 7]. Most thermally allowed pericyclic reactions take place through aromatic TSs, whereas TSs of thermally forbidden pericyclic reactions are usually less aromatic or antiaromatic. Other reactions, however, are driven by the deantiaromatization of the reactants or aromatization of the products.

The aim of this chapter is to illustrate with some representative examples the relationships between aromaticity and thermodynamic and kinetic stabilities. We focus mostly on classical organic aromatic molecules, although some examples from the realm of aromatic inorganic species are also discussed.

## 4.2 Aromaticity and Thermodynamic Stability

One of the simplest examples showing that the most aromatic compound among a series of isomers is, in many cases, the most stable is the phenanthrene and anthracene isomeric pair. Phenanthrene is about 4 kcal/mol more stable than anthracene [8], despite the H···H steric repulsion between H atoms in the bay region of phenanthrene (the H atoms of C4 and C5 shown in Figure 4.1). The simple Clar's model showing that phenanthrene has two $\pi$-sextets as compared to anthracene that has just one migrating $\pi$-sextet already provides an explanation for the higher stability of phenanthrene. A quantitative bond energy decomposition analysis [8] convincingly demonstrated the existence of better $\pi$-interactions in phenanthrene as compared to anthracene. Some authors consider that the interaction between H atoms in the bay region of phenanthrene is bonding (not repulsive) [9] and they justify the higher stability of phenanthrene as compared to anthracene through the existence of this H···H bond. Although this H···H interaction is controversial [10–12], two simple computational experiments prove that the H···H interaction in phenanthrene is repulsive. In the first one, one can remove the two bay hydrogens (H4 and H5) from phenanthrene and two equivalent H atoms in anthracene (H1 and H8) to generate two biradicals in their lowest-lying triplet states: 4,5-didehydrophenanthrene and 1,8-didehydroanthracene. The resulting phenanthrene biradical is more stable than the anthracene one by about 5 kcal/mol [8]. Therefore, without the bay H

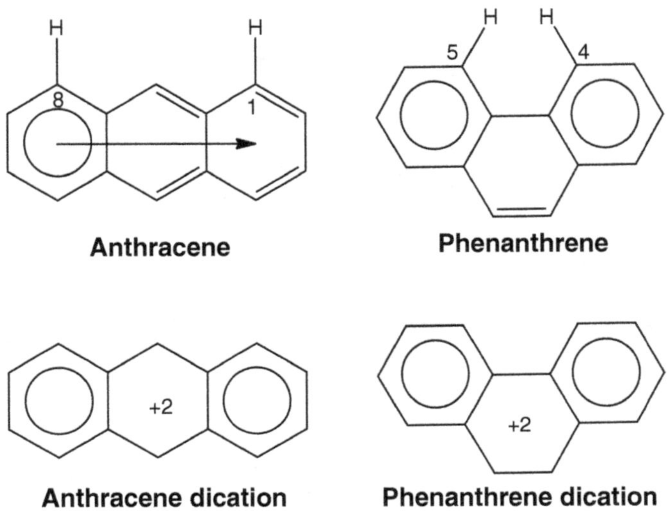


**Figure 4.1** The Clar structure of anthracene, phenanthrene, and their dications.

atoms, the phenanthrene topology becomes even more stable than the anthracene one and this cannot be attributed to H···H bonding interactions but to better π-interactions in phenanthrene [13]. The second experiment considers phenanthrene and anthracene dications. As can be seen in Figure 4.1, the Clar structures of phenanthrene and anthracene dications have both two π-sextets. Clar's model considers that the stabilization of the π-system in the two cases should be similar. And indeed, in the case of singlet state dicationic structures, Dominikowska and Palusiak [14] and Poater et al. [15] reported that the anthracene dication is more stable than the phenanthrene one by about 16 kcal/mol. The origin of the higher stability of dicationic anthracene is the same as in the neutral species, that is, better π-bonding interactions [15]. These two numerical experiments confirm the classical picture of better π-bonding causing phenanthrene's enhanced stability, despite the unfavorable H···H repulsion between bay hydrogens in phenanthrene. These results can be generalized to explain the enhanced stability of bent or kinked polycyclic benzenoids over linear ones [8]. The same trend applies to quinones, that is, 1,2-quinones are found to be more stable and more aromatic than their 2,3-counterparts [16].

A similar situation is found in the 1- and 2-indenone isomers and their aza derivatives (see Figure 4.2); 1-oxo compounds were reported to be more stable than the corresponding 2-oxo compounds by more than 20 kcal/mol [17]. The higher stability and aromaticity of the 1-one compared to the 2-one series was attributed to the higher aromaticity and better π-interactions of the former. This result was substantiated by energy decomposition analysis together with

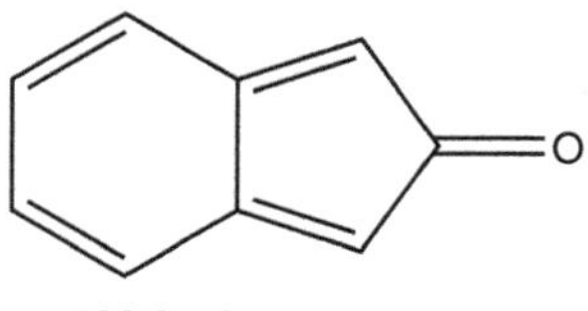

**2H-inden-2-one**

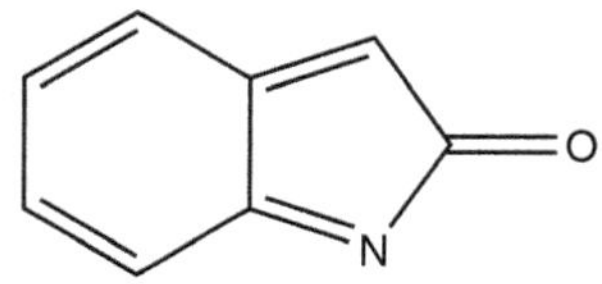

**2H-indol-2-one**

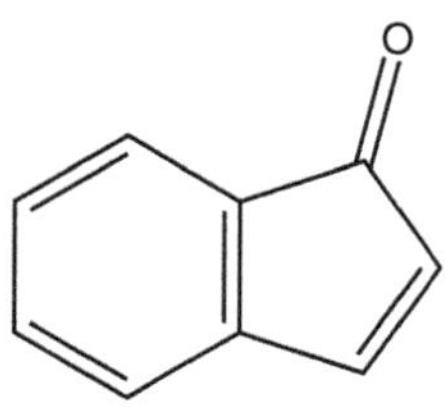

**1H-inden-1-one**

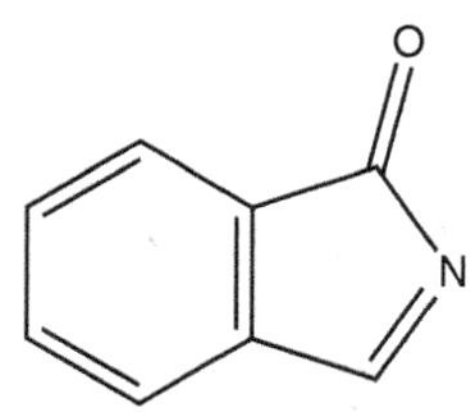

**1H-isoindol-1-one**

**Figure 4.2** The structure of 1- and 2-indenone isomers and their aza derivatives.

nucleus-independent chemical shift (NICS) and electronic *para*-delocalization index (PDI), aromatic fluctuation index (FLU), and multicenter electron delocalization index (MCI) calculations [17].

In this context, negatively charged fullerenes are another interesting species. Garcia-Borràs et al. found that there is an almost perfect correlation between the relative stability and aromaticity among the isomers of negatively charged $C_{2n}^{-q}$ ($n = 33 - 52$, $q = 2 - 6$) fullerenes [18, 19]. The authors concluded that, in fullerenes, there is an interplay between strain energy and aromaticity. In neutral empty fullerenes, reducing strain energy is the main factor for determining the stability of the isomeric carbon cages. This is reflected in the isolated pentagon rule (IPR), which states that all pentagons must be surrounded by hexagons to alleviate the strain produced by two fused pentagons [20]. On the contrary, in negatively charged fullerenes or endohedral metallofullerenes (EMFs), maximizing the total aromaticity is the main stabilizing element, and this is translated into the maximum aromaticity rule [18]. This rule affirms that the most stable anionic fullerene isomer is the one whose total aromaticity is maximized.

The last example we discuss in this section where aromaticity plays an essential role refers to the electronic ground state of [n]acenes. Bendikov et al. [21] reported that the restricted RB3LYP/6-31G(d) wave function of polyacenes longer than pentacene becomes unstable. For these species, a diradical open-shell singlet state obtained using the unrestricted broken symmetry UB3LYP/6-31G(d) method is found to be more stable than the closed-shell singlet state [22]. For these systems, the triplet state is found slightly higher in energy than the open-shell diradical singlet state, and the energy difference between these two states decreases when the

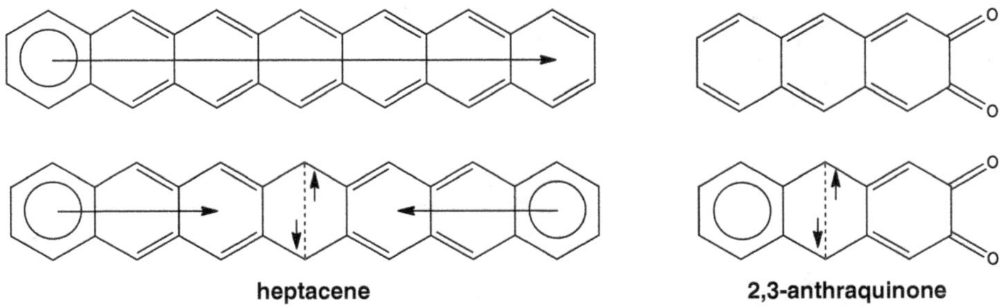


**Figure 4.3** The Clar structure of closed-shell singlet (top) and open-shell singlet (bottom) states for heptacene and 2,3-anthraquinone.

size of the acene increases. The change in the electronic ground state can also be understood using Clar's π-sextet model. As can be seen in Figure 4.3, when going from closed-shell to open-shell singlet state, a π-bond is lost and this is compensated by the formation on an extra π-sextet and some 1,4 interaction (Dewar-type resonance structure, Figure 4.3 left bottom).[1] This change is energetically unfavorable for [n]acenes with $n < 5$ or 6, but for larger acenes is thermodynamically favorable. The same situation applies to 2,3-quinones (but not to 1,2-quinones). As can be seen in Figure 4.3 the change in the ground state from singlet closed- to open-shell, which occurs at even lower number of six-membered rings (6-MRs) than in acenes, is driven by the same factors: the formation of an extra π-sextet and the development of a 1,4-interaction [16].

Despite the numerous examples confirming the link between aromaticity and thermodynamic stability, it is wrong, however, to conclude that among a series of isomers the most stable is always the most aromatic. Aromaticity is one of the many factors that affect the relative energies isomers but other factors can have a greater influence than aromaticity in determining the final relative energies of isomers. Many literature examples confirm this point. For instance, in a series of cyclopentafused pyrene congeners, Havenith et al. [23, 24] found that the relative order of stability does not follow the trend given by resonance energies. For instance, in the case of the dicyclopentapyrenes depicted in Figure 4.4, the least stable isomer (dicyclopenta[*cd,mn*]pyrene, by about 4 kcal/mol) had the highest aromatic character according to resonance energies, NICS, and diamagnetic susceptible exaltations [23].

Another example concerns the *ortho*-, *meta*-, and *para*-benzyne species, the three possible biradicals generated by removing two H atoms from benzene. For these systems, the order of aromaticity is *p*-benzyne > *m*-benzyne > *o*-benzyne,

1 It is worth noting that in the case of the open-shell diradical singlet state, there are other possible equivalent Clar structures with two π-sextets and the 1,4 interaction in rings different from the central one.

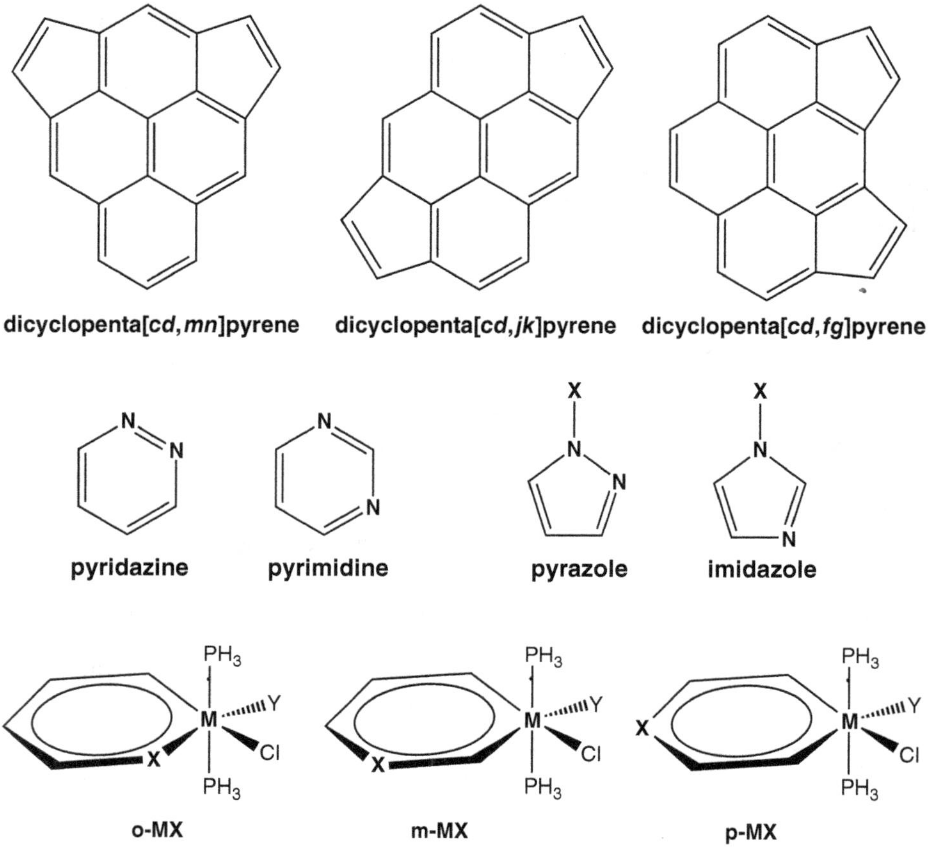


**Figure 4.4** The Lewis structure of three isomeric dicyclopentapyrenes, pyridazine, pyrimidine, pyrazole, imidazole, and some ortho, meta, and para heterometallabenzenes.

whereas the stability order of these biradicals is exactly the opposite: *o*-benzyne > *m*-benzyne > *p*-benzyne [13, 25]. Aromaticity and stability of aminomethylbenzoic acids also show opposite trends in some cases [26]. The authors of this study found that the more extensive conjugation between substituents in the isomers increases the stability but decreases the ring aromaticity. Moreover, Subramanian et al. [27] found no direct correlation between the thermodynamic stability and aromaticity in heterobicyclic isomers. Interestingly, N-substituted imidazoles are more stable than the corresponding pyrazoles (Figure 4.4) although the higher stability of imidazoles is not due to the higher aromaticity of imidazole rings but a weaker N—N bond found in the pyrazole rings [28]. Indeed, the pyrazole ring was found somewhat more aromatic than the imidazole one. For the same reason, pyrimidine is about 20 kcal/mol more stable than pyridazine (Figure 4.4) despite of similar aromatic character (in fact, pyridazine is slightly more aromatic than pyrimidine) [29].

In the group of metalla-aromatic species, the $C_{2v}$ isomers of $Al_2B_2^{2-}$ and $Ga_2B_2^{2-}$ were found to be more stable (by about 30 kcal/mol) than the more aromatic $D_{2h}$ isomers [30]. And for the species of $C_2Si_2^{2+}$ and $C_2Ge_2^{2+}$ [31], the linear isomer is more than 100 kcal/mol more stable than the $D_{2h}$ aromatic ring. An interesting example concerns the $B_6^{-2}$ and $Al_6^{-2}$ clusters. $Al_6^{-2}$ the octahedral and aromatic structure is more stable than the planar $D_{2h}$ antiaromatic structure. However, for $B_6^{-2}$ the situation is the opposite, and the $D_{2h}$ antiaromatic structure is favored [32]. The reason is the strong tendency of boron clusters to form 2-center-2-electron covalent localized bonds whereas delocalization is the dominant force in aluminium clusters [32, 33]. In the case of the *ortho*, *meta*, and *para* isomers of heterometallabenzenes such as those depicted in Figure 4.4, the relation between stability and aromaticity depends on the system considered. Thus, for M = Os, X = P, and Y = CO, the most aromatic *ortho* isomer according to $NICS(1)_{zz}$ and MCI values is also the most stable one. In contrast, for M = Ir, X = N, and Y = $Cl^-$, the most stable *meta* isomer is the least aromatic [34].

Finally, the most aromatic electronic state is not always the most stable. Thus, for instance, the antiaromatic singlet ground state of cyclobutadiene is more stable than its aromatic lowest-lying triplet state [35]. Another example corresponds to $Li_6^+$. In the shape of an octahedron, $Li_6^+$ has an electronic configuration of the type $1a_{1g}^2 1t_{1u}^3$ (S = 3/2, $^4A_{1g}$) with an atomization energy of 123.5 kcal/mol. This cluster was reported to be moderately aromatic [36]. However, the most stable isomer for $Li_6^+$ is not the octahedral one but a $C_{2v}$ structure with S = ½ and an atomization energy of 142.6 kcal/mol [37]. All indicators of aromaticity show that this $C_{2v}$ structure is not aromatic at all.

## 4.3 Aromaticity and Kinetic Stability

In 1825, when Michael Faraday [38] obtained benzene for the first time by distillation and named it dicarburet of hydrogen, he already noted that benzene was much less reactive than *trans*-2-butene. Such decreased reactivity has been considered characteristic of aromatic compounds ever since. However, not all aromatic compounds are unreactive. Moreover, some reactions are driven or facilitated by the formation of aromatic TSs, products or both. Therefore, the connection between aromaticity and reactivity has multiple faces. In this subsection, we discuss this relationship for some selected examples. Obviously, coverage is extensive but not exhaustive because there are plenty of reactions affected to some extent by the aromaticity of reactants, intermediates, TSs or products.

### 4.3.1 Acenes

The first example concerns the reactivity of acenes [39] that are polycyclic aromatic hydrocarbons (PAHs) consisting of linearly fused benzene rings.

The simplest acene is benzene. While benzene is aromatic in its ground state, it is antiaromatic in its lowest $\pi\pi^*$ excited state ($S_1$). This is the reason why benzene exhibits a rich photochemistry that provides access to complex molecular structures [40]. For instance, when benzene is irradiated with light at 254 nm, it rearranges in the $S_1$ state to benzvalene and fulvenes. The main driving force of this reaction is the relief of the antiaromatic character of the $S_1$ electronic state [41].

Acenes are of interest as versatile organic semiconductors [42, 43]. For higher acenes, spectroscopic data [44] seems to support a singlet open-shell ground state (antiferromagnetic) as suggested by computational analysis [21, 22]. Electrophilic substitution on these molecules is the most common reaction, whereas nucleophilic substitutions occurs only in strongly activated acenes [45]. As the number of rings increases, the members of the acene family become increasingly reactive, so that the higher members of acenes cannot be characterized experimentally [46]. Substitution by aryl- or alkylthio groups, or fluoro groups was reported to increase the stability of the larger members of the linear acenes [47, 48]. In this way, the longest synthesized member of the series corresponds to nonacene [44, 47, 48]. Therefore, the series of acenes constitute a good example that aromatic compounds are not necessarily kinetically stable. Indeed, the higher members of the acene series are extremely reactive despite being aromatic. Whereas benzene and naphthalene in their ground states are quite unreactive towards addition reactions [49], the central ring of anthracene is protonated [50], adds bromine [51], and undergoes Diels-Alder (DA) reactions [49] readily. Tetracene and pentacene participate in even more remarkable 1,4-cycloadditions [46, 52]. The successive reduction in the highest occupied molecular orbital-lowest unoccupied molecular orbital (HOMO-LUMO) gap and in the ionization potential, as well as the increasing proton affinities, are in line with the sequential loss of benzenoid character (aromaticity) predicted by Clar's qualitative $\pi$-sextet concept [53]. This result is in contrast with the NICS and resonance energy calculations of Schleyer et al. [46] indicating that the stabilization per $\pi$-electron is essentially constant along the series from benzene to heptacene.

In 2001, Schleyer and coworkers [46] studied the DA reaction of acetylene and acenes and found that the higher the acene, the lower the energy barrier and the more exothermic the reaction is. In addition, they reported that the more reactive inner rings are more aromatic than the less reactive outer rings. HOMO coefficients of acenes are consistent with the regioselectivity of DA reactions that prefer the middle rings, despite the greater aromaticity [54]. A study by Portella et al. [55, 56] confirmed that the more reactive inner rings are more aromatic than the outer rings by means of PDI, harmonic oscillator model of aromaticity (HOMA), and NICS descriptors of aromaticity. This trend is opposite to that found for the equivalent phenacenes (like phenanthrene, chrysene, picene,…), known experimentally to be more stable energetically than the corresponding isomeric acenes (like anthracene, tetracene, pentacene,…) and also the helicenes, in which the outer rings have the largest aromaticities [55, 56] and are the less reactive rings. The same trends of local aromaticity for acenes and phenacenes were reported by Cyrański et al. [57], who also found that the global aromaticity of both acenes and phenacenes decreases with the increase of the size, although with a greater decrease for the acenes series. Interestingly, however, the authors showed that the local aromaticity of the outer ring of phenacenes remains high and quite constant with increasing size. Therefore, the trends followed by the global and local aromaticities in phenacenes differ substantially with the increase in size.

The reaction mechanism of the addition of singlet oxygen ($^1\Delta_g$ $O_2$) with acenes was studied theoretically [54, 58]. For benzene and naphthalene, the reaction mechanism is concerted. However, starting from anthracene, acenes react with singlet oxygen via a biradical stepwise mechanism [58]. A computational study of the addition of HCl (which behaves as an electrophile) and water (which behaves as a nucleophile) to acenes show that the reactivity of acenes increases along the series up to hexacene and from hexacene and above remains basically constant [59]. As can be seen in Figure 4.5, the exothermicity of the addition of water and HCl are similar. Interestingly, the reactions of water and HCl with acenes are endothermic for acenes up to anthracene and exothermic for higher acenes. Despite similar exothermicities, the activation barriers for the addition of water (nucleophilic addition) are significantly higher than those for the addition of HCl. In both cases, the energy barriers decrease when the size of the acene increases up to hexacene and then remains constant.

In all cases, additions take place in the central ring of acenes. Therefore, addition reactions to acenes provide a clear example that the aromaticity of local rings in a PAH does not always control the regioselectivity of the addition reactions and that, in some cases, more aromatic rings can be more reactive than less aromatic rings. In a given addition reaction, the ground state energy of the PAH is the same whatever site reacts. The differences in energies of TSs (which closely parallel the energies of products) decide the preferential position of attack. In anthracene, for instance, the central ring is the most reactive, even though many methods have shown this ring to be the most aromatic ring. The reason for the higher reactivity of the inner ring is readily explained by using Clar's π-sextet rule [53]. As shown in Figure 4.6, the product of the DA cycloaddition of acetylene to the central ring of anthracene is more aromatic with two Clar π-sextet rings than the product resulting from the addition to the outer ring that contains only one migrating Clar π-sextet ring.

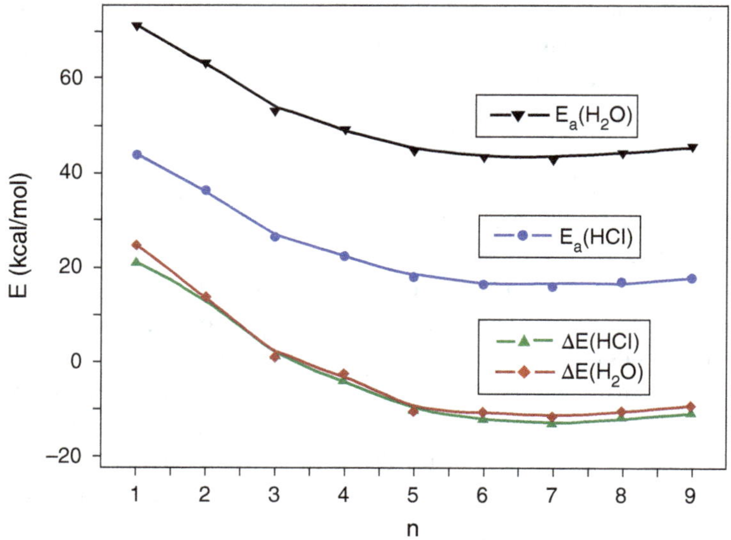


**Figure 4.5** Activation energies of the addition of water ($E_a(H_2O)$) and HCl ($E_a$(HCl)) to [n]acenes and reaction energies for the addition of water ($\Delta E(H_2O)$) and HCl ($\Delta E$(HCl)) to [n]acenes. All energies in kcal/mol. Source: Reprinted with permission from Ref. [59]. Copyright © 2007, American Chemical Society.

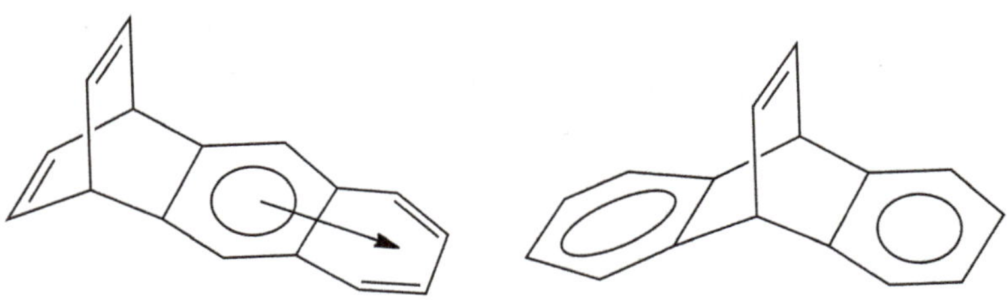

**Figure 4.6** The product of the addition to an outer ring of anthracene (left) and inner ring of anthracene (right).

Finally, referred to dimerization, theoretical [60] and experimental [61, 62] studies found that the preferred pathway leads to the formation of acene dimers via the central six-membered ring (6-MR). Higher acenes are predicted to self-react to form polymers rather than acene dimers.

### 4.3.2 Pericyclic Reactions

The pericyclic reactions are an important class of concerted reactions in which the breaking and making of bonds occurs simultaneously through a single cyclic TS.

These reactions were classified by Woodward and Hoffmann into five groups: sigmatropic shifts, cycloaddition, electrocyclic, chelotropic, and group transfer reactions [6, 7]. These reactions were rationalized by Woodward and Hoffmann [6, 7] with a series of rules based on the conservation of orbital symmetry. To predict whether a pericyclic reaction is symmetry forbidden or allowed, one has to analyze the symmetry of the molecular orbitals of the reactants and products and their relative energies to construct a molecular orbital correlation diagram. If the orbital symmetry is conserved along the reaction pathway the reaction is allowed and forbidden otherwise.

An alternative and simpler method to classify pericyclic reactions into allowed and forbidden reactions was developed by Zimmermann and Dewar [63, 64] one year after the publication of Woodward and Hoffmann, and it is called the Möbius–Hückel treatment. In contrast to the Woodward and Hoffmann approach, the Möbius–Hückel method does not need any symmetry element to be conserved during the reaction. This approach is based on the aromaticity of the TS. It is well known that TSs of pericyclic-allowed reactions are aromatic [65] despite being the least stable species along the reaction coordinate. The aromaticity of pericyclic TSs was first investigated systematically by Schleyer and coworkers [66–71] using geometric, energetic, and magnetic-based criteria of aromaticity, in all cases proving the electronic delocalization of these structures in allowed pericyclic reactions. More recently, Mandado et al. [72] further confirmed the aromaticity of TSs of allowed pericyclic reactions by means of multicenter electronic indices.

In the Möbius–Hückel approach, one has to depict the frontier molecular orbitals of the reactants (i.e., in a DA reaction the HOMO of the dienophile and the LUMO of the diene for instance, but one could take any other pair of orbitals), connect them, and count the number of phase changes around the orbital array. Where one has zero or an even number of sign inversions there is a Hückel array, whereas an odd number of sign inversions correspond to a Möbius array (Figure 4.7). A TS of a Hückel array of orbitals is stabilized and the reaction allowed if there are $4n+2$ electrons involved in the reaction and destabilized and the reaction forbidden with $4n$ electrons. Contrarily, in a Möbius system, the TS has aromatic stability with $4n$ electrons and it is destabilized with $4n+2$ electrons.

#### 4.3.2.1 Diels–Alder Cycloadditions

In 1938, Meredith G. Evans and Ernest Warhurst [73] noticed the analogy between the π electrons of benzene and the six delocalized electrons in the cyclic TS of the DA reaction of butadiene and ethylene (Figure 4.8). It is now well recognized that this reaction takes place via a synchronous and concerted mechanism through an aromatic boatlike TS [74]. The alternative stepwise diradical mechanism has an energy barrier that is about 2–7 kcal/mol higher [75]. Application of the

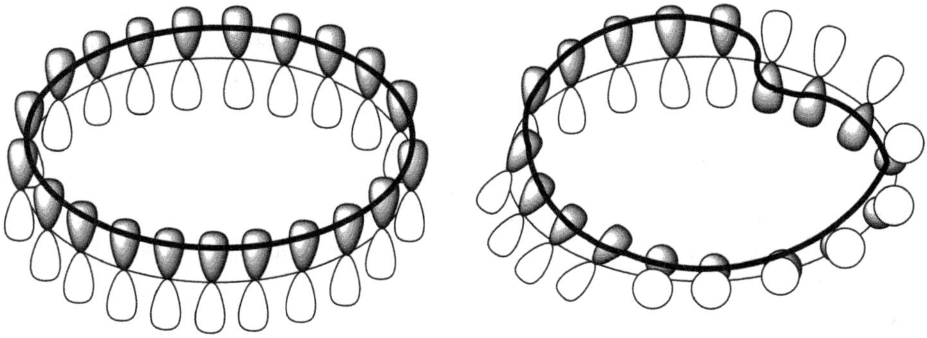

**Figure 4.7** Hypothetical $\pi$-systems with Hückel (left) and Möbius (right) topologies.

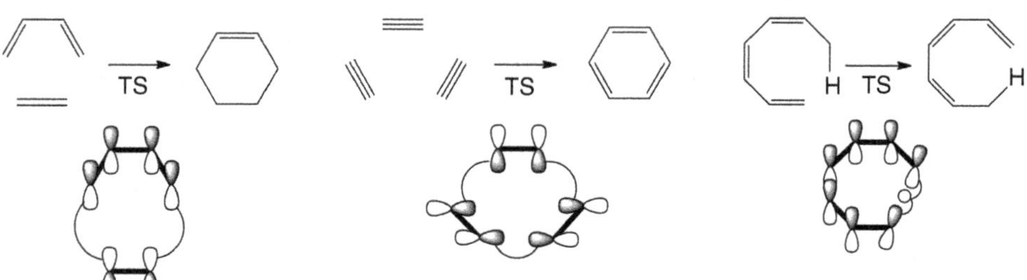


**Figure 4.8** Application of Möbius-Hückel treatment to [4 + 2] (left) and [2 + 2 + 2] (center) cycloadditions and [1,7]-sigmatropic shift (right).

Zimmerman–Dewar's rules to the TS of this reaction shows that the connection between molecular orbitals of butadiene and ethylene to yield cyclohexene occurs with an even number of phase changes (0 for the orbitals chosen in Figure 4.8). Since this reaction involves six electrons and the array of molecular orbitals in the transition follows a Hückel topology, then the TS is aromatic and the DA reaction is allowed. The cyclic delocalization of the six electrons in the TS involves $\sigma$ and $\pi$-orbitals and it is understood in terms of in-plane aromaticity [76–78].

The aromatic nature of this TS has been confirmed theoretically using magnetic-based indices such as NICS (NICS(0) = −19.4 ppm [65]) and the magnetic susceptibility exaltations [71] as well as electronic-based indices [79]. The performance of a series of aromatic descriptors was analyzed by applying them to the simplest DA between ethylene and 1,3-butadiene. The evolution of electronic indices (PDI, FLU, MCI, $I_{ring}$, $I_{NG}$, and $I_{NB}$) along the reaction path is depicted in Figure 4.9 [80]. All indicators of aromaticity correctly predict that a structure close to the TS is the most aromatic species along the reaction path, except FLU index. The failure of FLU is due to the fact that FLU determines variances of electronic patterns around the ring using as a reference electronic

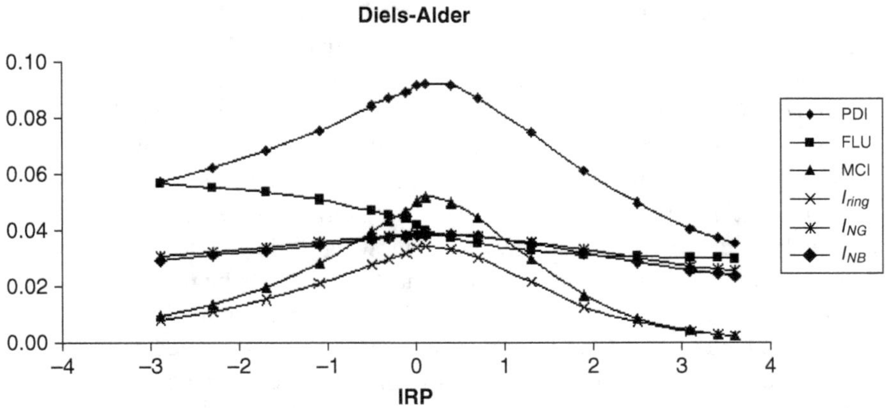


**Figure 4.9** Plot of PDI (electrons), $FLU^{1/2}$ (values divided by 10), MCI (electrons), $I_{ring}$ (electrons), $I_{NB}$, and $I_{NG}$ versus the reaction coordinate (IRP in $amu^{1/2}$ Bohr). Negative values of the IRP correspond to the reactants side of the reaction path and positive values to the product side of the butadiene-ethylene Diels-Alder cycloaddition. Source: Reprinted from Ref. [80] with permission. Copyright © 2009, CRC Press.

delocalization indices taken from standard aromatic species. Then, the higher the differences between the standard in aromatic systems and the actual electronic sharing, the lower aromaticity predicted by FLU (higher FLU values). For this reason, FLU cannot be used to discuss the aromaticity of species that are far from their equilibrium geometries.

Finally, it is worth mentioning that computations by Geerlings et al. [76] reveal that the DA cycloadditions involving various quinodimethanes gain aromaticity along the reaction coordinate and, consequently, these reactions are faster and thermodynamically more favorable that the prototype DA reaction between ethylene and 1,3-butadiene.

#### 4.3.2.2 [2+2+2] Cycloadditions

Zimmerman–Dewar rules applied to the TS of the [2+2+2] cycloaddition indicate that the connection among molecular orbitals of three acetylene molecules to yield benzene takes place with an even number of phase changes (0 for the orbitals chosen in Figure 4.8), and, therefore, with six electrons involved, it corresponds to an allowed reaction with an aromatic TS. Notably, even such a thermally allowed and strongly exothermic reaction with aromatic delocalization in cyclic TSs can have a substantial activation barrier. This reaction is highly exothermic (exp. $\Delta H_r^0 = -142.8$ kcal/mol) [81], but has a surprising large enthalpy barrier of about 80 kcal/mol for a thermally allowed process [82]. This high barrier results from the energetic cost of distorting the three acetylenes to the geometry they have in the TS as well as from the closed-shell repulsion between filled π-orbitals [82].

Aromaticity changes during the transformation from reactants to product has been analyzed in many studies [71, 81, 83–85] that found that this thermally allowed reaction goes through an aromatic TS. Most authors reported that the system evolves from localized σ and π electrons in the reactants to the well-known π-delocalization in benzene through TS that has mainly in-plane σ-electron delocalization with only minor π-electron delocalization. So, it seems reasonable to think that the aromaticity of the 6-MR being formed increases from reactants to the TS and it decreases slightly before growing again to reach the aromatic benzene product. This was indeed the trend observed [85] by means of a series of electronic-based aromaticity criteria. NICS gives a maximum of aromaticity at the TS [71], although, more importantly, when NICS is separated between the NICS(π) and NICS(σ) components, it is found that the π-component increases all the way from the reactant to the product, whereas the σ-component is highly diamagnetic at the TS but strongly paramagnetic for benzene. This result confirms that at the TS the in-plane contribution is larger, although the π-delocalization is also important.

#### 4.3.2.3 [1,7]-Sigmatropic Migrations

The [1,7]-sigmatropic migration of a hydrogen atom in (Z,Z)-1,3,5-heptatriene is a pericyclic transformation that occurs via a Möbius TS [86]. As can be seen in Figure 4.8, the array of molecular orbitals that connect the reactant with the product has an odd number of phase changes (1 for the orbitals chosen in Figure 4.8), and, therefore, with eight electrons (six π and two σ) involved, it corresponds to an allowed reaction with an aromatic TS of Möbius topology. The transition structure of this [1,7]-sigmatropic migration shows bond length equalization and the ring protons exhibit downfield $^{1}$H-NMR chemical shifts comparable to those of benzene [86]. These features support the aromatic character of this TS.

#### 4.3.2.4 Fullerene Additions

Finally, let us analyze two examples of reactions involving fullerenes. Fullerenes are a new carbon allotrope form in which carbon atoms are arranged into 12 pentagons and a variable number of hexagons in a more or less spherical shape. The most stable empty neutral fullerenes strictly obey the IPR [20]. DA cycloadditions are among the most employed reactions for the functionalization of fullerenes [87–91]. The electron withdrawing nature of fullerenes makes these molecules ideal dienophiles for DA reactions. $C_{60}$, which is the most abundant fullerene, readily undergoes [4+2] DA cycloaddition reactions with a variety of reactive dienes such as cyclopentadiene [92, 93]. $C_{60}$ has two types of bonds: [6,6]-bonds, where two 6-MRs are fused, and [5,6]-bonds corresponding to the ring junction between a 5- and a 6-MR. The [4+2] DA cycloaddition to neutral $C_{60}$ involves exclusively the [6,6]-bonds [94, 95]. The major factor controlling the observed regioselectivity in DA to $C_{60}$ is the more stabilizing interaction between the deformed reactants in the [6,6]-reaction pathway along the entire reaction coordinate [95] due to better HOMO(diene)-LUMO(dienophile) overlap (Figure 4.10).

Interestingly, for hexaanionic $C_{60}^{6-}$, the preference of reactivity is reversed and the [5,6]-bonds become more reactive [96]. This change in regioselectivity is gradual with the increase of the negative charge in fullerenes. Whereas, for neutral $C_{60}$, the DA addition of cyclopentadiene at the [6,6] bond is more exothermic than the [5,6] attack, the DA addition in negatively charged $C_{60}$ is more exothermic for the [5,6]-bond when the added number of electrons is larger than 4 (Figure 4.11). Energy barriers for the addition of fullerenes follow the same trend. In negatively charged fullerenes, there is an accumulation of the negative charge on 5-MRs that increases their aromatic character [19]. This change can be quantified with the MCI index that points out an increase in the aromaticity of the 5-MRs and a decrease of aromaticity in the 6-MRs upon reduction (Figure 4.11). When the reaction takes place on a pyracylenic [6,6] type bond, the aromaticity of two 5-MRs and two 6-MRs is lost. If the addition is on a corannulenic [5,6] bond, the aromaticity vanishes in three 6-MRs and one 5-MR. Then, the [6,6] attack is preferred for the neutral $C_{60}$ because it breaks the aromaticity of only two most aromatic 6-MRs as compared to the [5,6] addition that leads to the loss of aromaticity in three 6-MRs. On the other hand, for hexaanionic $C_{60}^{6-}$, the preferred attack is the [5,6] one because it involves the loss of aromaticity of only one most aromatic 5-MR as compared to the [6,6] addition that destroys the aromaticity of two 5-MRs [96].

The DA reaction between cyclopentadiene (Cp) and $C_{60}^{n-}$ ($n = 0$–$6$) provides an example of the influence of the aromatic properties of $C_{60}^{n-}$ in the regioselectivity of the reaction. A second example concerns the Bingel–Hirsch (BH) reaction [98] in IPR EMFs. EMFs are fullerenes that have metal clusters encapsulated into the carbon cage [99, 100]. In EMFs, a formal charge transfer up to six electrons from the metal cluster to the fullerene cage takes place, which accumulates mainly in the 5-MRs of the cage, increasing the aromaticity of these rings.

The BH addition is a cyclopropanation reaction where a fullerene and diethyl bromomalonate in the presence of a strong base such as 1,8-diazabicycloundec-7-ene (DBU) or sodium hydride react to produce a methanofullerene or a fulleroid [101]. Figure 4.12 illustrates the mechanism of the BH reaction. First, the base abstracts the acidic proton of the malonate derivative to generate a carbanion or enolate. Then this carbanion nucleophilically attacks the fullerene, thus generating a new carbanion with charge localized at the cage. In the final step, bromide is displaced in a nucleophilic substitution $S_N2$ reaction causing an intramolecular 3-MR closure. The attacked C—C bond in the fullerene at the end of the reaction can remain closed or it can be opened.

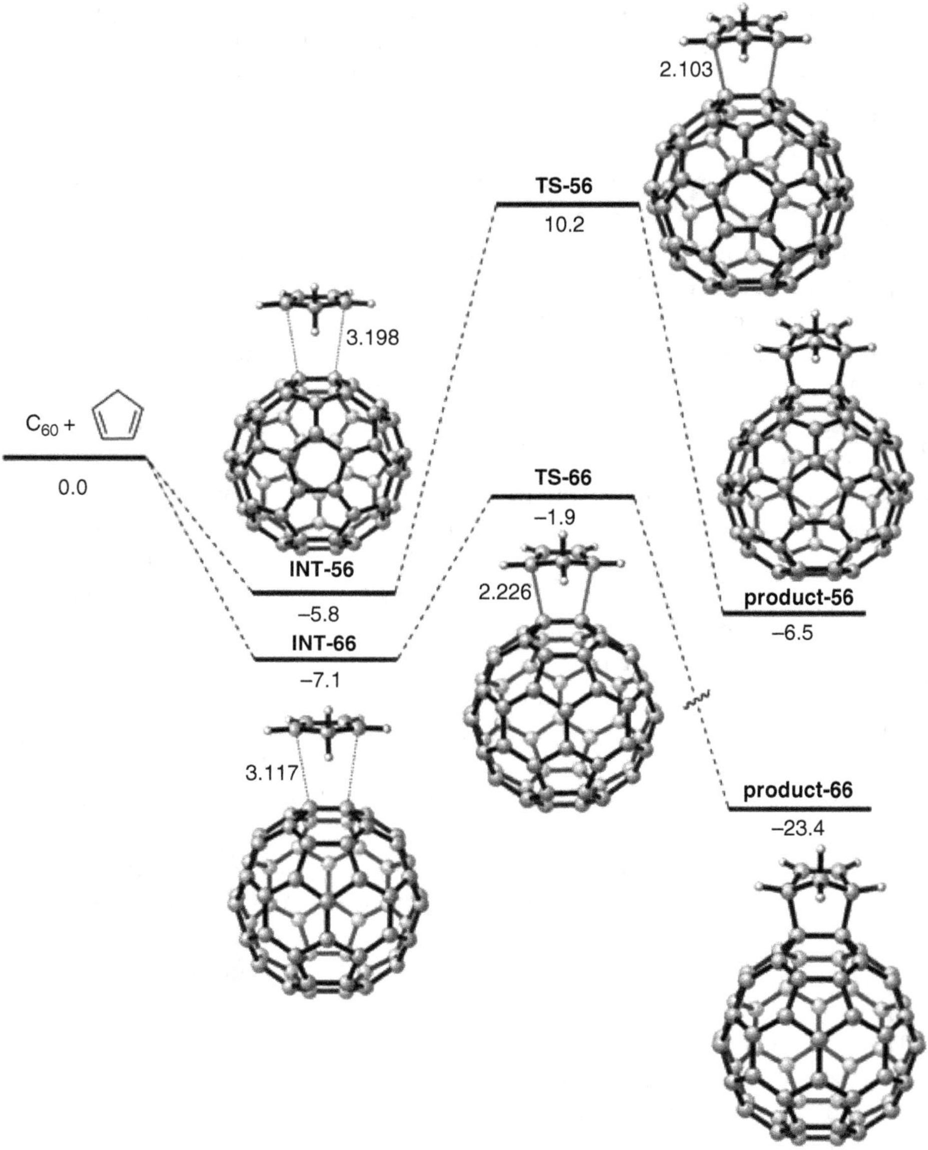


**Figure 4.10** Computed reaction profile of the reaction between cyclopentadiene and $C_{60}$. Energies are given in kcal/mol and bond distances in angstroms. All data have been computed at the BP86-D3/TZ2P+//RI-BP86-D3/def2-SVP level. Source: From Ref. [95] with permission. Copyright © 2013, Wiley.

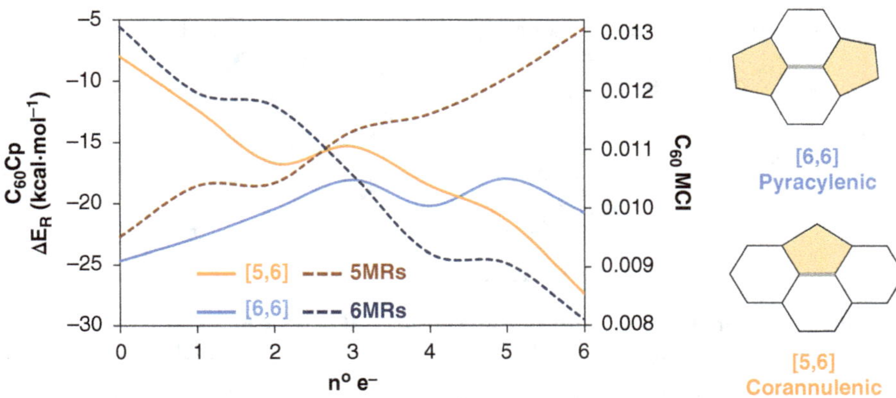


**Figure 4.11** Change in the reaction energy (in kcal/mol) of the Diels–Alder reaction between cyclopentadiene (Cp) and $C_{60}{}^{n-}$, and the evolution ofthe MCI aromaticity index (dashed lines, in electrons) of five- and six-membered rings of $C_{60}{}^{n-}$ as a function of the number of electrons added ($n = 1-6$). Source: Reprinted from Ref. [97] with permission. Copyright © 2017, Wiley.

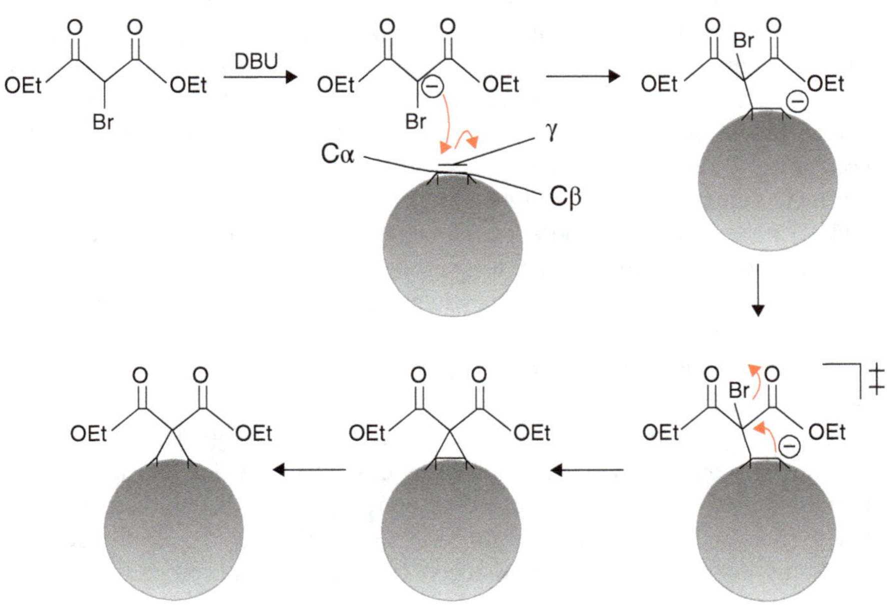


**Figure 4.12** General mechanism of the nucleophilic [2 + 1] Bingel–Hirsch reaction. Source: Adapted with permission from Ref. [102]. Copyright © 2016, Wiley.

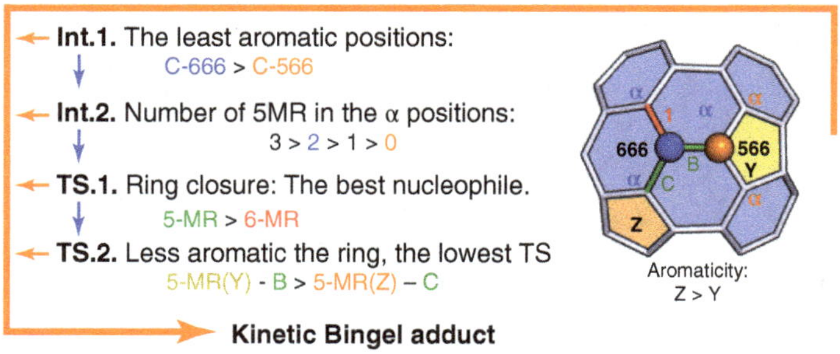


**Figure 4.13** Predictive Aromaticity-based criteria (PAC) for the Bingel–Hirsch addition to IPR EMFs. Source: Reprinted with permission from Ref. [103]. Copyright © 2016, Wiley.

The regioselectivity patterns for the BH addition on IPR EMFs can be predicted using a set of aromaticity-based criteria, the so-called Predictive Aromaticity Criteria (PAC) (see Figure 4.13). The PAC criteria identify the lowest energy intermediates and TSs for a given IPR EMF from the initial fullerene structure. The most aromatic (and therefore lowest in energy) intermediates lead to the lowest activation barriers for the BH addition. The PAC criteria states that: (i) the lowest energy intermediate should be located in a 666 region (C-666 in Figure 4.13) to avoid disrupting the aromaticity of 5-MRs, and (ii) among all possible C-666s, those presenting more pentagonal rings adjacent to the 666 region (denoted as α-positions in Figure 4.13) will likely be more favored to stabilize the negative charge placed on the cage. According to PAC, the lowest energy TSs for the cyclopropane ring closure will occur over: (i) 5-MRs because the C atoms of 5-MRs are more negatively charged and better suited for a nucleophilic attack, and (ii) among all 5-MRs, those located far from the metal cluster. These 5-MRs have less negative charge and are less aromatic (i.e., Y-5-MRs are preferred over Z-5-MR in Figure 4.13). These rules when applied to the fullerene fragment of Figure 4.13 predict that the bond B is the one that will be attacked. Application of PAC to the BH addition on $Sc_3N@D_{3h}$-$C_{78}$, $Sc_3N@I_h$-$C_{80}$, $Sc_3N@D_{5h}$-$C_{80}$ predicts correctly which are the most reactive bonds [103].

In summary, although the concept of aromaticity is usually related to those of stability and low reactivity, there are many examples in which the most stable isomer or electronic state does not correspond to the most aromatic situation, or where chemical reactivity is determined by the presence of chemical structures (reactants, intermediates, TSs or products) with (anti)aromatic properties.

## References

1 Pitt, M.P., Paskevicius, M., Brown, D.H. et al. (2013). *J. Am. Chem. Soc.* 135: 6930–6941.

2 Udovic, T.J., Matsuo, M., Unemoto, A. et al. (2014). *Chem. Commun.* 50: 3750–3752.
3 Hoffmann, R. (2015). *Am. Sci.* 103: 18–22.
4 Ritter, S.K. (2015). *Chem. Eng. News* 93: 37–38.
5 Balaban, A.T., Oniciu, D.C., and Katritzky, A.R. (2004). *Chem. Rev.* 104: 2777–2812.
6 Woodward, R.B. and Hoffmann, R. (1969). *Angew. Chem. Int. Ed.* 8: 781–853.
7 Woodward, R.B. and Hoffmann, R. (1970). *The Conservation of Orbital Symmetry*. Weinheim: Verlag Chemie.
8 Poater, J., Visser, R., Solà, M., and Bickelhaupt, F.M. (2007). *J. Org. Chem.* 72: 1134–1142.
9 Matta, C.F., Hernández-Trujillo, J., Tang, T.-H., and Bader, R.F.W. (2003). *Chem. Eur. J.* 9: 1940–1951.
10 Bader, R.F.W. (2006). *Chem. Eur. J.* 12: 2896–2901.
11 Poater, J., Solà, M., and Bickelhaupt, F.M. (2006). *Chem. Eur. J.* 12: 2889–2895.
12 Poater, J., Solà, M., and Bickelhaupt, F.M. (2006). *Chem. Eur. J.* 12: 2889–2905.
13 Poater, J., Bickelhaupt, F.M., and Solà, M. (2007). *J. Phys. Chem. A* 111: 5063–5070.
14 Dominikowska, J. and Palusiak, M. (2011). *Phys. Chem. Chem. Phys.* 13: 11976–11984.
15 Poater, J., Duran, M., and Solà, M. (2018). *Front. Chem.* 6: 561.
16 Szatyłowicz, H., Krygowski, T.M., Solà, M. et al. (2015). *Theor. Chem. Acc.* 134: 1–14.
17 Poater, J., Solà, M., Alkorta, I., and Elguero, J. (2014). *Eur. J. Org. Chem.* 5370–5377.
18 Garcia-Borràs, M., Osuna, S., Swart, M. et al. (2013). *Angew. Chem. Int. Ed.* 52: 9275–9278.
19 Garcia-Borràs, M., Osuna, S., Luis, J.M. et al. (2014). *Chem. Soc. Rev.* 43: 5089–5105.
20 Kroto, H.W. (1987). *Nature* 329: 529–531.
21 Bendikov, M., Duong, H.M., Starkey, K. et al. (2004). *J. Phys. Chem. A* 126: 7416–7417.
22 Poater, J., Bofill, J.M., Alemany, P., and Solà, M. (2005). *J. Phys. Chem. A* 109: 10629–10632.
23 Havenith, R.W.A., Jiao, H., Jenneskens, L.W. et al. (2002). *J. Am. Chem. Soc.* 124: 2363–2370.
24 Havenith, R.W.A., van Lenthe, J.H., Dijkstra, F., and Jenneskens, L.W. (2001). *J. Phys. Chem. A* 105: 3838–3845.

25 De Proft, F., Schleyer, P.v.R., van Lenthe, J.H. et al. (2002). *Chem. Eur. J.* 8: 3402–3410.
26 Lima, C.F.R.A.C., Gomes, L.R., and Santos, L.M.N.B.F. (2007). *J. Phys. Chem. A* 111: 10598–10603.
27 Subramanian, G., Schleyer, P.v.R., and Jiao, H. (1996). *Angew. Chem. Int. Ed. Engl.* 35: 2638–2641.
28 Curutchet, C., Poater, J., Solà, M., and Elguero, J. (2011). *J. Phys. Chem. A* 115: 8571–8577.
29 El-Hamdi, M., Tiznado, W., Poater, J., and Solà, M. (2011). *J. Org. Chem.* 76: 8913–8921.
30 Islas, R., Poater, J., Matito, E., and Solà, M. (2012). *Phys. Chem. Chem. Phys.* 14: 14850–14859.
31 Díaz-Cervantes, E., Poater, J., Robles, J. et al. (2013). *J. Phys. Chem. A* 117: 10462–10469.
32 El Bakouri, O., Solà, M., and Poater, J. (2016). *Phys. Chem. Chem. Phys.* 18: 21102–21110.
33 Huynh, M.T. and Alexandrova, A.N. (2011). *J. Phys. Chem. Lett.* 2: 2046–2051.
34 El-Hamdi, M., El Bakouri El Farri, O., Salvador, P. et al. (2013). *Organometallics* 32: 4892–4903.
35 Feixas, F., Vandenbussche, J., Bultinck, P. et al. (2011). *Phys. Chem. Chem. Phys.* 13: 20690–20703.
36 El Bakouri, O., Duran, M., Poater, J. et al. (2016). *Phys. Chem. Chem. Phys.* 18: 11700–11706.
37 Temelso, B. and Sherrill, C.D. (2005). *J. Chem. Phys.* 122: 064315.
38 Faraday, M. (1825). *Philos. Trans. R. Soc. London* 115: 440–446.
39 Zade, S.S. and Bendikov, M. (2012). *J. Phys. Org. Chem.* 25: 452–461.
40 Papadakis, R. and Ottosson, H. (2015). *Chem. Soc. Rev.* 44: 6472–6493.
41 Slanina, T., Ayub, R., Toldo, J. et al. (2020). *J. Am. Chem. Soc.* 142: 10942–10954.
42 Anthony, J.E. (2008). *Angew. Chem. Int. Ed.* 47: 452–483.
43 Anthony, J.E. (2006). *Chem. Rev.* 106: 5028–5048.
44 Tönshoff, C. and Bettinger, H.F. (2010). *Angew. Chem. Int. Ed.* 49: 4125–4128.
45 Payne, M.M., Parkin, S.R., Anthony, J.E. et al. (2005). *J. Am. Chem. Soc.* 127: 4986–4987.
46 Schleyer, P.v.R., Manoharan, M., Jiao, H.J., and Stahl, F. (2001). *Org. Lett.* 3: 3643–3646.
47 Kaur, I., Jazdzyk, M., Stein, N.N. et al. (2010). *J. Am. Chem. Soc.* 132: 1261–1263.
48 Purushothaman, B., Bruzek, M., Parkin, S.R. et al. (2011). *Angew. Chem. Int. Ed.* 50: 7013–7017.
49 Biermann, D. and Schmidt, W. (1980). *J. Am. Chem. Soc.* 102: 3163–3173.
50 Wang, D.Z. and Streitwieser, A. (1999). *Theor. Chem. Acc.* 102: 78–86.

**51** Wright, O.L. and Mura, L.E. (1966). *J. Chem. Educ.* 43: 150–150.

**52** Sarova, G.H. and Berberan-Santos, M.N. (2004). *Chem. Phys. Lett.* 397: 402–407.

**53** Solà, M. (2013). *Front. Chem.* 1: 22.

**54** Chien, S.-H., Cheng, M.-F., Lau, K.-C., and Li, W.-K. (2005). *J. Phys. Chem. A* 109: 7509–7518.

**55** Portella, G., Poater, J., Bofill, J.M. et al. (2005). *J. Org. Chem.* 70: 2509–2521.

**56** Portella, G., Poater, J., Bofill, J.M. et al. (2005). *J. Org. Chem.* 70: 4560–4560.

**57** Cyrański, M.K., Stępień, B.T., and Krygowski, T.M. (2000). *Tetrahedron* 56: 9663–9967.

**58** Reddy, A.R. and Bendikov, M. (2006). *Chem. Commun.* 1179–1181.

**59** Reddy, A.R., Fridman-Marueli, G., and Bendikov, M. (2007). *J. Org. Chem.* 72: 51–61.

**60** Zade, S.S., Zamoshchik, N., Reddy, A.R. et al. (2011). *J. Am. Chem. Soc.* 133: 10803–10816.

**61** Bouas-Laurent, H., Castellan, A., Desvergne, J.-P., and Lapouyade, R. (2000). *Chem. Soc. Rev.* 29: 43–55.

**62** Bouas-Laurent, H., Castellan, A., Desvergne, J.-P., and Lapouyade, R. (2001). *Chem. Soc. Rev.* 30: 248–263.

**63** Zimmermann, H. (1971). *Acc. Chem. Res.* 4: 272–280.

**64** Dewar, M.J.S. (1971). *Angew. Chem. Int. Ed. Eng.* 10: 761–776.

**65** Schleyer, P.v.R., Wu, J.I., Cossio, F.P., and Fernandez, I. (2014). *Chem. Soc. Rev.* 43: 4909–4921.

**66** Jiao, H.J. and Schleyer, P.v.R. (1993). *Angew. Chem. Int. Ed. Engl.* 32: 1763–1765.

**67** Jiao, H.J. and Schleyer, P.v.R. (1994). *J. Chem. Soc., Perkin Trans. 2* 407–410.

**68** Herges, R., Jiao, H., and Schleyer, P.v.R. (1994). *Angew. Chem. Int. Ed. Engl.* 33: 1376–1378.

**69** Jiao, H.J. and Schleyer, P.v.R. (1995). *J. Am. Chem. Soc.* 117: 11529–11535.

**70** Jiao, H.J. and Schleyer, P.v.R. (1995). *Angew. Chem. Int. Ed. Engl.* 34: 334–337.

**71** Jiao, H.J. and Schleyer, P.v.R. (1998). *J. Phys. Org. Chem.* 11: 655–662.

**72** Mandado, M., González-Moa, M.J., and Mosquera, R.A. (2007). *ChemPhysChem* 8: 696–702.

**73** Evans, M.G. and Warhust, E. (1938). *Trans. Faraday Soc.* 34: 614–624.

**74** Houk, K.N., Gonzalez, J., and Li, Y. (1995). *Acc. Chem. Res.* 28: 81–90.

**75** Beno, B.R., Houk, K.N., and Singleton, D.A. (1996). *J. Am. Chem. Soc.* 118: 9984–9985.

**76** Manoharan, M., De Proft, F., and Geerlings, P. (2000). *J. Chem. Soc., Perkin Trans. 2* 1767–1773.

**77** Cossío, F.P., Morao, I., Jiao, H.J., and Schleyer, P.v.R. (1999). *J. Am. Chem. Soc.* 121: 6737–6746.

**78** Morao, I., Lecea, B., and Cossio, F.P. (1997). *J. Org. Chem.* 62: 7033–7036.

**79** Matito, E., Poater, J., Duran, M., and Solà, M. (2005). *J. Mol. Struct. (Theochem)* 727: 165–171.

**80** Matito, E., Poater, J., Solà, M., and Schleyer, P.v.R. (2009). Aromaticity and chemical reactivity. In: *Chemical Reactivity Theory* (ed. P.K. Chattaraj), 419–438. Boca Ratón: Taylor and Francis/CRC Press.

**81** Morao, I. and Cossío, F.P. (1999). *J. Org. Chem.* 64: 1868–1874.

**82** Bach, R.D., Wolber, G.J., and Schlegel, H.B. (1985). *J. Am. Chem. Soc.* 107: 2837–2841.

**83** Havenith, R.W.A., Fowler, P.W., Jenneskens, L.W., and Steiner, E. (2003). *J. Phys. Chem. A* 107: 1867–1871.

**84** Santos, J.C., Polo, V., and Andrés, J. (2005). *Chem. Phys. Lett.* 406: 393–397.

**85** Feixas, F., Matito, E., Poater, J., and Solà, M. (2008). *J. Comput. Chem.* 29: 1543–1554.

**86** Rzepa, H.S. (2005). *Chem. Rev.* 105: 3697–3715.

**87** Sliwa, W. (1997). *Fullerene Sci. Technol.* 5: 1133–1175.

**88** Maynollo, J. and Krautler, B. (1996). *Fullerene Sci. Technol.* 4: 213–226.

**89** Hirsch, A. (1994). *The Chemistry of Fullerenes*. Stuttgart: Thieme.

**90** Hirsch, A. (1999). *Top. Curr. Chem.* 199: 1–65.

**91** Martin, N. (2006). *Chem. Commun.* 2093–2104.

**92** Pang, L.S.K. and Wilson, M.A. (1993). *J. Phys. Chem.* 97: 6761–6763.

**93** Giovane, L.M., Barco, J.W., Yadav, T. et al. (1993). *J. Phys. Chem.* 97: 8560–8561.

**94** Solà, M., Mestres, J., Martí, J., and Duran, M. (1994). *Chem. Phys. Lett.* 231: 325–330.

**95** Fernández, I., Solà, M., and Bickelhaupt, F.M. (2013). *Chem. Eur. J.* 19: 7416–7422.

**96** Garcia-Borràs, M., Osuna, S., Swart, M. et al. (2013). *Chem. Commun.* 49: 1220–1222.

**97** Solà, M. (2017). Aromaticity. In: *Encyclopedia of Physical Organic Chemistry*, vol. 1 (ed. Z.D. Wang and U. Wille), 511–542. Weinheim: Wiley.

**98** Bingel, C. (1993). *Chem. Ber.* 126: 1957–1959.

**99** Lu, X., Feng, L., Akasaka, T., and Nagase, S. (2012). *Chem. Soc. Rev.* 41: 7723–7760.

**100** Popov, A.A., Yang, S., and Dunsch, L. (2013). *Chem. Rev.* 113: 5989–6113.

**101** Kiely, A.F., Haddon, R.C., Meier, M.S. et al. (1999). *J. Am. Chem. Soc.* 121: 7971–7972.

**102** Martínez, J.P., Garcia-Borràs, M., Osuna, S. et al. (2016). *Chem. Eur. J.* 22: 5953–5962.

**103** Garcia-Borràs, M., Cerón, M.R., Osuna, S. et al. (2016). *Angew. Chem. Int. Ed.* 55: 2374–2377.

# 5

# Descriptors of Aromaticity: Geometric Criteria

*Motto: "There is no more basic enterprise in chemistry than the determination of the geometrical structure of a molecule. Such a determination, when it is well done ends all speculation as to the structure and provides us with the starting point for understanding of every physical, chemical and biological property of the molecule."*

Roald Hoffmann (Foreword to the monograph by L. V. Vilkov, V. S. Mastryukov, N. I. Sadova, Determination of the Geometrical Structure of Free Molecules, Mir Publishers, Moscow, **1993**)

## 5.1 Introduction

A rapid and versatile development in the determination of the molecular geometry by use of X-ray [1–3], gas phase diffraction [2, 4, 5], microwave wave and vibration-rotation, spectroscopies [2], as well as application of the quantum chemical modeling [6], and finally a great development of the methods directly describing the electronic structure of molecules [7, 8] have brought about a great explosion in new interpretations of chemical notions, processes, and phenomena.

As a result of the previously mentioned developments in experimental and theoretical techniques, chemical bonds are now recognized by both their geometrical and electronic characteristics. It is worth mentioning that, due to simple calorimetric measurements and theoretical approaches [9], particular bonds can be characterized not only by their lengths, but also by their energies [10]. Even if the relationships between these two characteristics are not very precise, it is important to take this aspect into account when molecular structure is subject to any deeper interpretation. For a particular bond, the deformations can be described by an empirical function (5.1) [11], which describes changes of the potential energy

*Aromaticity and Antiaromaticity: Concepts and Applications*, First Edition.
Miquel Solà, Alexander I. Boldyrev, Michał K. Cyrański, Tadeusz M. Krygowski, and Gabriel Merino.

E(R) of a diatomic molecule as a function of an interatomic distance, R [12].

$$E(R) = D\left\{e^{-2\alpha(R-R_e)} - 2e^{-\alpha(R-R_e)}\right\} \tag{5.1}$$

where $D$ is the energy of the system at equilibrium distance $R_e$ and $\alpha$ is a constant.

Another very popular approach describing deformation of molecular geometry that has been very fruitful in infra-red spectroscopy is the harmonic oscillator model [13]. In the simplest form applied to bond lengths, the energy of the bond deformation $E$ is proportional to the squared value of the difference between the equilibrium length $R_0$ and the deformed length $R_d$ multiplied by a force constant $k$ [14].

$$E = k(R_0 - R_d)^2 \tag{5.2}$$

Applications of the harmonic oscillator model in vibrational spectroscopy are associated with the fact that many potential energy surfaces are quite well approximated by a quadratic function for small displacements from their minima. However, in many cases more advanced approaches have to be applied [13, 15].

Since the harmonic oscillator model allows estimation of deformation energies described by changes in bond length, it has become natural to apply this model for cases in which, due to some intra- or inter- molecular forces, molecular geometry changes. This kind of approach was used in an approximate way, assuming some particular bond lengths as references, and then the deviations from them used to answer the question how far on an energetic scale is a given bond from the reference one fulfilling some particular chemical meaning. The idea of stabilization energy and the Kekulé structure contributions derived from bond lengths, named as HOSE (harmonic oscillator stabilization energy) [16] took into account deviations from the lengths of the single and double bonds and then estimated deformation energy with reference to Kekulé structures with single and double bonds; for a review see [17]. The energies obtained from the HOSE model correlated very well (cc = 0.991) with the Hess and Schaad resonance energies [18–20] for 22 molecules of alternant π-electron hydrocarbons. Even more striking has been the application of the HOSE model for estimating the contributions of the Kekulé structures. Estimated values by the HOSE model and by the Randić approach [21] gave cc = 0.985 with a slope 0.998. Even more convincing is a linear regression between the weights of the canonical structures derived by use of a graph-topological approach [22] and the contributions estimated by HOSE, cc = 0.997 for 150 data points.

## 5.2 Geometry-Based Estimation of the Molecular Energy

More than a half century ago Pauling [23] introduced a concept of a fractional bond order, $n$, as a difference between a given bond length, R($n$) and a length of the standard single bond, R(1), defined by the formula:

$$R(n) - R(1) = -c\ln(n) \tag{5.3}$$

where $c$ is an empirical constant. The concept was found to be very fruitful in structural chemistry. It has been even applied to identify and analyze reaction paths [1, 24, 25]. In 1967, Johnston and Parr [26] introduced an empirical rule relating bond energy E($n$) to the bond order n:

$$E(n) = E(1)n^p \tag{5.4}$$

E(1) and E($n$) in Eq. (5.4) represent the energies of bonds with bond numbers 1 and n, respectively, whereas p is an empirical constant. Combining (5.3) and (5.4) leads to a formula for bond energy E($n$) [27]:

$$E(n) = E(1)\exp\{\alpha[R(1) - R(n)]\} \tag{5.5}$$

Taking into account experimental data [28, 29] it was possible to derive a formula for the heat of formation of a molecule from atoms (HtFfA) directly from bond lengths:

$$HtFfA = -100.53n - 87.99\sum_{i=1}^{N}\exp\{2.255(1.533 - R_i)\} \tag{5.6}$$

Comparison of the calculated and experimental [29] data for nine benzenoid hydrocarbons (benzene, naphthalene, anthracene, phenanthrene, tetracene, chrysene, triphenylene, 3,4-benzophenanthrene, and pyrene) gave an estimated error less than 1% of the mean calculated HtFfA values. Further, the parameterization of expression (5.5) led to an empirical formula for an individual C—C bond energy, BE, which is (5.7):

$$BE = 87.99\exp\{2.255(1.533 - R)\} \tag{5.7}$$

Summation of BE values for all bonds leads to the energy content (EC) of the whole molecules, or of the ring (ring energy content, REC). Table 5.1 presents the REC values for selected benzenoid hydrocarbons (for more examples see Ref. [30]).

There are a few interesting findings presented in Table 5.1. The REC, but also harmonic oscillator model of aromaticity (HOMA) (for details of this aromaticity index see later in this chapter) descriptors based on experimental geometry are different for symmetrically equivalent rings. This is due to the specific conditions of X-ray diffraction determination of molecular geometry. If a molecule in a crystal lattice lies in a special position, that is, on some element of symmetry, then its symmetrical property is maintained. If, however, it lies in a general position, that is, not on any symmetry element, then its molecular environment in the crystal is no more symmetrical and hence different intermolecular interactions act on a molecule which, in the gas-phase, would be symmetrical [35]. Note also that REC and HOMA values for the external rings are not far from the value for benzene,

**Table 5.1** REC and HOMA values for individual rings of four benzenoid hydrocarbons, calculated using experimental bond lengths taken from [30].

| System | HOMA | REC |
|---|---|---|
| Benzene[a)] |  |  |
| Phenanthrene[b)] | 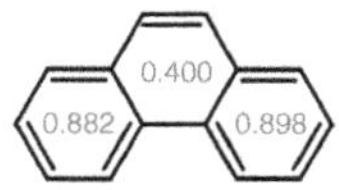 | 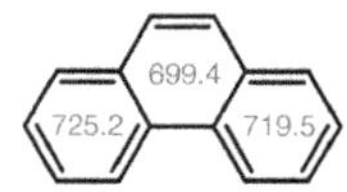 |
| Triphenylene[c)] | 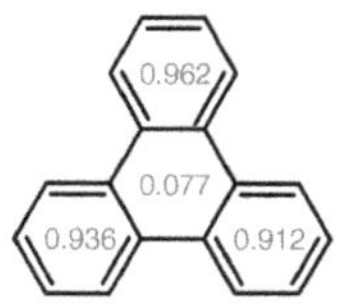 |  |
| Perylene[d)] | 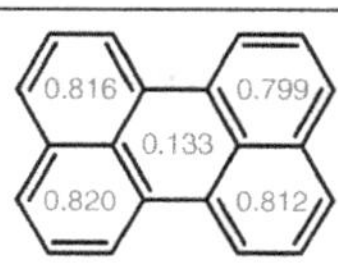 | 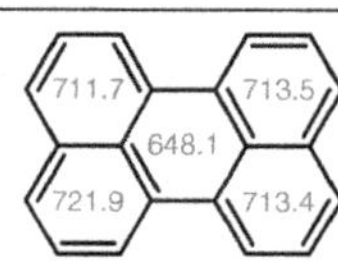 |

a) Ref. [31].
b) Ref. [32].
c) Ref. [33].
d) Ref. [34].

whereas those for central rings are dramatically smaller, indicating much lower aromaticity. Following the Clar classification [36, 37] the external rings in phenanthrene or triphenylene are so-called rings with an aromatic π-sextet, whereas the central rings in triphenylene and perylene belong to the category of "empty" rings, that is, with a lower participation of π-electrons in their structure. In the case of linear polycyclic acenes, the aromatic π-sextet is assumed to be a "migrating" π-sextet. Finally, the central ring in phenanthrene is a ring with a double bond, that is, with a more significant contribution of localized single and double bonds.

Molecular geometry has provided several consistent relationships between bond lengths and bond energies [38–43] that were applied to both planar [40] and nonplanar systems [41].

## 5.3 Bond Length Alternation as a Basis for Defining Aromaticity Indices

Access to the experimental geometries of molecules has allowed the early discovery that, in aromatic systems, there is an observed equalization of bond lengths. Contrary to this, non- or antiaromatic systems have a significant bond length alternation. Experimental data for benzene and 1,3,5,7-cyclooctatetraene illustrate this point very well. The high resolution Raman spectroscopy structure determination gave CC bond lengths in benzene equal to 1.397(3) Å for all bonds [44], very close to the low temperature (10K) X-ray determination with 1.3972(5) Å [31]; whereas gas phase electron diffraction geometry determination of 1,3,5,7-cyclooctatetraene finds two kinds of bonds with lengths 1.334(1) Å and 1.462(1) Å, for double and single bonds, respectively [45]. The early observed lack of the bond length alternation in the archetypal aromatic system – benzene – and a very large one observed in the typically cyclo-olefinic system 1,3,5,7-cyclooctatetraene, led to the idea of applying the degree of bond length alternation as a quantitative descriptor of aromatic character in molecules.

Unlike other aromaticity characteristics, geometry-based aromaticity indices may be applied in a much wider way than any other descriptors. They may be used on individual rings; on any fragment of π-electron system; on the whole molecule; or on only the peripheral (perimeter) bonds, as recommended by Julg [46]. If applied to noncyclic systems they estimates the extent of π-electron delocalization, rather than aromaticity. It is worth mentioning that peripheral bonds are characterized by a much greater π-electron delocalization than those inside the molecules [47].

### 5.3.1 The Julg Aromaticity Index $A_J$

The first attempt to make use of bond length alternation for estimating aromaticity [46, 48] relied on defining an aromaticity index $A_J$ being a normalized function of variance for perimeter CC bond lengths:

$$A_j = 1 - \frac{255}{n} \sum_{i=1}^{n} \left(1 - \frac{R_i}{R_{av}}\right)^2 \tag{5.8}$$

where $n$ is the number of peripheral bonds of length $R_i$ whereas $R_{av}$ is the mean value of all bond lengths (in the original paper the bond length was denoted by $d$). The constant 225, which results from the normalization condition, was chosen to preserve zero aromaticity for the Kekulé structure of benzene with bond lengths as in 1,3-butadiene (a model nonaromatic system). The Julg index returns unity for any system with all bonds of the same length (fully aromatic one).

This was a very good idea for quantifying the degree of bond length alternation of a molecule; but a serious inconvenience has appeared: what to do in the case of systems with heteroatoms? In such a case, $A_J$ was used for hetero-π-electron systems applied only to the CC bonds as carried out by Schleyer et al. for five-membered heterocyclic systems [49].

### 5.3.2 The Harmonic Oscillator Model of Aromaticity, HOMA (1972 and 1993)

The HOMA is a modification of the Julg idea. The mean bond length $R_{av}$ in Eq. (5.8) was replaced [50] by a conceptual parameter, the optimal bond length, $R_{opt}$, assuming this quantity being the closest to that realized in a fully aromatic system. The basic idea of the HOMA model is that energy cost of extension or compression of the bond depends on the force constants, which, in turn, are dependent on bond lengths. Assuming a 2:1 ratio between the force constants for double and single bonds [15, 51, 52] taking as reference CC bond the values for CC bond length in ethane ($R_{C\text{-}C}$ = 1.524 Å) and ethene ($R_{C=C}$ = 1.334 Å) [53] and assuming that energies of extension of $R_{opt}$ to the length of single bond and of compression to the length of the double bond are equal, the reference bond length is given by

$$R_{opt} = \frac{1}{3}(R_{C-C} + 2R_{C=C}) \tag{5.9}$$

in good agreement with the experimental values for benzene 1.397(3) Å for the CC bond determined by Raman spectroscopy [44] and by means of low-temperature solid state X-ray diffraction, 1.3972(5) Å [31]. The formula for HOMA reads:

$$HOMA = 1 - \frac{\alpha}{n}\sum_{i=1}^{n}(R_{opt} - R_i)^2 \tag{5.10}$$

where $\alpha = 98.89$ is an empirical normalization constant, chosen to give HOMA = 0 for a model nonaromatic system and HOMA = 1 for a system where all bonds are equal to $R_{opt}$ = 1.397 Å; $n$ is the number of CC bonds included in the summation; $R_{opt}$ is the optimal bond length for aromatic systems; and the $R_i$ are the experimental or computed bond lengths.

The modification of HOMA presented in 1993 [54] introduces two important changes. First, it extended the model to hetero-π-electron systems with the following bonds: CN, CO, CP, CS, NN, NO. Second, it introduced an adjusted $R_{opt}$ value, changing from the average value based on the CC bonds in ethane and ethene (1.397 Å) to the value based on the CC bond lengths from electron diffraction measurements of butadiene-1,3 (1.467 and 1.349 Å [55], for a single and double bonds, respectively) resulting in $R_{opt}$ = 1.388 Å. These changes also resulted also in a change of the normalization constant, which for this optimal bond length is $\alpha = 257.7$. The motivation for this change was that the reference systems for estimation of resonance energy were acyclic olefins [56, 57]. For a recent review see Cyrański [58].

Consequently, application of the new reference bond lengths, $R_s$ and $R_d$ for single and double ones led to HOMA parameters in the forms presented in

Eqs. (5.11) and (5.12):

$$R_{opt} = \frac{R_s + wR_d}{1 + w} \tag{5.11}$$

$$\alpha = 2\left[(R_s - R_{opt})^2 + (R_d - R_{opt})^2\right]^{-1} \tag{5.12}$$

where $w$ denotes the ratio of force constants for the double and single bonds, and, as previously, this was assumed to be $w = 2$ [15, 51]. Accepting these changes, the HOMA index can be expressed as in Eq. (5.13) [54]:

$$HOMA = 1 - \frac{1}{n}\left\{\alpha(CC)\sum(R(CC)_{opt} - R_i)^2 + \alpha(CX)\sum(R(CX)_{opt} - R_i)^2 + \alpha(CY)\sum(R(CY)_{opt} - R_i)^2 + \alpha(XY)\sum(R(XY)_{opt} - R_i)^2\right\} \tag{5.13}$$

where α and R values are specified by the type of bonds located in parentheses and have the meanings presented earlier in the text; $n$ is the number of bonds taken into account.

Unfortunately, the modification introduced some kind of inconsistency, and parameters of HOMA (collected in Table 5.2) consist of two kinds of the previously mentioned models, and also some of them in between. One is, if the bonds used

**Table 5.2** Reference bond lengths $R_s$ and $R_d$, and appropriate $R_{opt}$, c and α values used in the HOMA calculation and its components.

| Type of bond | $R_s$ (Å) | $R_d$ (Å) | $R_{opt}$ (Å) | α | c | References |
|---|---|---|---|---|---|---|
| BB | 1.6474 | 1.5260 | 1.5665 | 244.147 | 0.1752 | [59] |
| BB | 1.6474 | 1.5260 | 1.5693 | 250.544 | 0.1750 | [59] |
| BC | 1.5472 | 1.3616 | 1.4235 | 104.507 | 0.2678 | [60] |
| BC | 1.5542 | 1.3796 | 1.4378 | 118.009 | 0.2520 | [60] |
| BC | 1.5542 | 1.3796 | 1.4386 | 118.618 | 0.2520 | [60] |
| BN | 1.564 | 1.363 | 1.402 | 72.03 | | [61] |
| CC | 1.467 | 1.349 | 1.388 | 257.7 | 0.1702 | [54] |
| CN | 1.465 | 1.269 | 1.334 | 93.52 | 0.2828 | [54] |
| CO | 1.367 | 1.217 | 1.265 | 157.38 | 0.2164 | [54] |
| CP | 1.814 | 1.640 | 1.698 | 118.91 | 0.2510 | [54] |
| CS | 1.807 | 1.611 | 1.677 | 94.09 | 0.2828 | [54] |
| CSe | 1.959 | 1.7591 | 1.8217 | 84.9144 | 0.2970 | [62] |
| NN | 1.420 | 1.254 | 1.309 | 130.33 | 0.2395 | [54] |
| NO | 1.415 | 1.164 | 1.248 | 57.21 | 0.3621 | [54] |

to estimate $R_{opt}$ represent single and double bonds with some kind of π-electron delocalization, for example, the case with CC bonds taken from butadiene or CO bonds taken from the monomer of formic acid, then we obtain a greater value of normalization constant α (257.7 and 157.38 for CC and CO bonds, respectively). Opposite to this is the case where the reference single and double bonds are typically single and double ones as in the case of CC bonds in ethane and ethene, or CS bonds in $S(CH_3)_2$ and $H_2C{=}S$ or CN in $H_2N{-}CH_2$ and in $HN{=}CH_2$. For these cases, the normalization constant α in each is much lower, 98.89, 94.09, and 93.52, respectively. This inconsistency of the HOMA model and the differences in α values have to be taken into consideration when aromaticity is quantified by means of the HOMA models. The subject of differences between various definition of $R_{opt}$ and, in consequence, of the normalization constant α are discussed in another part of the chapter. Table 5.2 summarizes the various parameters of the HOMA models.

One more modification came in 1998 [61], when the way of estimating $R_{opt}$ by use of Eq. (5.9) and a strongly simplified approximation assuming $w = 2$, was replaced by a much better justified method based on *ab initio* computation for ethane, ethene, $BH_3NH_3$ and $BH_2{=}NH_2$ with the aim of getting force constants for the stretching motions of C-C, C=C, B-N, and B=N bonds.

As a result, the obtained ratio $w$ for BN bonds amounts to 4.2, dramatically greater than the formerly accepted value of $w = 2.0$. Undoubtedly this is a problem to be thoroughly considered in future works on geometry-based aromaticity indices. The result obtained for BN bond is $R_s = 1.564$ Å, $R_d = 1.363$ Å, resulting in $R_{opt} = 1.402$ Å and $\alpha = 72.03$. Application of these HOMA parameters gave for 15 systems with BN bonds acceptable agreement with the data obtained by means of the Bird $I_6$ index of aromaticity [63].

To describe aromaticity of systems with CSe bonds, Zborowski and Proniewicz [62] obtained parameters of the HOMA index for the CSe bond. Reference lengths of the single and double were taken from $CH_3SeH$ and $CH_2Se$, respectively, from experimental data. Following the procedure of Madura et al. [61], a value of $w = 1.79$ was obtained, that is, quite close to 2.0. In the case of CC and CO bonds, corresponding values were 2.22 and 2.60, respectively. For this reason, the carbon-selenium bond was included in the group of bonds for which the approximation $w = 2$ is valid.

Implementation of the HOMA index for π-electron systems with BB [59] and CB bonds [60] was carried out in two ways. In the case of π-electron systems containing BC bonds, the basic parameters of HOMA (i.e., $R_{opt}$, α) were estimated in three ways: (i) from experimental BC single and double bonds, (ii) from optimized computations on model molecules with this bond with the ratio w = 1.962, and (iii) the same as (ii) but assuming the ratio w of force constants of the double to single bond not as 2:1. The differences between final results of HOMA calculated for 10 model molecules containing C—B bonds are usually much less than 5 percent, and for $HOMA_{exp}$ and NICS(0) and NICS(1) [64] good correlations were found. In the case of π-electron systems containing BB bonds, the HOMA parameters were obtained in two ways: within the approximation that k(double bond)/k(single bond) = 2 and from the estimated force constants ratio. The new estimated parameters for BB bonds [59] and the earlier estimated ones for CB bonds [60] allowed comparison of the HOMA values for hydrocarbons and their derivatives with BB and BC bonds. For 19 π-electron systems a substantial role of the replacement of CC by BB bonds was found, particularly in cases where BB is located inside the molecule. As a rule, insertion of BB bonds dramatically decreases aromaticity of the rings with this bond. This kind of effect on aromaticity has also been studied for incorporation of boron and several other heteroatoms into pyrene and other, similar systems [65].

## 5.4 Separation of HOMA into Two Components EN and GEO (1996)

Formula (5.10) can be analytically transferred into Eq. (5.14) [66]:

$$HOMA = 1 - \frac{\alpha}{n}\sum_{i=1}^{n}(R_{opt} - R_i)^2 = 1 - EN - GEO \tag{5.14}$$

where *EN* and *GEO* are defined as:

$$GEO = \frac{1}{n}\sum_{i=1}^{n}\alpha(R_{av} - R_i)^2 \tag{5.15}$$

and

$$EN = \alpha(R_{opt} - R_{av})^2 \tag{5.16}$$

The term *GEO* represents a decrease of aromaticity due to an increase of bond length alternation, whereas EN reflects an increase of the average bond length in a given object. Note that the increase of EN (i.e., elongation of mean bond length) denotes a decrease of aromaticity because the value of HOMA diminishes. The separation carried out in this way works only for carbocyclic systems. To apply the preceding separation of HOMA for systems with hetero-atoms [67, 68], one has to use the Pauling bond order [23], Eq. (5.3). The value of the constant *c* in Eq. (5.17) can be estimated applying bond lengths for typical single, $R_s$, and double, $R_d$, bonds with bond orders $n = 1$ and 2, respectively:

$$c = \frac{R_s - R_d}{\ln 2} \tag{5.17}$$

Therefore, for any bond its bond order can be calculated using the Eq. (5.18) and the appropriate parameters ($R_s$ and $c$):

$$n = \exp\left[\frac{(R_s - R_n)}{c}\right] \tag{5.18}$$

An optimal bond order, $n_{opt}$, can be evaluated using Eq. (5.18) for $R_n = R_{opt}$. It was found, that $n_{opt}$ has practically the same value for all bonds taken into consideration (in the range between 1.584 and 1.602), therefore, $n_{opt} = 1.590$ was proposed for all bonds.

Finally, by replacing bond lengths by their respective bond orders in Eqs. (5.14)–(5.16), the formulae enabling the separation of HOMA into its energetic and geometric terms for heterosystems were introduced. The values of the constant $c$ for particular bonds are collected in Table 5.2. This formula has been used, for example, to study aza-analogues of benzenoid hydrocarbons [68].

## 5.5 Harmonic Oscillator Model of Electron Delocalization, HOMED (2007)

Raczyńska et al. introduced a modification of the HOMA concept (Eq. (5.14)) named HOMED [69–71], which is assumed to be a descriptor of all kinds of the resonance effects (any type of π-electron delocalization):

$$HOMED = 1 - \frac{\alpha}{n}\sum (R_{opt} - R_i)^2 \tag{5.19}$$

where $\alpha$ is a normalization constant, $n$ is the number of bonds forming a system, $R_{opt}$ is the optimum bond length and $R_i$ is the $i$th bond length.

The HOMED formula, Eq. (5.19), is formally the same as HOMA of Eq. (5.10) and the difference is due to another definition of reference systems for single and double bonds, and also, due to a different definition of the normalization constant. The normalization constant for HOMED is the same as for HOMA, Eq. (5.10), in the case of systems that consist of an even number of bonds; but for systems with an odd number of bonds, the normalization constant can be determined using the formulae [71]:

$$\alpha = \frac{2i + 1}{(i + 1)(R_{opt} - R_s)^2 + i(R_{opt} - R_d)^2} \tag{5.20}$$

for systems with $i + 1$ single bonds and $i$ double bonds, and

$$\alpha = \frac{2i + 1}{i(R_{opt} - R_s)^2 + (i + 1)(R_{opt} - R_d)^2} \tag{5.21}$$

for systems that consist of $i$ single and $i + 1$ double bonds. $R_s$ and $R_d$ are the single and double bond lengths in reference systems.

The most important change in comparison with HOMA (1993) is the way that the reference systems for single bonds as simple saturated species were chosen.

**Table 5.3** Parameters for the HOMED index taken from [69, 71].

| Type of bond | $R_s$ (Å) | $R_d$ (Å) | $R_{opt}$ (Å) | α |
|---|---|---|---|---|
| CC | 1.5300 | 1.3288 | 1.397 | 98.89 |
| CN | 1.4658 | 1.2670 | 1.334 | 93.52 |
| CO | 1.4238 | 1.2017 | 1.273 | 70.80 |

Table 5.3 gives for the reference the bond lengths of ethane, methylamine, and methanol for single CC, CN, and CO bonds. The same procedure was applied to unsaturated molecules to determine the reference double bond lengths. The reference systems for the double CC, CN, and CO bond lengths are ethene, methylamine, and formaldehyde, respectively. The optimal bond lengths in the table come from benzene, 1,3,5-triazine, and protonated carbonic acid [71].

Comparison of HOMA (1993) and HOMED data for a wide set of acyclic π-electron systems [71] revealed that these two approaches are frequently not equivalent. It was also found that HOMED values obtained using different levels of theory are comparable even when $\alpha$ and bond lengths differ when changing the method/basis set. For acyclic π-electron systems, where only σ–π delocalization is possible, HOMED values lie in the interval 0.0–0.4 that proves some weak delocalization. For heteroallyl systems, where n-π conjugation occurs, HOMED values are between 0.4 and 0.8, which suggests moderate delocalization in heteroallyl systems.

HOMED has also been tested for a few cases of cyclic conjugated π-electron systems [71]. In the case of azines, HOMA (1993) and HOMED gave almost identical results. In the case of six-membered rings containing nitrogen and oxygen atoms, numerical values of HOMED and HOMA (1993) are different but the direction of variability is similar. The most significant differences are observed for five-membered systems, where mixed $n$-π and π-π conjugations are possible. This may lead to nonequivalent resonance structures, with or without the charge separation.

In the case of compounds that form a structurally close family, the HOMED and HOMA (1993) descriptions are well related [72].

## 5.6 Harmonic Oscillator Model for Heterocycles with π Electron and/or *n*-Electron Delocalization: The HOMHED Index (2012)

Extension of the HOMA-concept to heterocyclic systems with π electron and/or $n$-electron delocalization was undertaken by Frizzo et al. [73], who applied slightly different reference bond lengths for evaluating HOMA index parameters

**Table 5.4** Parameters for HOMHED index [73].

| Type of bond | $R_s$ (Å) | $R_d$ (Å) | $R_{opt}$ (Å) | α |
|---|---|---|---|---|
| CC | 1.530 | 1.316 | 1.387 | 78.6 |
| CN | 1.474 | 1.271 | 1.339 | 87.4 |
| CO | 1.426 | 1.210 | 1.282 | 77.2 |
| CS | 1.819 | 1.559 | 1.672 | 74.4 |
| NN | 1.454 | 1.240 | 1.311 | 78.6 |
| NO | 1.463 | 1.218 | 1.300 | 60.0 |
| NS | 1.765 | 1.541 | 1.616 | 71.7 |

for systems with CC, CN, CO, CS, NN, NO, and NS bonds. The experimental bond lengths applied for estimation of the new HOMA parameters ($R_s$, $R_d$, $R_{opt}$, and α) were taken from tables of bond lengths determined by X-ray and neutron diffraction [74]. The resulting parameters were obtained using the mean values for a given type of bond and are collected in Table 5.4.

The new obtained parameters were compared with those of Raczyńska HOMED [71] and of HOMA (1993) [54], and applied to 16 molecular systems. Differences between HOMA (1993) and HOMHED were observed mostly for systems with CO and CN bonds. They result from the difference in the model bonds applied for estimation of $R_{opt}$ and α in HOMA (1993) and in HOMHED, since for HOMA (1993) the reference systems for CC, CO, and CN were butadiene, COOH in the monomer of HCOOH and amidine – all systems with some portion of resonance effect operating in them. In the case of Frizzo and Martins' approach, similarly as in the case of HOMED [71] single bonds, $R_s$, contain atoms of $sp^3$ hybridization and hence the results are in better agreement.

## 5.7 Applications of the Bond Orders for Estimating Aromaticity

Simultaneously with the introduction of the HOMA index based on the bond lengths, [50], its analogue based on the Hückel molecular orbital (HMO) bond order was proposed. Taking into account that there is a well-known relation between HMO bond orders and bond lengths [75], the HOMA index in terms of bond orders, $p_{rs}$, reads:

$$HOMA = 1 - \frac{3.60}{n} \sum \left(\frac{2}{3} - p_{rs}\right)^2 \tag{5.22}$$

Application of SCF bond orders for benzenoid and nonbenzenoid hydrocarbons [76, 77] gave much greater differentiation of the data than those calculated using the HMO approach.

Another attempt to quantitatively describe aromaticity using bond orders was made by Fringuelli et al. [78]. They proposed using bond orders instead of the bond lengths because bonds between different atoms can have the same length, but differ in the bond order. The bond orders $N$ were calculated by means of the Gordy equation [79].

$$N = a \cdot R^{-2} - b, \tag{5.23}$$

where $a$ and $b$ denote constants (see Table 5.5). The sum of the differences of the bond orders, $\Delta N$, of a ring was taken as a descriptor of its aromaticity. The smaller

**Table 5.5** Values of constants $a$ and $b$ used in the calculation of bond orders with Eq. (5.23).

| Type of bond | a | b | References |
|---|---|---|---|
| BB | 9.12 | 1.94[a)] | [79] |
| BC | 8.05 | 2.11[a)] | [79] |
| BN | 7.15 | 2.10[a)] | [79] |
| BN | 7.72 | 2.16[a)] | [61] |
| BO | 6.75 | 2.14[a)] | [79] |
| CC | 6.80 | 1.71[a)] | [79] |
| | 7.366 | 2.115 | [80] |
| CN | 6.48 | 2.00[a)] | [79] |
| | 6.941 | 2.205 | [80] |
| CP | 13.54 | 3.02 | [63] |
| CO | 5.75 | 1.85[a)] | [79] |
| | 5.315 | 1.616 | [80] |
| CS | 11.9 | 2.59[a)] | [79] |
| | 11.250 | 2.400 | [80] |
| CSe | 15.24 | 3.09 | [78] |
| CTe | 21.41 | 3.81 | [78] |
| NN | 5.28 | 1.41[a)] | [79] |
| | 6.132 | 1.916 | [80] |
| NO | 4.98 | 1.45[a)] | [79] |
| | 4.634 | 1.165 | [80] |
| NS | 10.53 | 2.50 | [63] |
| NSe | 13.31 | 2.86 | [63] |
| OO | 4.73 | 1.22[a)] | [79] |
| OS | 17.05 | 5.58 | [63] |
| SS | 19.30 | 3.46 | [63] |

a) The sign of original parameter has been changed for consistency.

$\Sigma\Delta N$ value indicates a more aromatic system (for benzene $\Sigma\Delta N = 0$). In this way, they established a sequence of decreasing aromaticity: benzene > thiophene > selenophene > tellurophene > furan. The sequence was consistent with reactivity- and magnetism-based criteria of aromaticity.

## 5.8 Bird's Aromaticity Indices $I_5$ and $I_6$ (1985)

Bird [63, 81, 82] applied the same idea as Julg and François [46] replacing bond lengths, $R$, by the Gordy bond orders, [77] Eq. (5.23). As a consequence, the coefficient of variation for the bond orders of a heterocyclic ring is given by Eq. (5.24):

$$V = \frac{100}{N_{av}}\sqrt{\frac{1}{n}\sum_{i=1}^{n}(N_i - N_{av})^2} \tag{5.24}$$

and aromaticity index:

$$I_A = 100 \cdot \left(1 - \frac{V}{V_k}\right) \tag{5.25}$$

In Eq. (5.24) $N_i$ is the individual bond order, $N_{av}$ denotes the mean bond order, and $n$ is the number of bonds. In the case of a fully aromatic ring, $V$ has a value of 0, whereas for a nondelocalized Kekulé structure with alternating single and double bonds, the value depends on the type of ring system. Thus in Eq. (5.25) for a six-membered heterocyclic ring, the constant $V_k = 33.3$, whereas for a five-membered ring, the $V_k = 35$. The aromaticity index $I$ depends on a ring size, therefore subscripts were added. So $I_5$, $I_6$, $I_{5,6,}$ and $I_{5,5}$ denote Bird index for five-membered [63, 83], six-membered [81], five- and six-membered or five- and five-membered rings fused together [84], respectively.

## 5.9 Pozharskii Criterion of Aromaticity, $\Delta\overline{N}$ (1985) and Bond Alternation Coefficient, BAC (1995)

Pozharskii [85, 86] introduced in 1985 another criterion of aromaticity ($\Delta\overline{N}$), which is the average fluctuation of the bond orders in the cycle:

$$\Delta\overline{N} = \frac{\sum \Delta N}{n} \tag{5.26}$$

$$\sum \Delta N = \sum_{i,j} |N_i - N_j| \tag{5.27}$$

where $N$ denotes bond order and $n$, the number of bonds.

The extremal values are for benzene ($\Delta\overline{N} = 0$) and for cyclopentadiene ($\Delta\overline{N} = 0.49$). These values can be pictured using a percentage scale. Thus, at the beginning of the scale will be cyclopentadiene (0 percent) and at the end, benzene (100 percent).

To evaluate the aromaticity, this index allows use of both the experimental and theoretical bond lengths. It may be applied for the estimation of the aromaticity of individual rings in polycyclic aromatic systems. The Pozharskii index gives satisfactory agreement between calculations and experiments for six-membered heterocycles with one or more heteroatoms and for five-membered heterocycles with one heteroatom.

Another approach, which bases solely on bond length alternation has been the BAC (bond alternation coefficient) index introduced by Krygowski, Ciesielski et al. in 1995 [27]. It bases on the sum of squared differences of the sequential bond lengths as given by Eq. (5.28):

$$BAC = \sqrt{\sum_n (R_n - R_{n+1})^2} \tag{5.28}$$

The summation runs over a defined cycle. It can be applied to account aromaticity in the case of monocyclic systems; however, in the case of polycyclics, the problem becomes much more complex. It must be stressed that the Pozharskii index (but also BAC) has a major disadvantage: It cannot be applied in the case of planar antiaromatic systems and for estimating the aromatic character of the highly symmetric systems such as, for example, 1,3,5-triazines where, due to cyclic conjugation, an equalization of bond lengths is observed. Moreover, the BAC can be applied only to systems with homogenous type of bonds, what is a limitation for most of heteroaromatics.

## 5.10 Applications

Geometry-based aromaticity indices are widely used as very easily accessible characteristics of π-electron delocalization in both carbo- and heterocyclic systems. Extensive analyses of many systems and also chemical problems with use of this criterion of aromaticity have been discussed in two Chemical Reviews devoted to "Structural Aspects of Aromaticity" [87] and "Aromaticity from the Viewpoint of Molecular Geometry: Application to Planar Systems" [88]. However, application for heterocyclic systems is associated with some discomfort due to different reference systems in both – the modifications of HOMA (see earlier parts of the chapter) – and various definitions of bond orders in the Bird-type approaches. Their application is discussed in Chapter 9 devoted to heteroaromaticity. Despite the quantitative nonequivalence of the data obtained by different indices,

**Table 5.6** HOMA and HOMED indices for simple heteroaromatic systems [71].

| System | HOMA | HOMED | System | HOMA | HOMED | System | HOMA | HOMED |
|---|---|---|---|---|---|---|---|---|
| 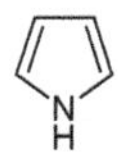 | 0.854 | 0.921 | 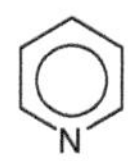 | 0.995 | 0.999 | 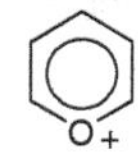 | 0.748 | 0.926 |
| 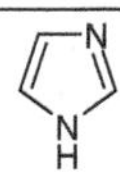 | 0.882 | 0.903 | 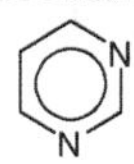 | 0.999 | 0.999 | 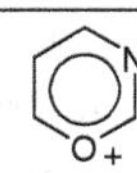 | 0.699 | 0.893 |
|  | 0.189 | 0.749 | 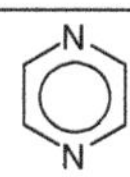 | 0.996 | 0.999 | 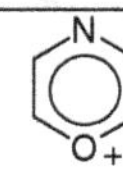 | 0.816 | 0.951 |
| 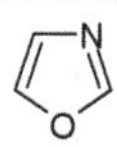 | 0.224 | 0.702 | 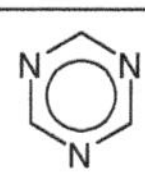 | 1.000 | 1.00 | 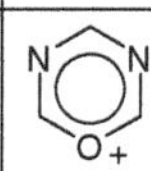 | 0.661 | 0.870 |

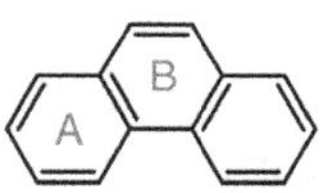

**1**

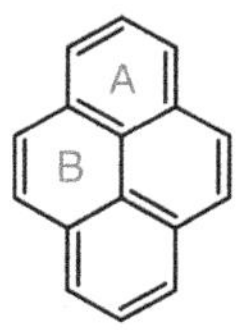

**2**

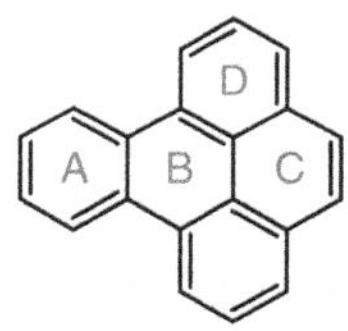

**3**

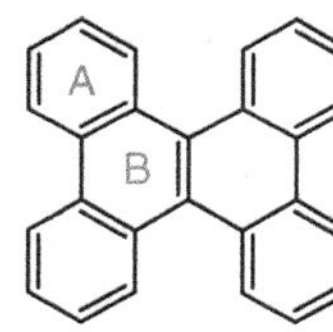

**4**

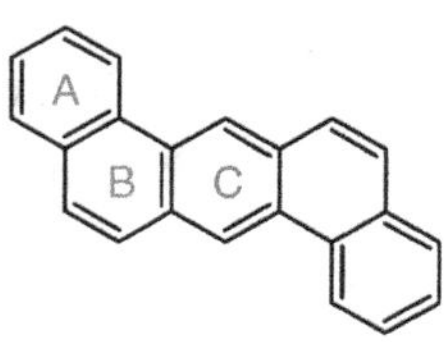

**5**

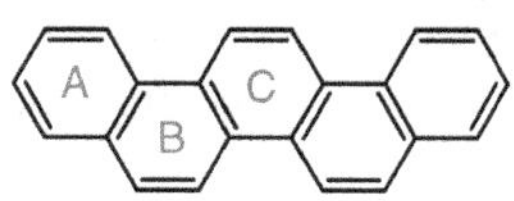

**6**

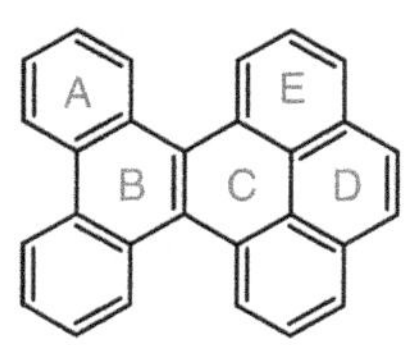

**7**

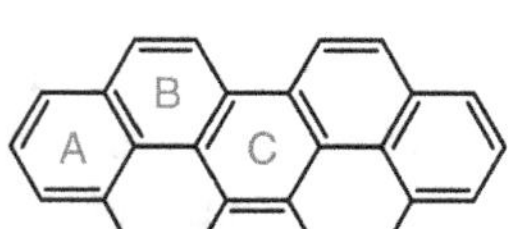

**8**

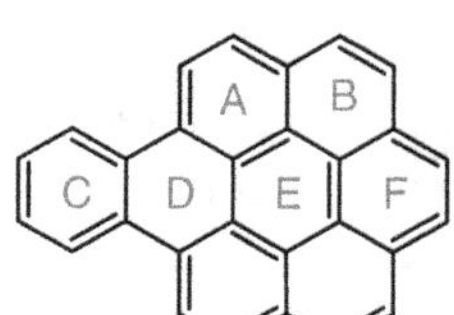

**9**

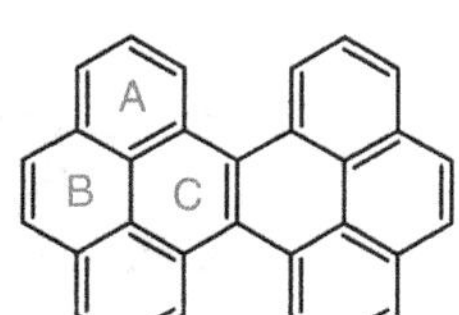

**10**

**Figure 5.1** Selected polycyclic hydrocarbons with ring labels [90].

**Table 5.7** PDI (in electrons), HOMA, and NICS (in ppm) values for the polycyclic hydrocarbons listed in Figure 5.1 [90].

| | | Polycyclic hydrocarbon | | | | | | | | | |
|---|---|---|---|---|---|---|---|---|---|---|---|
| **Descriptor** | **Ring** | **1** | **2** | **3** | **4** | **5** | **6** | **7** | **8** | **9** | **10** |
| **PDI** | **A** | 0.080 | 0.069 | 0.086 | 0.084 | 0.083 | 0.080 | 0.084 | 0.068 | 0.058 | 0.071 |
| | **B** | 0.047 | 0.043 | 0.026 | 0.034 | 0.041 | 0.051 | 0.034 | 0.044 | 0.051 | 0.045 |
| | **C** | | | 0.044 | | 0.068 | 0.059 | 0.031 | 0.045 | 0.082 | 0.031 |
| | **D** | | | 0.073 | | | | 0.045 | | 0.033 | |
| | **E** | | | | | | | 0.071 | | 0.029 | |
| | **F** | | | | | | | | | 0.055 | |
| **HOMA** | **A** | 0.856 | 0.834 | 0.889 | 0.811 | 0.872 | 0.847 | 0.807 | 0.840 | 0.749 | 0.721 |
| | **B** | 0.435 | 0.553 | −0.030 | 0.383 | 0.356 | 0.520 | −0.518 | 0.550 | 0.686 | 0.477 |
| | **C** | | | 0.518 | | 0.788 | 0.640 | −0.496 | 0.570 | 0.853 | −0.511 |
| | **D** | | | 0.838 | | | | 0.479 | | 0.341 | |
| | **E** | | | | | | | 0.723 | | 0.510 | |
| | **F** | | | | | | | | | 0.732 | |
| **NICS** | **A** | −10.06 | −12.74 | −8.63 | −9.44 | −9.93 | −9.80 | −9.30 | −12.63 | −10.97 | −12.22 |
| | **B** | −6.82 | −5.07 | −1.18 | −4.13 | −5.38 | −7.15 | −3.74 | −4.82 | −9.33 | −5.88 |
| | **C** | | | −5.47 | | −11.27 | −8.35 | −2.29 | −13.04 | −9.77 | −1.83 |
| | **D** | | | −11.58 | | | | −6.04 | | −5.86 | |
| | **E** | | | | | | | −12.38 | | 0.05 | |
| | **F** | | | | | | | | | −11.39 | |

if a chosen set of homostructural systems is considered, then the various aromaticity criteria lead to comparable results. A good perspective on the previously presented remarks is shown by HOMED and HOMA indices for a few typical heterocyclic systems, see Table 5.6 [71].

This problem is less important in the case of carbocyclic systems but still exists. However, the HOMA index has mostly been applied to these kind of systems. This index was used in a critical analysis of the local aromaticity concept in polycyclic aromatic compounds [89] as well as for assessment of the Clar's aromatic π-sextet rule for polycyclic aromatic compounds [90]. In both cases, HOMA data were confronted with a collection of other aromaticity characteristics. As a rule, no good correlation between them was found, in agreement with an earlier documented concept of multidimensional character of aromaticity indices [91–93]. A good illustration of the extent of aromaticity variation in rings of polycyclic hydrocarbons is shown in Figure 5.1 and Table 5.7.

It is important to note that the aromaticity of a ring depends dramatically on its topological environment.

## 5.11 Impact of the Electric Field on Aromaticity

It is a well-known fact that electrophiles or nucleophiles, which are often electrically charged units, cause chemical reactions known as electrophilic or nucleophilic substitution [94, 95]. As a result, the new chemical species still maintain their π-electron structure – that is, aromatic π-electron delocalization, but may be substantially changed. However, we should be aware that aromatic molecules necessarily exist in some chemical environment and may interact with charged or dipolar species. The question is posed: How does the electric field affect π-electron delocalization in aromatic compounds? The answer comes from studies of polycalicenes [96], where it was shown that an increase in electric field induces dramatic changes in aromaticity (see Figure 5.2). This effect is

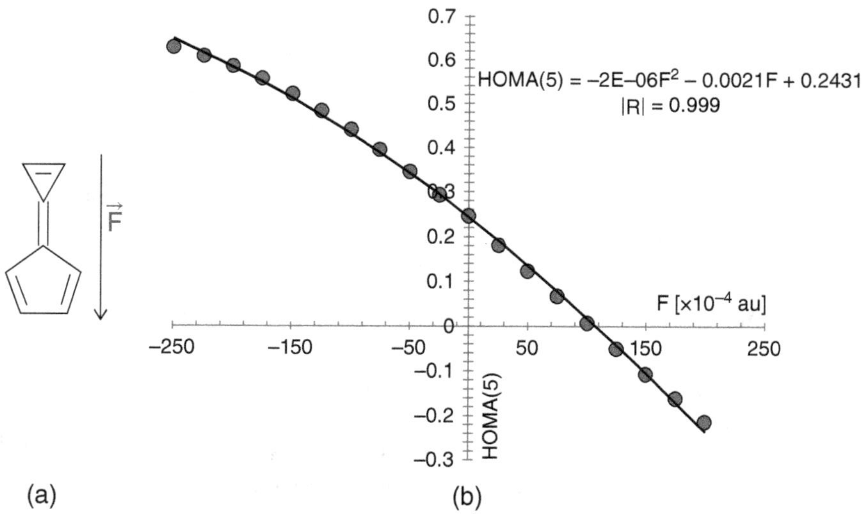


**Figure 5.2** (a) Orientation of a calicene molecule in an external electric field ($\vec{F}$). (b) Dependence of the HOMA index and the value of external electric field (a.u.) for a five-membered ring. Source: Reprinted with permission from Ref. [96]. Copyright © 2015, American Chemical Society.

also observed in terms of changes of aromaticity in polycalicenes, where dipolar calicene units making up the polycalicenes clearly affect the aromaticity of three- or five-membered rings.

## 5.12 Stacking Interactions versus H-Bonding in Nucleobases

In the three-dimensional structure of DNA helixes, two very important interactions determine their stability: hydrogen bonding and stacking interactions. In both cases, these interactions affect the π-electron delocalization of their participants. It has recently been shown [97] how the aromaticity of rings in nucleobases is changed as a result of stacking interactions. The estimation of aromatic character used four aromaticity indices: PDI (*para*-delocalization index) [98], ATI (average two-center index) [99], FLU (aromatic fluctuation index) [100, 101], and HOMA [54]. It appears that the variability of HOMA values due to stacking interactions depends on the kind of ring. The five-membered rings in adenine and guanine are relatively aromatic with HOMA ~0.9 and the range of changes are ~0.009 and ~0.018 for adenine and guanine, respectively. The variability of HOMA for six-membered rings is greater in two ways. The HOMA values of six-membered rings in adenine, guanine, thymine, and uracyl for noninteracting rings are 0.891, 0.754, 0.538, and 0.570, respectively. The ranges of HOMA values due to these interactions are at the level 0.05–0.07 units of HOMA and usually stacking increases the aromaticity of these rings. On the other hand, the aromatic characters of five-membered rings is much more weakly dependent on stacking interactions – the ranges of HOMA value due to these interactions are at the level 0.01 units of HOMA.

In a similar way, some attention has been paid to the consequences of H-bonding on the aromatic character of nucleobases. Their aromatic characters, both as individual molecules and as in the Watson-Crick pairs, were studied computationally at the B3LYP/6-311+G** and MP2/6-311+G* levels of theory [102]. Analysis of the changes in aromaticity of the bases involved in Watson–Crick pairs led to the conclusion that, for bases containing one or two carbonyl groups, their aromatic character increases due to H-bond formation. The double bonds become longer, resulting in higher aromatic character of the ring to which the CO bond is attached. The observed increase is between 0.03 and 0.13 units of HOMA.

## 5.13 Showing the Interaction Path for Substituent Effects

Since geometry-based characteristics of π-electron delocalization may be applied to any π-electron fragment of a molecule, HOMA has been applied for estimating the interaction path of substituent effects. The paths of π-electron delocalization in four-substituted 1,2-benzoquinone were studied [103] using four aromaticity indices: HOMA [54], MCI (multicenter index) [99, 116], DI (delocalization index) [104], and FLU [100, 101, 104] at the B3LYP/6-311+G(d,p) level of theory. The qualitative picture of interactions between electron donating/attracting substituents with carbonyl groups of the quinone is shown in Figure 5.3.

The changes in π-electron delocalization (as determined by HOMA) for transmitting paths involving OC2C3C4 and OC1C6C5C4 on the substituent constants are presented in Figure 5.4. In the former case, the slope is negative, indicating a decrease of the π-electron delocalization with a decrease of π-electron donating power of the substituent. In the latter case, the relationship is similar, but the slope is less steep and positive. A similar picture was found for the dependences of MCI on the substituent constants.

The differentiation between interactions are well shown by dependences of CO bond lengths and DI values of C2O and C1O on substituent constants (see Figure 5.5). Evidently C2O interacts with substituents much more strongly than C1O.

A similar but more comprehensive analysis was carried out for the conjugation paths in monosubstituted 1,2 and 2,3-naphthoquinones [104]. Optimization of 10 monosubstituted 1,2 and 2,3-naphthoquinones at the B3LYP/6-311+G** level allowed characterization of possible conjugation paths using HOMA, DI, and FLU.

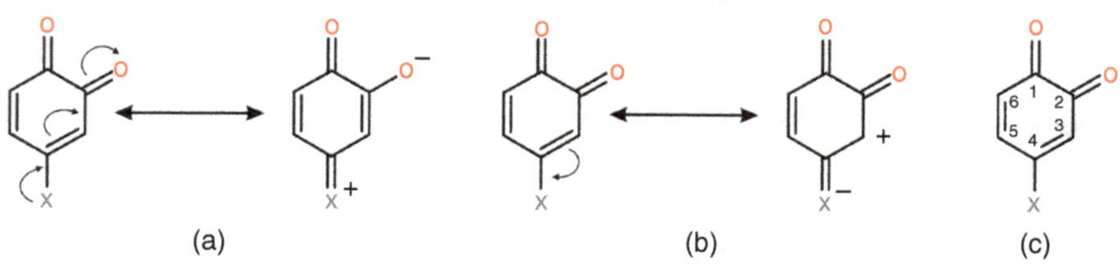


**Figure 5.3** (a) Interaction between π-electron-donating substituents and carbonyl groups for *meta*-type substitution in 4-X-1,2-benzoquinone; (b) Interaction between π-electron-attracting substituents and carbonyl groups for *para*-type substitution in 4-X-1,2-benzoquinone; (c) labeling of atoms in 4-X-substituted of 1,2-benzoquinone. X stands for NO, $NO_2$, CN, CHO, H, Me, OMe, OH, $NH_2$, NHMe or $NMe_2$.

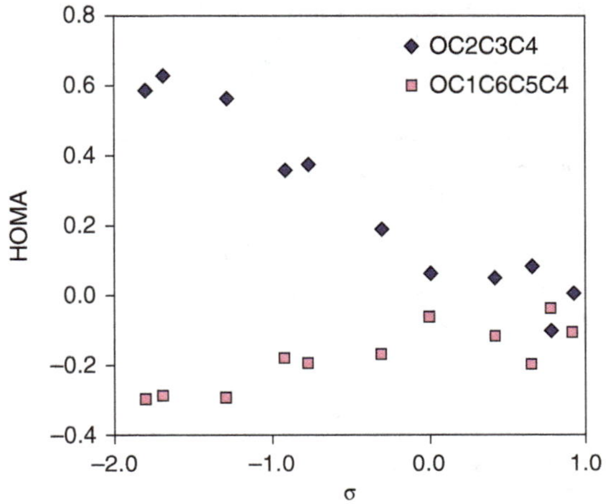


**Figure 5.4** Dependence of HOMA for transmission paths OC2C3C4 (*meta*-type way) and OC1C6C5C4 (*para*-type way) on substituent constants σ ($\sigma_p^+$ for electron-donating substituents and for others $\sigma_p$). Source: Reprinted with permission from Ref. [103]. Copyright © 2011, American Chemical Society.

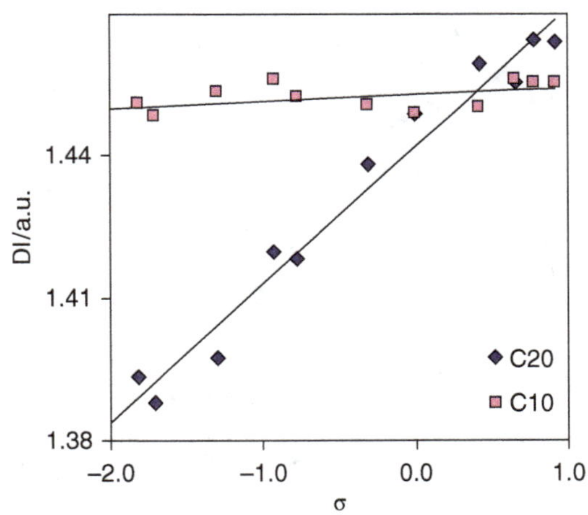


**Figure 5.5** Correlation between the bonding DIs of C10 and C20 with substituent constants σ ($\sigma_p^+$ for electron-donating substituents and for others $\sigma_p$); for C20 bond cc = 0.985. Source: Reprinted with permission from Ref. [103]. Copyright © 2011, American Chemical Society.

All these indices presented a similar picture of interaction paths. Tables 5.8 and 5.9 present the results of regression between the HOMA values for a given kind of interaction path and the Hammett substituent constants.

Two important conclusion can be drawn. First, the interaction paths that consist of even numbers of bonds (including C-X) usually have much better correlation coefficients than those containing odd numbers of bonds. Second, the former usually have greater (as absolute values) slopes, indicating stronger interactions.

**Table 5.8** Dependences of HOMA for Conjugation Paths on Substituent Constants for 5- and 6-Substituted 2,3-Naphthoquinone Derivatives. *N* stands for the number of bonds between X and oxygen atoms. The correlation coefficients and regression equations are given [104].

| System | Conjugation pathway | N | **HOMA path** vs. σ |
|---|---|---|---|
| (1) | 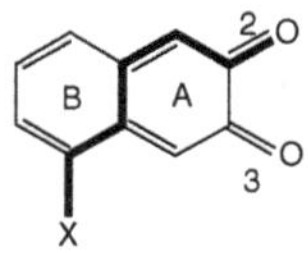 | **6** | **R = 0.056**<br>y = 0.0029x – 0.2469 |
| (2) | 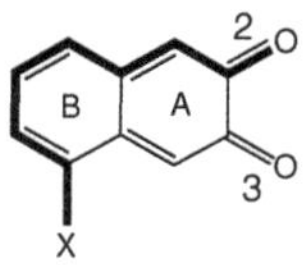 | **8** | **R = 0.80**<br>y = –0.0456x + 0.4346 |
| (3) | 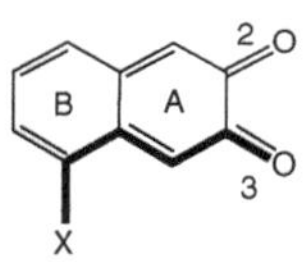 | **5** | **R = 0.57**<br>y = 0.0385x + 0.0361 |
| (4) | 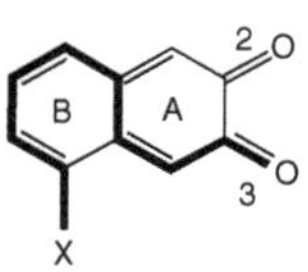 | **9** | **R = 0.74**<br>y = –0.0192x + 0.1205 |
| (5) | 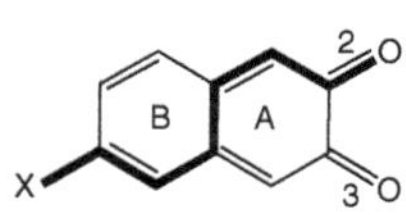 | **7** | **R = 0.73**<br>y = –0.0227x + 0.0559 |
| (6) | 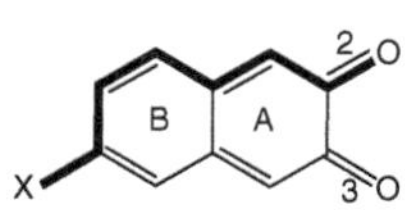 | **7** | **R = 0.88**<br>y = 0.0467x + 0.2141 |
| (7) | 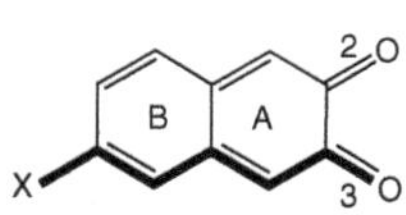 | **6** | **R = 0.97**<br>y = –0.1299x + 0.4132 |
| (8) | 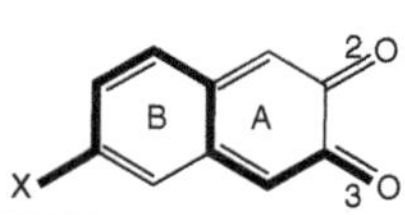 | **8** | **R = 0.89**<br>y = –0.033x – 0.0043 |

**Table 5.9** Dependences of HOMA on Substituent Constant for 4-, 5-, 6-, and 7-Substituted 1,2-Naphthoquinone Derivatives. *N* stands for the number of bonds between X and oxygen atoms. The correlation coefficients and regression equations are given [104].

| System | Conjugation pathway | N | **HOMA path** vs. σ |
|---|---|---|---|
| (9) | 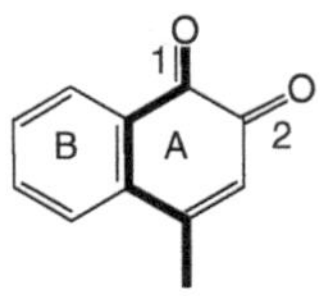 | **5** | **R = 0.88**<br>$y = 0.0954x - 0.3176$ |
| (10) | 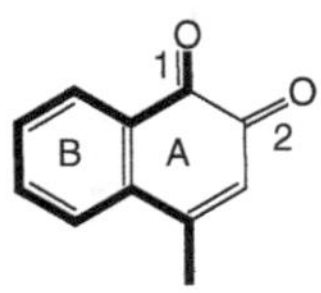 | **9** | **R = 0.89**<br>$y = 0.0503x + 0.3531$ |
| (11) | 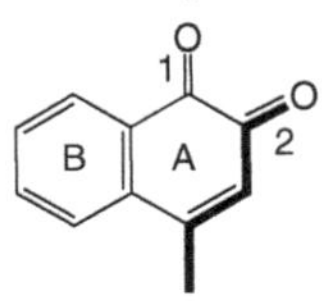 | **4** | **R = 0.97**<br>$y = -0.1834x + 0.2726$ |
| (12) | 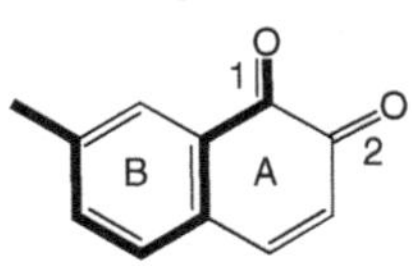 | **7** | **R = 0.44**<br>$y = 0.0087x + 0.4118$ |
| (13) | 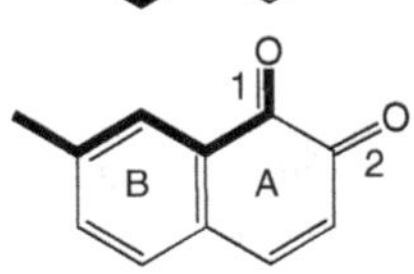 | **5** | **R = 0.53**<br>$y = 0.0153x + 0.1699$ |
| (14) | 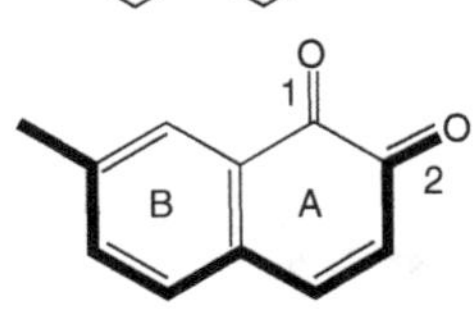 | **8** | **R = 0.99**<br>$y = -0.0489x + 0.4411$ |
| (15) | 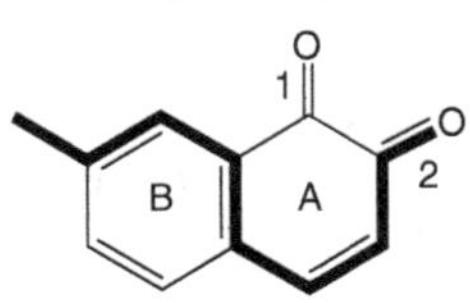 | **8** | **R = 0.99**<br>$y = -0.0493x + 0.4302$ |
| (16) | 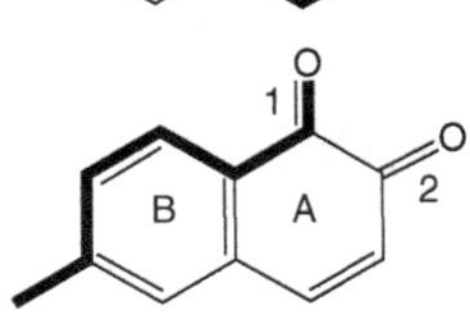 | **6** | **R = 0.99**<br>$y = -0.0833x + 0.3732$ |
| (17) | 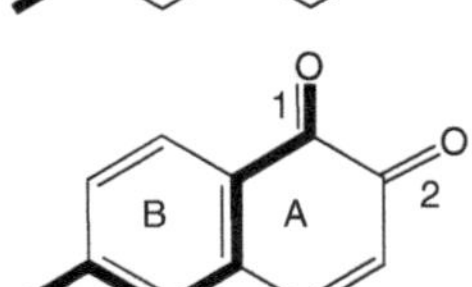 | **6** | **R = 0.99**<br>$y = -0.0876x + 0.3492$ |

**Table 5.9** (Continued)

| | | | |
|---|---|---|---|
| (18) | 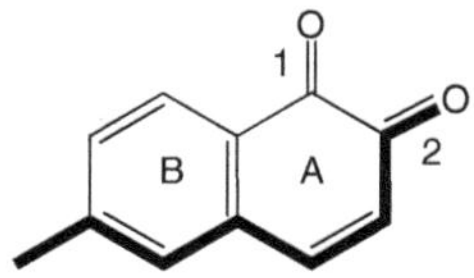 | **7** | **R = 0.79**<br>y = 0.0157x + 0.3133 |
| (19) | 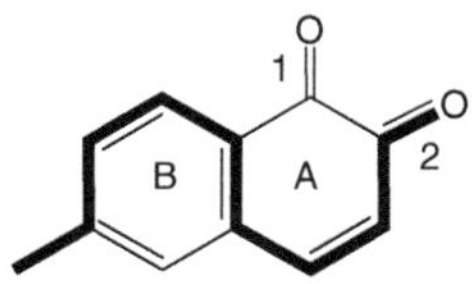 | **9** | **R = 0.79**<br>y = 0.0146x + 0.4645 |
| (20) | 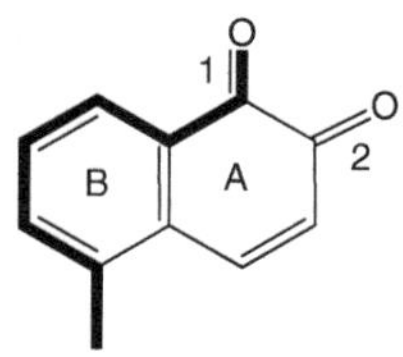 | **7** | **R = 0.11**<br>y = −0.0014x + 0.4294 |
| (21) | 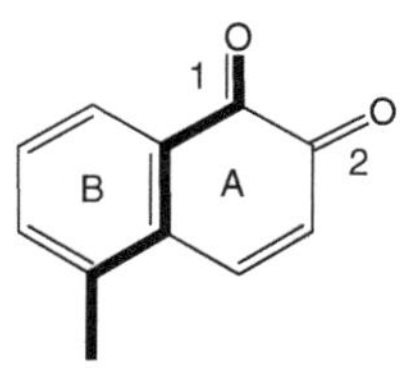 | **5** | **R = 0.21**<br>y = 0.0056x + 0.0826 |
| (22) | 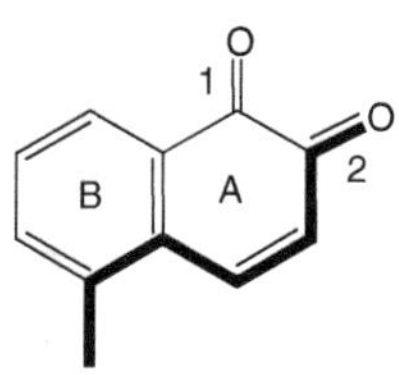 | **6** | **R = 0.90**<br>y = −0.0461x + 0.1997 |
| (23) | 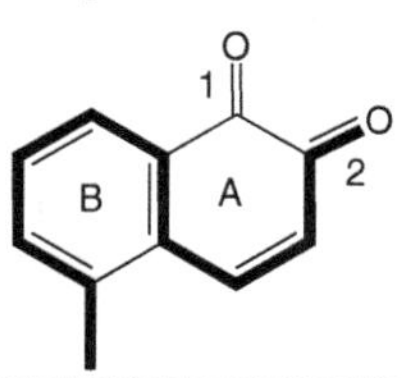 | **10** | **R = 0.95**<br>y = −0.0275x + 0.5479 |

## 5.14 Applications in the Field of Quasiaromatic Systems

Quasi-aromatic systems may be well illustrated by a simple scheme as follows (Figure 5.6) [105].

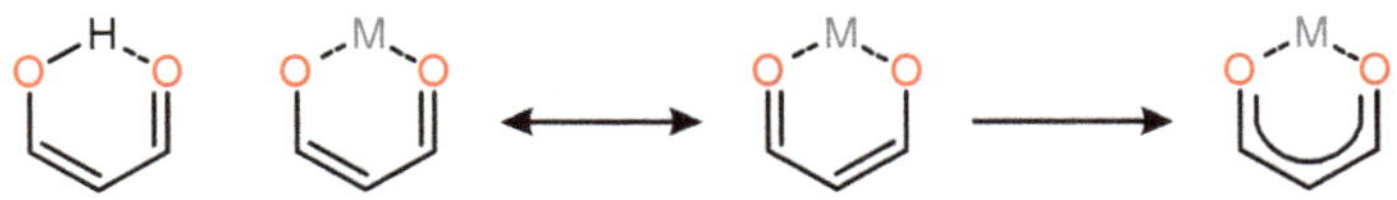


**Figure 5.6** Schemes of quasi-aromatic systems with chelating proton (I) and metal cation (II). CH bonds are not shown.

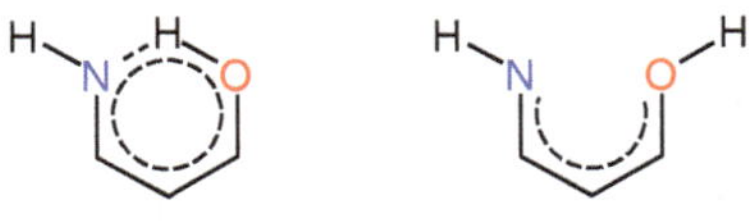


**Figure 5.7** HOMA and NICS values for two conformations of (1Z,3E)-3-iminoprop-1-ol [105].

A chelating chain consisting of π-electron bonds changes its electron delocalization depending on the kind of the chelated cation. The definition was presented a half century ago: "molecules should be called quasi-aromatic only if they contain an acyclic conjugated π-electron system and show chemical properties typical of aromatic compounds, especially reaction by substitution with retention of type. The significant mesomeric stability is implied" [106]. It is important to stress the difference between quasiaromaticity and metalla-aromaticity [107]. In quasi-aromatic systems, coordinated hydrogen or metal atoms do not participate in π-electron delocalization that operate in the chelate part of the system. Figure 5.7 nicely illustrates to what extent chelation may affect π-electron delocalization in open and closed conformations of (1Z,3E)-3-iminoprop-1-ol. Interestingly, the NICS (nucleus independent chemical shift) [64, 108] characteristics in these kind of cases is disappointing, for a clear reason. In the quasi "ring" no effective ring current can be formed and hence the lack of changes in NICS characteristics [105].

## 5.15 Extension of HOMA to Noncyclic and Non-π-electron Systems

In 2014, the idea of HOMA was applied to cyclic, non-π-electron systems [109]. In the original formula of HOMA [54] (Eq. (5.10)) bond lengths were replaced by the electron densities at bond critical points. The value of the normalization coefficient α was also changed appropriately:

$$HOMA(\rho) = 1 - \frac{\alpha_{BCP}}{n}\sum_{i=1}^{n}\left(\rho_i^{BCP} - \rho_{opt}^{BCP}\right)^2 \tag{5.29}$$

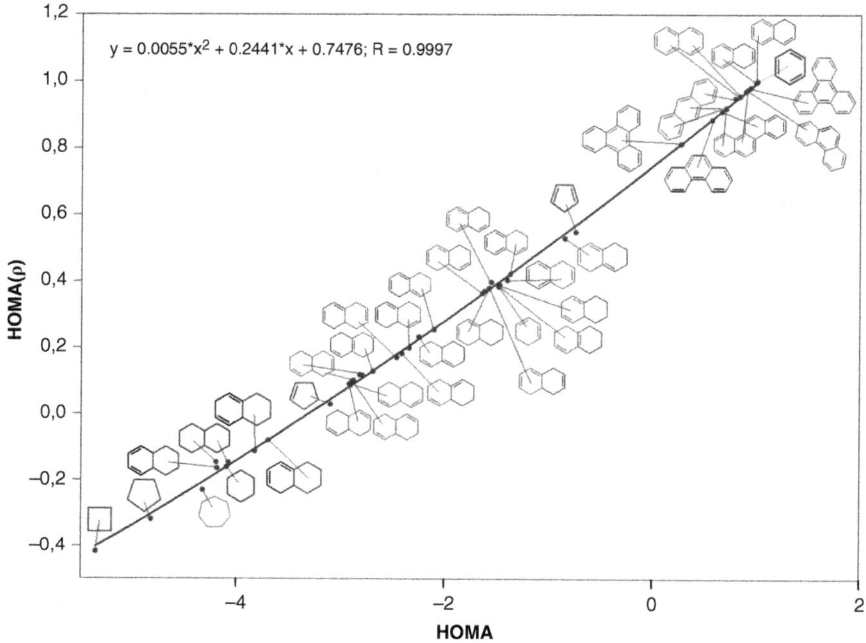


**Figure 5.8** The correlation between the HOMA geometrical index and HOMA(ρ) index based on electron densities at bond critical points for series of saturated, unsaturated, and aromatic hydrocarbon rings. Source: Reprinted with permission from Ref. [109]. Copyright © 2014, Royal Society of Chemistry.

ρ stands for the electron density at the BCP of the $i$th bond in the analyzed structure and the optimal value in the reference (benzene); and $n$ is the number of CC bonds in the considered structure. Regressions of these BCP properties against the original HOMA characteristics were found as quadratic fits with R greater than 0.999 as shown in Figure 5.8.

However, when this kind of treatment was applied to series of saturated and unsaturated hydrocarbons, no clear correlation was found. Howard and Krygowski showed that topological charge density parameters correlate very well with bond lengths as so aromaticity parameters [110]. Based on the HOMA philosophy of estimation of aromatic character, Dominikowska and Palusiak also proposed a similar approach based on ellipticity of a chemical bond within the QTAIM (quantum theory of atoms in molecules) [111]. This approach can also be used for both theoretical and experimental data and shows good agreement with other indicators of aromaticity for hydrocarbons. Another approach, CTED (corrected total electron density) based on ellipticity at C—C bond critical points and bond length alternations has been proposed by Tokatli and Ucun [112] to estimate π-electron density distribution in cyclic systems. They showed a rather

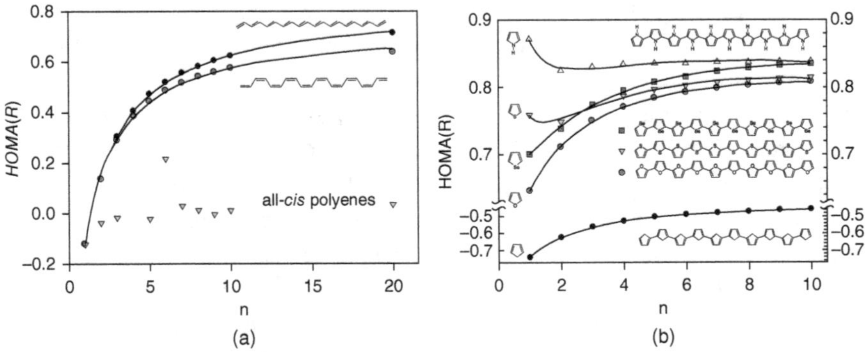

**Figure 5.9** Relationships between the aromaticity index HOMA of (a) conjugated all-*trans*, *cis–trans*, and all-*cis* polyenes and the number of conjugated double bonds in the polyene molecule. (b) polycyclopentadiene, polyfuran, polythiophene, and polypyrrole and the number of cyclic moieties repeated in the structure. Source: Reprinted with permission from Ref. [113]. Copyright © 2015, Royal Society of Chemistry.

comprehensive description of aromaticity by CTED [112, 113] and HOMA [54] (but also PDI [98], FLU [100, 101], and NICS [64, 108]) for a wide range of aromatic, nonaromatic, and antiaromatic compounds.

The original HOMA [54] has also been applied to conjugated all-*trans*, *cis–trans* and all-*cis* polyenes as well as to model polymers: polycyclopentadiene, polyfuran, polythiophene, and polypyrrole (see Figure 5.9) [113].

The plot of HOMA values all-*trans* and *cis–trans* polyenes as a function of *n* (the number of conjugated double bonds in the structure) demonstrates a very significant increase of π-electron delocalization to a horizontal asymptote. The trends show that the electron delocalization in the former system is more efficient than in the *cis–trans* ones. It is also clear that, in the case of all-*cis* polyenes, it is practically zero and does not change with elongation of the polyene chain. Horizontal asymptotes are also observed in the case of polycyclopentadiene, polyfuran, polythiophene, and polypyrrole. Only in the latter case is a decrease of aromaticity observed in comparison to the parent mono-ring system. In most cases, HOMA reaches the value 0.8–0.9, which is typical for six-membered aromatic rings in polycyclic aromatic hydrocarbons such as naphthalene or pyrene, indicating the presence of significant electron delocalization in the system [113].

## 5.16 Conclusions

What are the pros? Geometry-based aromaticity indices are easily accessible characteristics of π-electron delocalization. Usually they are mutually correlated, particularly for set of structurally similar systems. The great advantage of using them is that, in most cases, they may be applied in many different situations: to the whole molecules, to individual rings, and also to any π-electron fragment of the molecule in question. Also the size of a system does not cause any problem providing that a reliable geometry is obtained. The estimations can be made for planar, distorted, heterocyclic [114], and also acyclic systems. They can be also successfully made for analysis of electron delocalization in excited states [115] and in the presence of an external electric field [96]. Recently, application of the HOMA concept appeared useful in description of some acyclic and conducting systems. It's also important to note that the HOMA concept in which bond lengths are replaced by bond critical point properties correlate very well with HOMA based on bond lengths. As has been shown, geometry-based aromaticity models can be very easily parameterized and anyone can obtain a valuable tool for a given problem. What are the cons? The geometry-based evaluations depend on the quality of the structural information available. As has been shown, the estimations also depend somewhat on the type of parameterization, so the chemical problem has to be investigated using a consistent model. Although the estimations are very fast and efficient, some other effects unrelated to π-electron delocalization may interfere with the final findings – one of the obvious factors is molecular strain, present especially in highly distorted systems. Despite the great opportunities that geometry-based indices provide, an appropriate level of caution is always advised.

## References


1 Dunitz, J.D. (1978). *X-ray Analysis and the Structure Determination of Organic Molecules*. Ithaca: Cornell University Press.

2 Domenicano, A. and Hargittai, I. (ed.) (1992). *Accurate Molecular Structures, their Determination and Importance*, International Union of Crystallography, 127–169. Oxford: Oxford University Press.

3 Gilli, G. and Gilli, P. (2009). *The Nature of the Hydrogen Bond, Outline of a Comprehensive Hydrogen Theory*, International Union of Crystallography. Oxford: Oxford University Press.

4 Vilkov, I.V., Mastryukov, V.S., and Sadova, N.I. (1983). *Determination of the Geometrical Structure of Free Molecules*. Moscow: Mir Publishers.

5 Hargittai, I., Orville-Thomas, W., and Edts, J. (1981). *Diffraction Studies on Non-crystalline Substances*. Budapest: Akademiai Kiado.

6 Bachrach, S.M. (2014). *Computational Organic Chemistry*. Hoboken, New Jersey: Wiley.

7 Matta, C.F., Boyd, R.J., and Becke, A. (2007). *The Quantum Theory of Atoms in Molecules: From Solid State to DNA and Drug Design*. Weinheim: Wiley-VCH Verlag GmbH & Co. KGaA.

8 Glendening, E.D., Landis, C.R., and Weinhold, F. (2012). *WIREs Comput. Mol. Sci.* 2: 1–42.

9 Yu-Ran, L. (2007). *Comprehensive Handbook of Chemical Bond Energies*. Boca Raton: CRC Press.

10 Dewar, M.J.S. and Schmeising, H.N. (1959). *Tetrahedron* 5: 166–178.

11 Morse, P.M. (1929). II. Vibrational levels. *Phys. Rev.* 34: 57–64.

12 Coulson, C.A. (1961). *Valence*. Oxford: Oxford University Press.

13 Chalmers, J.M. and Griffiths, P.R. (2001). *Handbook of Vibrational Spectroscopy*. Chichester: Wiley.

14 Wojtkowiak, B. and Chabanel, M. (1977). *Spectrochemie Moleculaire*. Paris: Techniqe et Documentation.

15 Wilson, E.B., Decius, J.C., and Cross, P.C. (1955). *Molecular Vibrations. The Theory of Infrared and Raman Vibrational Spectra*. New York: McGraw-Hill.

16 Krygowski, T.M., Anulewicz, R., and Kruszewski, J. (1983). *Acta Crystallogr. B* 39: 732–739.

17 Krygowski, T.M. (1988). Resonance structure contributions derived from the experimental geometry of molecules. In: *Structure and Reactivity* (ed. J.L. Liebman and A. Greenberg), 231–254. Weinheim: VCH.

18 Hess, B.A. and Schaad, L.J. (1971). *J. Am. Chem. Soc.* 93: 305–310.

19 Hess, B.A. and Schaad, L.J. (1971). *J. Org. Chem.* 36: 3418–3423.

20 Hess, B.A. and Schaad, L.J. (1971). *J. Am. Chem. Soc.* 93: 2413–2416.

21 Randić, M. (1977). *Tetrahedron* 33: 1905–1920.

22 Ciesielski, A., Krygowski, T.M., Cyrański, M.K., and Balaban, A.T. (2011). *Phys. Chem. Chem. Phys.* 13: 3737–3747.

23 Pauling, L. (1947). *J. Am. Chem. Soc.* 69: 542–553.

24 Bürgi, H.B. (1975). *Chem. Int. Ed. Eng.* 14: 460–473.

25 Bürgi, H.B. and Dunitz, J.D. (1994). *Structure Correlation*. Weinheim: VCH.

26 Johnston, H.S. and Parr, C. (1963). Hydrogen transfer reactions. *J. Am. Chem. Soc.* 84: 2544–2551.

27 Krygowski, T.M., Ciesielski, A., Bird, C.W., and Kotschy, A. (1995). *J. Chem. Inf. Comput. Sci.* 35: 203–210.

28 George, P. (1975). *Chem. Rev.* 75: 85–111.

29 Pedley, J.B., Taylor, R.D., and Kirby, S.P. (1986). *Thermochemical Data of Organic Compounds*. London: Chapman & Hall.

30 Krygowski, T.M. and Cyrański, M.K. (1998). Molecular geometry as a source of chemical information for π-electron compounds. In: *Theoretical Organic Chemistry – Theoretical and Computational Chemistry* (ed. C. Parkanyi), 153–187. Amsterdam: Elsevier Science.

**31** Jeffrey, G.A., Ruble, J.R., McMullan, R.K., and Pople, J.A. (1987). *Proc. Royal Soc.* London, Series A 414: 47–57.

**32** Kay, M.I., Okaya, Y., and Cox, D.E. (1971). *Acta Crystallogr. B* 27: 26–33.

**33** Ferraris, G., Jones, D.W., and Yerkess, J. (1973). *Z. Kristallogr.* 138: 113–128.

**34** Krygowski, T.M., Ciesielski, A., Świrska, B., and Leszczyński, P. (1994). *Pol. J. Chem.* 68: 2097–2107.

**35** Bunn, C.W. (1961). *Chemical Crystallography*. Oxford: Clarendon Press.

**36** Clar, E. (1964). *Polycyclic Hydrocarbons*, vol. 1 and 2. London, New York: Academic Press and Springer Verlag, Berlin-Göttingen.

**37** Clar, E. (1972). *The Aromatic Sextet*. Chichester: Wiley.

**38** Exner, K. and Schleyer, P.v.R. (2001). *J. Phys. Chem. A* 105: 3407–3416.

**39** Grimme, S. (1996). *J. Am. Chem. Soc.* 118: 1529–1534.

**40** Howard, S.T., Cyrański, M.K., and Stolarczyk, L.Z. (2001). *Chem. Commun.* 2: 197–198.

**41** Cyrański, M.K., Howard, S.T., and Chodkiewicz, M.L. (2004). *Chem. Commun.* 21: 2458–2459.

**42** Howard, S.T. (2003). *Phys. Chem. Chem. Phys.* 5: 3113–3119.

**43** Howard, S.T. and Lamarche, O. (2003). *J. Phys. Org. Chem.* 16: 133–141.

**44** Stoicheff, B.P. (1954). *Can. J. Phys.* 32: 339–346.

**45** Bastiansen, O., Hedberg, L., and Hedberg, K. (1957). *J. Chem. Phys.* 27: 1311–1317.

**46** Julg, A. and François, P. (1967). *Theor. Chim. Acta* 7: 249–259.

**47** Zborowski, K.K. and Krygowski, T.M. (2014). *Tetrahedron Lett.* 55: 6359–6361.

**48** Julg, A. (1971). A new definition of degree of aromaticity in aromaticity, pseudo-.aromaticity, anti-aromaticity. In: *Proceedings of an International Symposium held in Jerusalem* 31 March – 3 April 1970 (ed. E.D. Bergmann and B. Pullman), 383–385. Jerusalem: Israel Academy of Science and Humanities.

**49** Schleyer, P.v.R., Jiao, H., Goldfuss, B., and Freeman, P.K. (1995). *Angew. Chem. Int. Ed.* 34: 337–340.

**50** Kruszewski, J. and Krygowski, T.M. (1972). *Tetrahedron Lett.* 13: 3839–3842.

**51** Schwartz, J.C.P. (1964). *Physical methods in Organic Chemistry*. Edinburgh, London: Oliver & Boyd.

**52** Krygowski, T.M. and Cyrański, M. (1997). Aromatic character of carbocyclic pi-electron systems deduced from molecular geometry. In: *Advances in Molecular Structure Research*, vol. 3 (ed. I.M. Hargittai). London: JAI Press.

**53** Bastiansen, O. and Traetteberg, M. (1962). *Tetrahedron* 17: 147–154.

**54** Krygowski, T.M. (1993). *J. Chem. Inf. Comput. Sci.* 33: 70–78.

**55** Kveseth, K., Seip, R., and Kohl, D.A. (1980). *Acta Chem. Scand. A* 34: 31–42.

56 Pauling, L. and Sherman, J. (1933). *J. Chem. Phys.* 1: 606–617.
57 Jug, K. and Köster, A.M. (1990). *J. Am. Chem. Soc.* 112: 6772–6777.
58 Cyrański, M.K. (2005). *Chem. Rev.* 105: 3773–3811.
59 Zborowski, K.K., Alkorta, I., Elguero, J., and Proniewicz, L.M. (2013). *Struct. Chem.* 24: 543–548.
60 Zborowski, K.K., Alkorta, I., Elguero, J., and Proniewicz, L.M. (2012). *Struct. Chem.* 23: 595–600.
61 Madura, I.D., Krygowski, T.M., and Cyrański, M.K. (1998). *Tetrahedron* 54: 14913–14918.
62 Zborowski, K.K. and Proniewicz, L.M. (2009). *Pol. J. Chem.* 83: 477–484.
63 Bird, C.W. (1985). *Tetrahedron* 41: 1409–1414.
64 Chen, Z.F., Wannere, C.S., Corminboeuf, C. et al. (2005). *Chem. Rev.* 105: 3842–3888.
65 Mazurek, A. and Dobrowolski, J.C. (2015). *J. Phys. Org. Chem.* 28: 290–297.
66 Krygowski, T.M. and Cyrański, M. (1996). *Tetrahedron* 52: 1713–1722.
67 Krygowski, T.M. and Cyrański, M. (1996). Part IV. A general model for the π-electron systems. *Tetrahedron* 52: 10255–10264.
68 Cyrański, M. and Krygowski, T.M. (1996). Part V. Analysis of the aromatic character of aza-analogues of benzenoid hydrocarbons. *Tetrahedron* 52: 13795–13802.
69 Raczyńska, E. D. Duczmal, K., Hallmann, M. (2007). 81: 1655–1666.
70 Raczyńska, E.D., Hallmann, M., and Duczmal, K. (2008). *Pol. J. Chem.* 82: 1077–1090.
71 Raczyńska, E.D., Hallman, M., Kolczyńska, K., and Stępniewski, T.M. (2010). *Symmetry* 2: 1485–1509.
72 Raczyńska, E.D. and Kamińska, B. (2013). *J. Mol. Model.* 19: 3947–3960.
73 Frizzo, C.P. and Martins, M.A.P. (2012). *Struct. Chem.* 23: 375–380.
74 Allen, F.H., Kennard, O., Watson, D.G. et al. (1987). *J. Chem. Soc. - Perkin Trans. 2* 12: S1–S19.
75 Coulson, C.A. and Gołębiewski, A. (1960). *Tetrahedron* 11: 125–130.
76 Dewar, M.J.S. and Trinajstić, N. (1970). *Collect. Czechoslov. Chem. Commun.* 35: 3136–3189.
77 Dewar, M.J.S. and Trinajstic, N. (1970). *Coll. Czech. Chem. Comm.* 35: 3483–3518.
78 Fringuelli, F., Marino, G., Taticchi, A., and Grandoli, G. (1974). *J. Chem. Soc. Perkin Trans.* 2: 332–337.
79 Gordy, W. (1947). *J. Chem. Phys.* 15: 305–310.
80 Kotelevskii, S.I. and Prezhdo, O.V. (2001). *Tetrahedron* 57: 5715–5729.
81 Bird, C.W. (1986). *Tetrahedron* 42: 89–92.
82 Bird, C.W. (1993). *Tetrahedron* 49: 8441–8448.

**83** Bird, C.W. (1990). *Tetrahedron* 46: 5697–5702.
**84** Bird, C.W. (1987). *Tetrahedron* 43: 4725–4730.
**85** Pozharskii, A.F. (1985). *Teoreticheskie osnovy khimii geterotsiklov* (Theoretical principles of Heterocyclic Chemistry), in Russian, 22–30. Moscow: Khimia.
**86** Pozharskii, A.F. (1985). *Chem. Heterocycl. Compd.* 21: 717–749.
**87** Krygowski, T.M. and Cyrański, M.K. (2001). *Chem. Rev.* 101: 1385–1419.
**88** Krygowski, T.M., Szatyłowicz, H., Stasyuk, O.A. et al. (2014). *Chem. Rev.* 114: 6383–6422.
**89** Bultinck, P. (2007). *Faraday Discuss.* 135: 347–365.
**90** Portella, G., Poater, J., and Sola, M. (2005). *J. Phys. Org. Chem.* 18: 785–791.
**91** Katritzky, A.R., Barczyński, P., Mussumura, G. et al. (1989). *J. Am. Chem. Soc.* 111: 7–15.
**92** Katritzky, A.R., Karelson, M., Sild, S. et al. (1998). *J. Org. Chem.* 63: 5228–5231.
**93** Cyrański, M.K., Krygowski, T.M., Katritzky, A.R., and Schleyer, P.v.R. (2002). *J. Org. Chem.* 67: 1333–1338.
**94** Taylor, R. (1990). *Electrophylic Aromatic Substitution.* New York: Wiley.
**95** Mąkosza, M. and Winiarski, J. (1987). *Acc. Chem. Res.* 20: 282–289.
**96** Dominikowska, J., Krygowski, T.M., Oziminski, W.P., and Palusiak, M. (2015). *J. Org. Chem.* 80: 9091–9101.
**97** Mohajeri, A. and Davari, N. (2010). *Struct. Chem.* 21: 1069–1078.
**98** Poater, J., Fradera, X., Duran, M., and Solà, M. (2003). *Chem. Eur. J.* 9: 400–406.
**99** Bultinck, P., Ponec, R., and Van Damme, S. (2005). *J. Phys. Org. Chem.* 18: 706–718.
**100** Matito, E., Duran, M., and Solà, M. (2005). *J. Chem. Phys.* 122: 014109.
**101** Matito, E., Duran, M., and Solà, M. (2006). *J. Chem. Phys.* 125: 059901.
**102** Cyrański, M.K., Gilski, M., Jaskólski, M., and Krygowski, T.M. (2003). *J. Org. Chem.* 68: 8607–8613.
**103** Szatyłowicz, H., Krygowski, T.M., Palusiak, M. et al. (2011). *J. Org. Chem.* 76: 550–556.
**104** Shahamirian, M., Cyrański, M.K., and Krygowski, T.M. (2011). *J. Phys. Chem. A* 115: 12688–12694.
**105** Krygowski, T.M., Bankiewicz, B., Czarnocki, Z., and Palusiak, M. (2015). *Tetrahedron* 71: 4895–4908.
**106** Lloyd, D.M.G. and Marshall, D.R. (1964). *Chem. Ind.* 42: 1760–1761.
**107** Masui, H. (2001). Metalloaromaticity. *Coord. Chem. Rev.* 219–221: 957–992.
**108** Schleyer, P.v.R., Maerker, C., Dransfeld, A. et al. (1996). *J. Am. Chem. Soc.* 118: 6317–6318.
**109** Ostrowski, S. and Dobrowolski, J.C. (2014). *RSC Adv.* 4: 44158–44161.

**110** Howard, S.T. and Krygowski, T.M. (1997). *Can. J. Chem.* 75: 1174–1181.

**111** Dominikowska, J. and Palusiak, M. (2012). *Struct. Chem.* 23: 1173–1181.

**112** Tokatli, A. and Ucun, F. (2014). *J. Phys. Org. Chem.* 27: 380–386.

**113** Ostrowski, S. and Dobrowolski, J.C. (2015). *RSC Adv.* 5: 9467–9471.

**114** Szatyłowicz, H., Stasyuk, O.A., and Krygowski, T.M. (2016). Calculating the aromaticity of heterocycles. In: *Advances in Heterocyclic Chemistry: Heterocyclic Chemistry in the 21st Century: A Tribute to Alan Katritzky* (ed. E.F.V. Scriven and C.A. Ramsden), 301–327. San Diego: Elsevier Academic Press.

**115** Krygowski, T.M. and Cyrański, M.K. (1999). *Tetrahedron* 55: 11142–11148.

**116** Bultinck, P., Rafat, M., Ponec, R. et al. (2006). *J. Phys. Chem. A* 110: 7642–7648.

# 6

# Descriptors of Aromaticity: Energetic Criteria

*A madman trails a boot tied with a string addressing the boot from time to time: "Come on, you stubborn dog." The physician arrives and to humour his patient says: "A nice little dog you have here!" "Are you crazy?" replies the madman, "Don't you see it's a boot?" "You're recovering finely," the physician says, "tomorrow I'll allow you to return home." After the physician is out of sight, the madman winks to the boot and says: "You see, dog, I've told you we would fool him, haven't I?" Sometimes the discussions about aromaticity take place in the same mood as above, replacing* dog *by* aromatic.

Alexandru T. Balaban (*Pure & Appl. Chem.* **1980**, *52*, 1409)

## 6.1 Introduction

Energetic description of molecules is accessible by both experimental (calorimetric measurements) [1, 2] as well as theoretical calculations by use of quantum chemistry methods [3]. In both cases, the obtained results are related to the stability of molecular systems. This stability, however, may be understood in two ways: thermodynamic or kinetic stability. The mixture of hydrogen and oxygen is thermodynamically unstable, but kinetics of the reaction depends strongly on the external conditions: pressure, concentrations of substrates, temperature, and so on, and at room temperature the mixture is safe. At the beginning, stability of chemical species was understood in both senses together. The archetypal aromatic system, benzene, was identified early on as being resistant to chemical reactions, and so efforts were made to quantify this molecular characteristic. The first successful attempt for this purpose was done by Pauling and Sherman [4] and then by Kistiakovsky et al. [5] introducing the concept of resonance energy (RE). The general idea of RE is based on a comparison of the energy of a given molecule

*Aromaticity and Antiaromaticity: Concepts and Applications*, First Edition.
Miquel Solà, Alexander I. Boldyrev, Michał K. Cyrański, Tadeusz M. Krygowski, and Gabriel Merino.

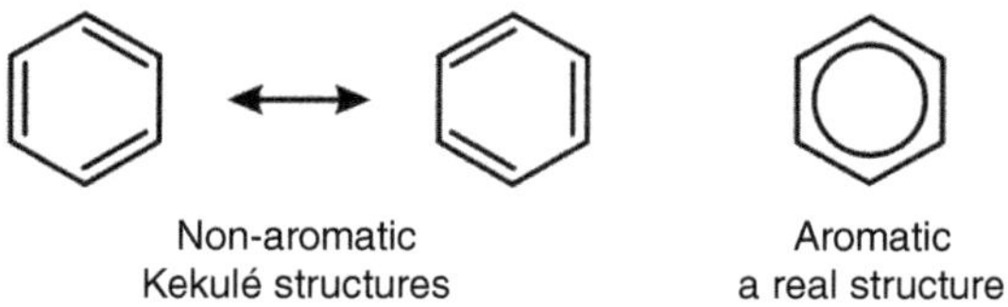


**Figure 6.1** Nonaromatic Kekulé structures and real aromatic structure of benzene.

and its analog with olefinic structure of the π-electron structure, the so-called Kekulé structure. This is exemplified by Figure 6.1 [6]. The RE values quantify thermodynamic meaning of the stability, and actually the term when referred to aromaticity is often named as aromatic stabilization energy, ASE [6], and is associated with π-electron delocalization in the molecule in question.

For clarity, it is useful to define what the electron delocalization is. The IUPAC definition reads [7, 8]:

> "delocalization is a redistribution of the valence-shell electron density throughout a molecular entity as compared with some localized models (individual atoms in their valence states, separated bonds or separated fragments). Different topological modes of the electron delocalization include: (i) ribbon delocalization of either π- or σ-electrons (i.e., electrons occupying respectively π- and σ-orbitals); (ii) surface delocalization of σ-electrons occurring through an overlap of radially oriented σ-orbitals of a cyclic molecule as is the case of cyclopropane and (iii) volume delocalization of σ-electrons through an overlap of σ-orbitals directed inside a molecular polyhedron." In organic chemistry, very popular is an explanation of delocalization presented in advanced organic chemistry textbooks (e.g., [9]): "localized chemical bonding is the bonding in which the electrons are shared by two and only two nuclei, while delocalized bonding is considered when electrons are shared by more than two nuclei."

The idea of RE is based on a difference between the actual π-electron energy of a molecule (benzene) and the π-electron energy of an analogous hypothetical species with a localized π-system (Kekulé structure of benzene). This may be lead out from the valence bond (VB) theory [10] in which the overall wave function $\psi$ describing the molecule is a linear combination of N wave functions $\psi_i$ describing N canonical structures $i = 1, 2, 3, \ldots N$. Figure 6.2 illustrates nicely this problem exemplified by canonical structures of benzene and showing as well a high level of complexity in describing the electron structure of benzene. Thus the concept and the values of RE introduced by Pauling et al. [4, 10] represent some kind of approximation for quantifying energetics of aromaticity. Nevertheless, the great diversity of canonical structures (two the most stable Kekulé structures, three

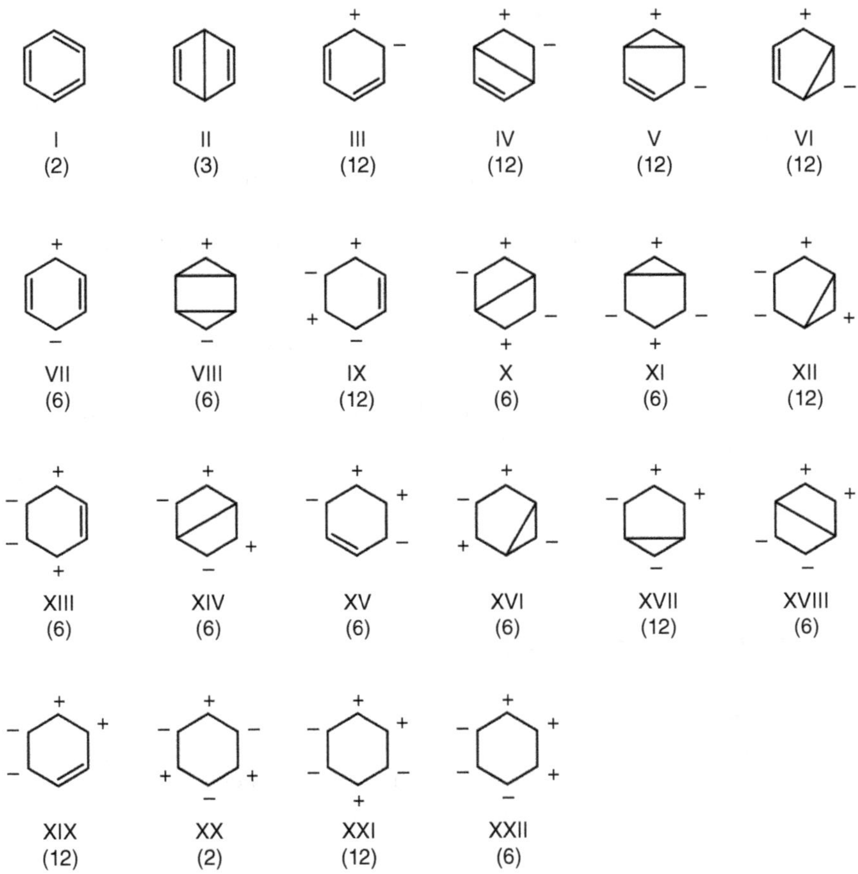


**Figure 6.2** Valence bond resonant structures of benzene [11].

the Dewar ones and additional 170 with single, double, and triple separations of charge) disclose sensitivity of the electron structure of benzene on potential intramolecular interactions when its ring is a part of greater π-electron system. Hence, serious complications for polycyclic systems and, in particular, for heterocyclic ones. However, despite of this, the concept of RE, actually named as ASE, may be regarded as the cornerstone of modern organic chemistry [12]. In the last few decades, the VB theory has not been often applied for estimation of RE, yet the idea based on the difference between the real energy of the molecule and of its "nonaromatic," that is, localized structure has been subject of numerous approaches, which is the subject of this chapter.

Before going into detailed problems and interpretations associated with energetic aspects of π-electron delocalization, it is important to realize that these energies are very small numbers in comparison with the total energy of the molecules in question. This may be nicely visualized by the example as follows [6]:

> We might liken this problem to a meal where some spice was added: although nominally the weight of the spice is small, it is essential for the taste of the food, and significantly influences its (other) properties. Despite its leading role, aromatic effects are not an easy task to separate out and quantify, for two main reasons: (i) the "spice-free" meal is unavailable, so any comparison has to rely on arbitrary chosen approximate model, which is never perfect and changes from case to case (read: from meal to meal) and (ii) the weights of the spicy meal and its "spice-free" model are very similar, which makes comparisons difficult. Continuing with the food analogy, sceptics sometimes argue that in the literature "a meal" has also been made of the importance of this topic(!) [13]. However, aficionados can (and do) frequently point out that cyclic π-electron delocalization, commonly called aromaticity, is a concept of immense practical importance in chemistry [14].

## 6.2 Thermochemical Approaches

The first quantitative criteria of aromaticity introduced by Pauling et al. [4] and Kistiakowsky et al. [5] were based on heats of atomization and hydrogenation, respectively. Practically, it was done by comparison of the energy of atomization of benzene ($\Delta H_a$) with the sum of bond energies of a hypothetical reference built up by the CC bonds taken from ethane and ethene, and the CH bonds from methane, resulting in RE = 36 kcal/mol:

$$RE = -\Delta H_a^0(\text{benzene})_g - [3E(C{=}C) + 3E(C{-}C) + 6E(C{-}H)] \tag{6.1}$$

The Kistiakowsky approach was based on a different scheme, that is, on the hydrogenation enthalpies of benzene, cyclohexane, and cyclohexene. The energetic scheme is presented in Figure. 6.3. The hydrogenation energy of benzene (to cyclohexane) is equal to 49.8 kcal/mol, whereas the hydrogenation energy of an "isolated" C=C bond in cyclohexene is 28.6 kcal/mol. The Kekule structure for benzene contains three "isolated" double bonds. Thus, the hydrogenation energy of three cyclohexene molecule (85.8 kcal/mol) should be compared with that of benzene, resulting in RE = 36.0 kcal/mol. However, this excellent agreement of the results for these two different schemes seems to be accidental. In both approaches, many approximations are introduced since the obtained REs are subject of limited precision and accuracy of experimental data as well as different perturbations due to important factors like unbalanced strain, changes of hybridization, conjugation,

Figure 6.3 Heats of hydrogenation to assess the aromatic stabilization energy of benzene [6, 15, 16].

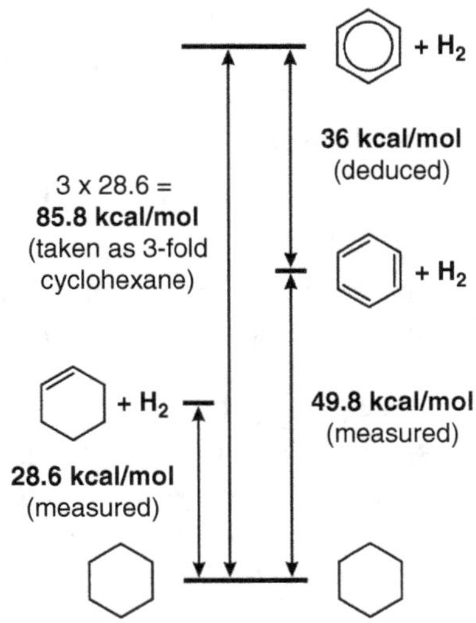


hyperconjugation effects, and so on. In the case of heterocyclic systems, additional problems appear since the combustion products of many heterocycles are ill-defined and hence the thermochemical measurements are less credible.

Aromatic compounds have long been recognized [4, 5] as systems where the nonadditivity of collective properties determine mostly their physicochemical behavior [10]. The RE values may be understood as numerical estimates of the nonadditivity since they are the differences between energy of a given system and that of its analog built up by addition of bond energies taken from localized $\pi$-electron systems.

The most effective studies of aromatic compounds and aromaticity by means of thermochemistry methods have been carried out mostly in the beginning of the second half of the 20th century. At that time, most important contributions (summarized in Ref. [17]) were reported dealing with estimation and application of energies for particular types of bonds, which could be later applied for construction of appropriate reference $\pi$-electron systems with localized bonds. Then, various chemical structures could be built up leading to many different schemes for estimations of RE values. Application of various schemes of termochemical approaches for describing aromatic compounds are well presented in Ref. [6].

## 6.3 Energetic Approaches Based on Molecular Geometry

It was shown that there exist some correlation between bond length and bond energy [18] that results from a combination of the Johnston and Parr bond order – bond energy formula [19] with the Pauling bond order – bond length relationship [20] that leads to an empirical formula given by Eq. (6.2) for the bond energy (BE), which reproduces well the experimental heats of formation of hydrocarbons [18].

$$BE_i = 87.99 \cdot \exp[2.255 \cdot (1.553 - R_i)] \tag{6.2}$$

A similar conclusion came later to Exner and Schleyer [21] based on the Atoms-in-Molecules (AIM) model [22], and finally to Howard et al. [23]. These results are an acceptable ground for the application of bond lengths to the energetic analysis of molecular systems.

To this category of energetic-scaling aromatic properties belong the so-called HOSE model [24–30]. By definition HOSE (Harmonic Oscillator Stabilization Energy) is the negative value of the energy necessary to deform the real molecule into its Kekulé (or resonance/canonical) structure with localized single and double bonds of typical lengths:

$$E_{def} = -\frac{1}{2} \cdot \left[ \sum_{r=1}^{n_1} (R'_r - R^s_o)^2 \cdot k'_r + \sum_{r=1}^{n_2} (R''_r - R^d_o)^2 \cdot k''_r \right] \tag{6.3}$$

where $R'_r$ and $R''_r$ stand for the lengths of $\pi$ bonds in the real molecule, whereas $n_1$ and $n_2$ are the numbers of the corresponding formal single and double bonds in the ith canonical structure, respectively. As a result of the deformation process, the $n_1$ bonds corresponding to the single bonds in the ith canonical structure are lengthened, whereas the $n_2$ bonds corresponding to the double bonds in the ith canonical structure are shortened to the reference bond lengths $R^s_o$ and $R^d_o$, respectively. The deformations follow the harmonic oscillator model with the force constants ($k_r$) fulfilling the relation: $k_r = a + bR_r$. Some improvement for this approximated viewpoint was presented by the application of *ab initio* calculations for the stretching motion of C—C, C=C, B—N, and B=N bonds leading to the ratio k(2)/k(1) = 2.35 for CC bonds and a dramatically larger 4.2 for BN bonds. This treatment has been applied for the harmonic oscillator model of aromaticity (HOMA) [31]. Since that time, this approach is sometimes used as more accurate, particularly for polar bonds [32, 33].

The HOSE model [27] allows us not only to obtain the stabilization energies well correlating with the Hess and Schaad RE values [34, 35], but also to estimate the weights of canonical structures well correlating with the Randić estimations [36]. It was found that the weights of the canonical structures derived by use of graph-topological approach [37] correlate excellently with the HOSE estimates (correlation coefficient R = 0.997) for 150 data points of $\pi$-electron rings.

On a similar basis, Bird introduced a model defining aromaticity indices $I_5$ or $I_6$ depending on size of the ring [38, 39] and applicable to a wide range of heterocyclic systems with heteroatoms such as, P, N, or O. In some cases, the Bird approach allows us to estimate weights of the canonical structures [40].

## 6.4 Theoretical Approaches

The first theoretical approach for estimating RE comes from Erich Hückel [41], who did not publish it, but presented this idea only in scientific discussions. Applying the Hückel Molecular Orbital (HMO) theory, Streiwieser et al. [42] utilized this idea for the description properties of small ring hydrocarbons and radicals and then popularized it in his famous monograph [43]. The formula for the Hückel resonance energy (HRE) reads:

$$HRE = -[E_\pi - n_{C=C}(2\alpha + 2\beta)] \tag{6.4}$$

where $n_{C=C}$ stands for number of double bonds in the reference system, whereas α and β stand for the Coulomb and resonance integrals of the HMO theory, respectively. When HRE values are calculated per one electron, then the obtained values estimated for all annulenes (except cyclobutadiene, for which HRE = 0) revealed them as aromatic systems. Hence, the HRE was found as a highly unreliable estimation of molecular stability. Additionally, very unstable pentalene has the RE value of 2.46β, whereas for benzene HRE, equals to 2.00β. Undoubtedly, the incorrect results are due to two factors: (i) unsuitable reference structures as discussed by Schaad and Hess [44] and (ii) too primitive methodology and model of calculations. Some substantial improvement within HMO theory was achieved by Hess and Schaad [34, 35, 45–48]. To get more proper RE values, it was sufficient to replace the HMO bond energy for a double C=C bond (2β) and for a single bond (zero) as in the original Hückel approach by the bond energies for topologically different types of bonds, as presented in Table 6.1. The appropriate REs (here, named as HSRE) were estimated for more than 150 molecules of alternant and nonalternant hydrocarbons and even for some heterocyclic systems [34].

Interestingly, when HSRE are calculated per π-electrons (HSRE/π), the results, shown in Table 6.2, seem to be rational.

Hess and Schaad RE per π-electron (REPE) well represent changes in stabilization of annulenes when plotted against the number of carbon atoms clearly following the Hückel's $4n + 2$ rule [49, 50] (Figure 6.4).

**Table 6.1** Hess-Schaad π-electron energies (in β units) for eight topologically different types of bonds.

| Type of bond | $E^\pi$ | Type of bond | $E^\pi$ |
|---|---|---|---|
| $H_2C{=}CH$ | 2.0000 | C=C | 2.1716 |
| HC=CH | 2.0699 | HC—CH | 0.4660 |
| $H_2C{=}C$ | 2.0000 | HC—C | 0.4362 |
| HC=C | 2.1083 | C—C | 0.4358 |

**Table 6.2** Hess-Schaad Resonance Energies (in β units) for benzenoid hydrocarbons [34].

| System | HSRE | HSRE/$\pi$ | System | HSRE | HSRE/$\pi$ |
|---|---|---|---|---|---|
| Benzene | 0.39 | 0.065 | Chrysene | 0.96 | 0.053 |
| Naphthalene | 0.55 | 0.055 | Picene | 1.17 | 0.053 |
| Anthracene | 0.66 | 0.047 | Triphenylene | 1.01 | 0.056 |
| Tetracene | 0.75 | 0.042 | Pyrene | 0.81 | 0.051 |
| Pentacene | 0.84 | 0.038 | Perylene | 0.97 | 0.048 |
| Phenanthrene | 0.77 | 0.055 | Coronene | 1.28 | 0.053 |

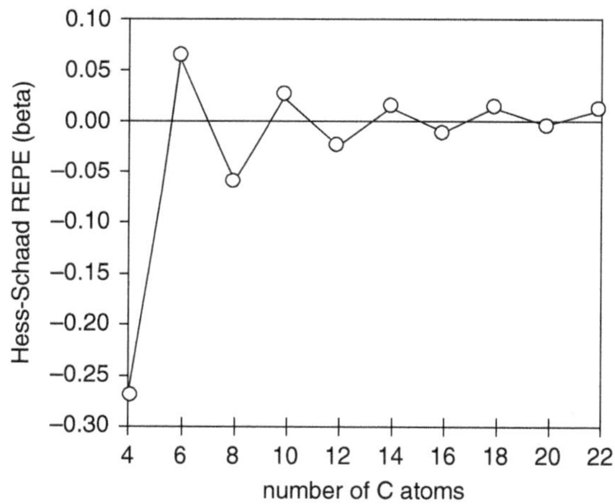


**Figure 6.4** Dependence of REPE on number of $\pi$-electrons in annulenes. Source: Reprinted with permission from Ref. [44]. Copyright 2001, American Chemical Society.

A real improvement was achieved when Dewar et al. [51–53], applying the Parr-Pariser $\pi$-electron method [54, 55], demonstrated that the bond energies in acyclic polyenes are additive. Then the Dewar RE was defined as:

$$\mathrm{DRE} = \Delta H_{\mathrm{a}} - \Delta H_{\mathrm{a}}{}^{\mathrm{add}} \tag{6.5}$$

where $\Delta H_a$ is the estimated heat of atomization of a given conjugated system, and $\Delta H_a{}^{add}$ is the heat of atomization for the reference structure. For hydrocarbons the formula reads:

$$DRE = \Delta H_a^M - (n_1 E_{C-C} + n_2 E_{C=C} + n_3 E_{C-H}) \tag{6.6}$$

where $n_i$ (i = 1, 2, 3) is the number of a given bond type; see the complete review on this topic by Schaad and Hess [44]. The Dewar approach has become a ground for the concept of isodesmic [56–58] and homodesmotic [59, 60] reactions.

Isodesmic reactions demand an equal number of formal single and double bonds in products and reactants. Homodesmotic reactions are a particular case of isodesmic reactions in which there is the same number of bonds between given atoms in each hybridization state in products and reactants. The number of hydrogen atoms joined to the atoms in given hybridization states must also coincide in products and reactants.

Homodesmotic reactions are preferred because they minimize different sources of errors, like lack of compensation in strain, hyperconjugation, anomeric effects, and so on [61, 62]. They are, till now, regarded as the most valuable description of some "extra" stabilization over given reference system(s). If possible, only cyclic reference compounds should be used in ASE calculations from homodesmotic reactions [61, 62]. In principle, the energies of substrates and products in these reactions can be taken either from the quantum chemical calculations or from thermochemical (calorimetric) measurements. Moreover, for the same purpose – estimation of stabilization – many isodesmic or homodesmotic reactions can be constructed, as shown in Table 6.3. To some extent, this is a short review of most often schemes applied to get information about ASE/RE. In a similar way, attempts may be undertaken for more complex $\pi$-electron systems [73].

The main problem in ASE calculations is the choice of appropriate reference molecules. The failure to take reasonable molecules of reference leads to discrepancies as large as ca. 50 kcal/mol in the calculation of ASEs for benzene. Variety of the results, presented in Table 6.3, allows us to enlighten and understand a complex problem of various factors influencing the final result, that is, impact of various intramolecular interactions in the reference structure that decide about the final result. The broadly diverging evaluations of ASE of benzene reported can be refined to c. 29 kcal/mol when corrections for conjugation, hyperconjugation, and protobranching are included [74, 75]. Moreover, based on substituted methylene derivatives (which arguably from a chemical perspective localize $\pi$-electron structure of benzenoids to highest extent, and have similar $\sigma$ frameworks), well-balanced steric CH$\cdots$CH repulsions and considering appropriate *cis-trans* topology, Cyrański and coworkers showed that extremely convergent values of ASE of benzenoid hydrocarbons can be obtained, provided that the reaction coefficients are optimally derived [75]. The exampling reaction for pyrene is given by Eq. (6.7). The values presented in Table 6.4 show that the lowering of stability highly depends on topology and is accompanied by an increase in the size of the molecule.

In line with chemical expectations, the angular system is more stable than the linear one, as exemplified by phenanthrene and anthracene, whereas the relative aromatic stabilization of coronene is lower than the value for pyrene. This approach can easily be extended to nonbenzenoid systems or nonplanar aromatics (bent or twisted, see the following) or heterocyclics, however it needs careful involvement of many reference systems, analyses of their $\pi$-electron topologies, and other (like steric) effects, which might be troublesome.

**Table 6.3** Fifteen different aromatic stabilization energies (in kcal/mol) for benzene at B3LYP/6-311+G** (+ZPE) [6].

| N | Reaction Scheme | ASE |
|---|---|---|
| 1 | 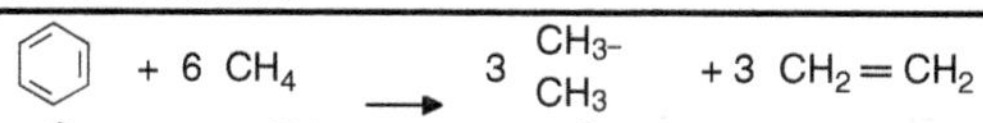 | 66.9[a] |
| 2 |  | 55.3[b] |
| 3 | 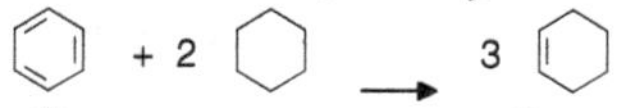 | 37.5[c] |
| 4 | 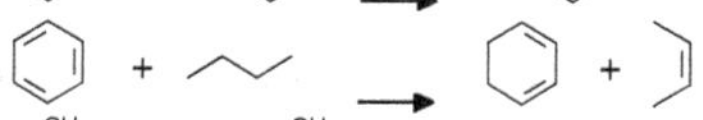 | 36.5[d] |
| 5 | 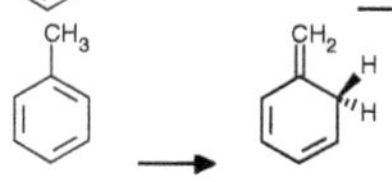 | 33.2[e] |
| 6 |  | 28.3[f] |
| 7 | 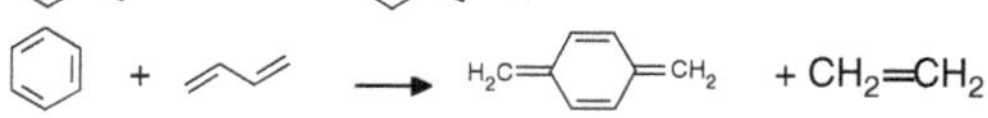 | 32.2[g] |
| 8 | 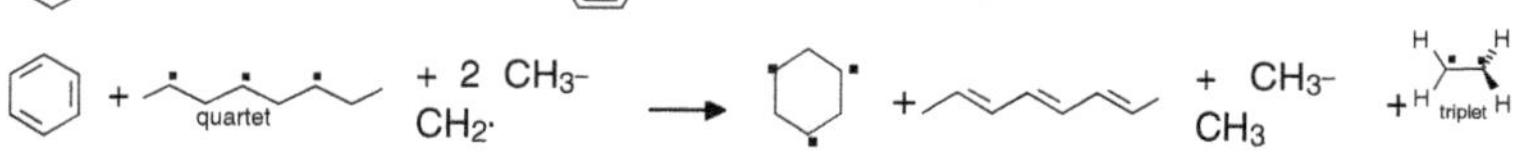 | 28.9[h] |
| 9 | 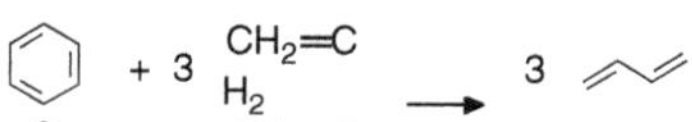 | 23.2[i] |
| 10 | 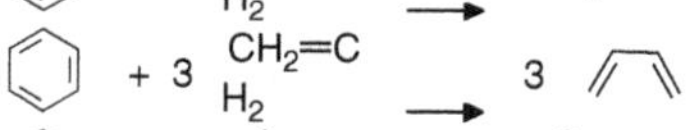 | 33.6[j] |
| 11 | 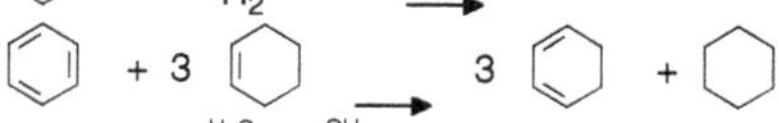 | 32.4[k] |
| 12 | 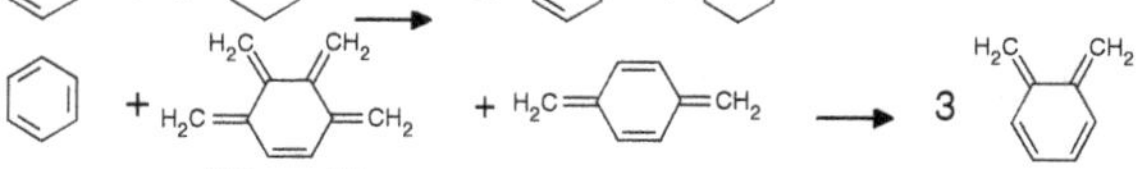 | 28.7 |
| 13 | 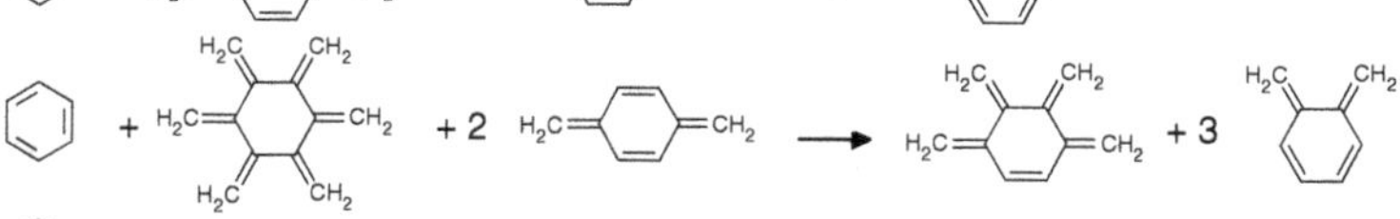 | 28.6 |
| 14 | 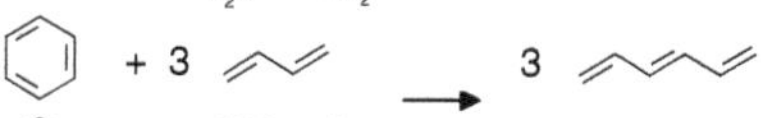 | 19.3[l] |
| 15 |  | 18.4[m] |

a) Experimental values: 61.1 kcal/mol [56], 64.2 kcal/mol [63], 64.7 kcal/mol [64].
b) Experimental values: 48.7 kcal/mol [65], 48.5 kcal/mol [64].
c) Experimental values: 35.6 kcal/mol [66, 67], 35.9 kcal/mol [64].
d) Experimental value: 33.4 kcal/mol [64].
e) Data corrected for *anti-syn* diene mismatch (3.6 kcal/mol) [67], experimental values are 38 and 23.1 kcal/mol [68, 69].
f) Data corrected for *anti-syn* diene mismatch (2 × 3.6 kcal/mol) [67], experimental value 28.8 kcal/mol [2, 64, 70].
g) Data corrected for *anti-syn* diene mismatch (3 × 3.6 kcal/mol) [67], experimental value (uncorrected for the *anti-syn* diene mismatch): 16.7 kcal/mol [64].
h) Taken from Ref. [71] at B3LYP/6-31G(d,p).
i) Experimental value: 20.6 kcal/mol [64].
j) 34.1 kcal/mol for planar *cis*-butadiene [72].
k) Experimental values: 30.5 kcal/mol [2], 28.8 kcal/mol [64, 67].
l) Experimental value: 22.5 kcal/mol [64].
m) Experimental value: 21.2 kcal/mol [64].

**Table 6.4** The mean values of Aromatic Stabilization Energies of selected benzenoid hydrocarbons in kcal/mol based on 15 reactions of the type shown by Eq. (6.7), with estimated standard deviations. The reference ASE of benzene equals to 34.1 kcal/mol [72].

| System | ASE | System | ASE |
|---|---|---|---|
| Naphthalene | 39 (1) | Benz[a]anthracene | 54 (1) |
| Anthracene | 42 (2) | Chrysene | 56 (3) |
| Phenanthrene | 51 (3) | [4]helicene | 52 (2) |
| Pyrene | 49 (2) | Anthanthrene | 46 (2) |
| Tetracene | 39 (2) | Coronene | 58 (2) |

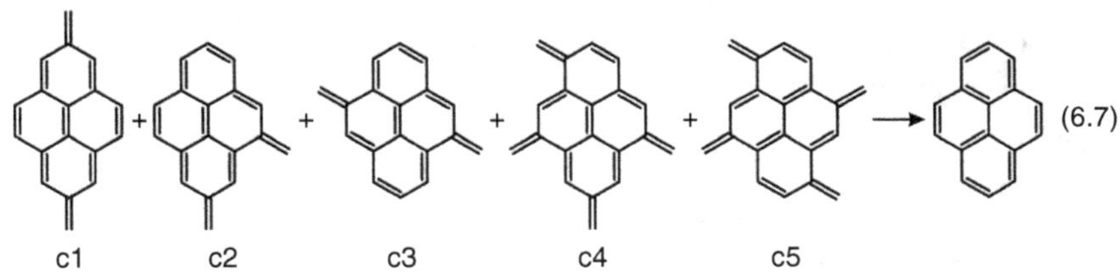


(6.7)

In general, a single reaction scheme is used to determine the ASE. For example, for five-membered heterocycles, a single reaction that balances most efficiently strain, hyperconjugation, change of hybridization, and heteroatom-heteroatom interactions is given as follows:

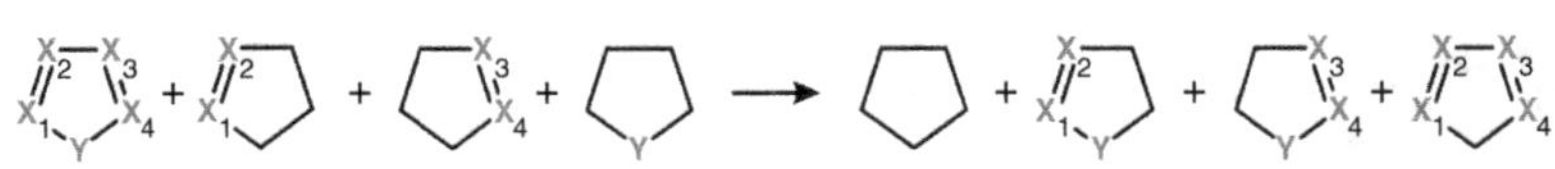

(6.8)

Equation (6.8) was applied for accurate ASE determination of all aza and phospha derivatives of furane, tiophene, pyrrole, and phosphole to solve the problem whether aromaticity is statistically a multidimensional phenomenon [62].

In an attempt to eliminate the stoichiometric arbitrariness in the evaluation of ASEs, Fishtik et al. [76–78] proposed to use a complete set of possible reaction schemes with an approach based on group additivity methods. Their best estimates of the ASE of benzene and pyridine were 31.8 and 29.6 kcal/mol, respectively [78].

The block-localized wave function (BLW) method [79] has also been used to compute ASEs. In BLW, electron-localized states are defined by restricting the expansion of molecular orbitals (MOs) within preset specific physical zones with coefficients optimized self-consistently. With this method, we can optimize 1,3,5-cyclohexatriene molecules by localizing the double bonds in the

1,3,5-positions. The difference in energy between optimal 1,3,5-cyclohexatriene and fully optimized benzene yields the adiabatic resonance energy (ARE) of benzene. The difference between the ARE of benzene and three times the ARE of butadiene yields the ASE of benzene, which is 27.5 kcal/mol at the HF/6-311+G(d) level of theory [80].

In 2002, Schleyer and Pühlhofer [67] introduced the isomerization stabilization energies (ISEs) as a particular way to obtain ASEs. For a given (anti)aromatic compound, ISE is the energy difference between a methyl derivative of an annulene and an isomeric species with acyclic conjugation and an exocyclic methylene group. For instance, for benzene, the ISE can be derived from toluene (which has an ASE practically identical to benzene) and a nonaromatic methylenecyclohexadiene (see isomerization reaction 6.9 or reaction 5 in Table 6.3). In most cases, we can design several nonaromatic methylenecyclohexadiene isomers. If so, then we have to do an average of all possible ISEs. For naphthalene, the ISE will be given by the isomerization reaction 6.10. ISE values have been also used to validate the Baird's rule of aromaticity [81].

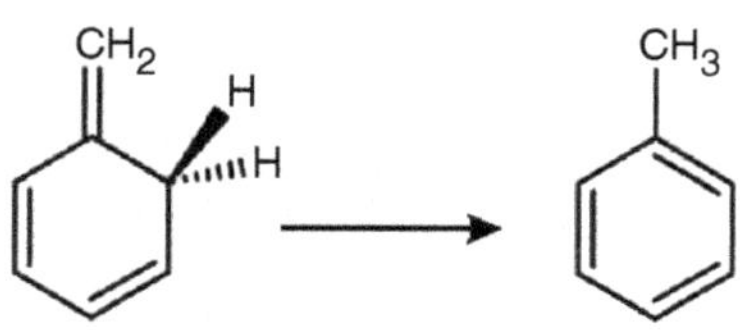


(6.9)

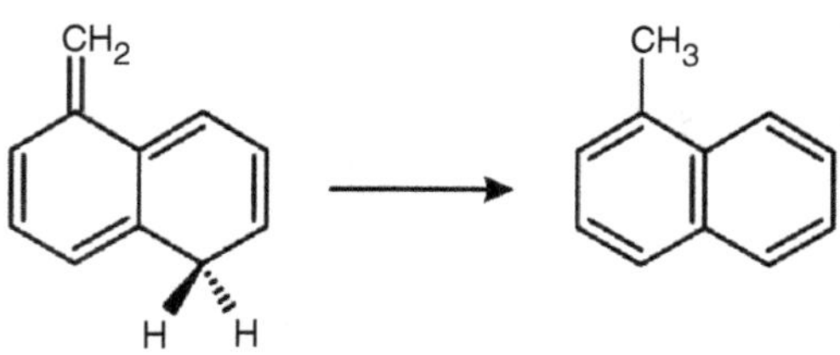


(6.10)

A quite different way for the estimation of ASEs is based on graph-topological approaches [82–84]. Two kinds of them are most frequently present in literature, one derived by Trinajstić et al. [85, 86], and another one by Aihara [87–90]. Both have introduced and then developed the concept of topological resonance energy (TRE). In the case of TRE, the characteristic polynomial is built up by the reference structures containing only the acyclic Sachs graphs for a given molecule taken into consideration [91, 92]:

$$TRE = \sum_{j=1}^{N} g_j \left( x_j - x_j^{ac} \right) \tag{6.11}$$

where N is the number of vertices in a graph, $x_j$ and $x_j^{ac}$ are the roots of the characteristic polynomials of the aromatic system and of the acyclic polyene-like reference structure, respectively. And $g_j$ stands for the orbital occupation of the jth MO. TRE, defined graph-theoretically, represents an ASE arising from cyclic conjugation in the π-system. In principle, these kind of approaches are closely related to the typical HMO procedures and a good equivalence between HSRE and TRE was found [93].

The concept of conjugated circuits of Randić [94] is based on a completely different idea. The *conjugated circuit* is defined as "a cycle within an individual Kekulé valence structures in which there is regular alternation of formal single and double bonds [95]. The circuits are necessarily of even length. While (4*n*+2) stabilize the system, the (4*n*) conjugated ones lead to its destabilization" [6]. According to Randić [94, 96], the conjugated circuits resonance energy (CCRE) of a polycyclic π-electron system may be considered as the total contribution from all conjugated circuits in the molecule:

$$CCRE = \frac{1}{k}\sum_{n}(r_n R_n + q_n Q_n) \tag{6.12}$$

where k is the number of Kekulé valence structures for a given system, $r_n$ and $q_n$ define the numbers of the conjugated circuits with ring size 4*n*+2 and 4*n*, respectively, in a given molecular system. $R_n$ and $Q_n$ describe the empirically determined contributions to the RE coming from $4n+2$ and $4n$ kinds of circuits. Graph-topological methods allowed to derive [37] the Hückel's $4n+2$ rule [49, 97], the Clar's rule of the classification rings in benzenoid hydrocarbons [98, 99], and the Randić concept of conjugated circuits [94]. It was also shown that for polycyclic benzenoid hydrocarbons some correlation exist between the thermodynamic and kinetic stabilities [100]. Solà, Bickelhaupt, and coworkers [101] showed for a few particular cases that kinked polycyclic benzenoid hydrocarbons are more stable than the straight ones. By use of graph-topological methods, it was possible to formulate a general rule stating that the kinked polycyclic benzenoid hydrocarbons are more stable than the straight ones [102].

Finally, Fernández and Frenking [103] found that it is possible to use the $\Delta E_\pi$ term of the orbital interaction component in a variational energy decomposition analysis (EDA) [104–107] of cyclic and acyclic conjugated systems to estimate the strength of π-conjugation in (anti)aromatic and (anti)homoaromatic species. In this EDA method, the dissociation energy in a molecule into two or more fragments is decomposed into:

$$-D_e = \Delta E_{\text{prep}} + \Delta E_{\text{int}} \tag{6.13}$$

In this formula, the preparation energy $\Delta E_{\text{prep}}$ (also referred as deformation or strain energies) is the amount of energy required to deform the individual (isolated) fragments from their equilibrium structure to the geometry that they acquire in the molecule and to bring them to their reference electronic states. The interaction energy $\Delta E_{\text{int}}$ corresponds to the actual energy change when these geometrically deformed and electronically prepared fragments are combined to form the molecule. $\Delta E_{int}$ is analyzed in the framework of the Kohn-Sham MO model using a quantitative decomposition of the interaction into electrostatic, Pauli repulsion (or exchange repulsion), and orbital interactions:

$$\Delta E_{int} = \Delta V_{elstat} + \Delta E_{Pauli} + \Delta E_{oi} \; (+\Delta E_{disp}) \tag{6.14}$$

The instantaneous interaction energy $\Delta E_{int}$ among the fragments in a molecule is partitioned into three terms: first, the quasiclassical electrostatic interaction $\Delta V_{elstat}$ between the fragments; second, the repulsive exchange (Pauli) interaction $\Delta E_{Pauli}$ between electrons of the two fragments having the same spin; and third, the orbital (covalent) interaction $\Delta E_{oi}$, which comes from polarization and orbital mixing between the fragments (see Figure 6.5).

The latter term can be decomposed into contributions of orbitals with different symmetry $\Gamma$, which makes possible to distinguish between $\sigma$, $\pi$, and $\delta$ contributions to bonding ($\Delta E_{oi} = \Sigma\Delta E_{\Gamma}$). If the density functional used in the calculations contains dispersion corrections, then in Eq. (6.14), there is another term, $\Delta E_{disp}$. As an example, Table 6.5 contains the EDA results for benzene and 1,3,5,7-*all-cis*-octatetraene. The first two rows show the EDA of benzene dissociated into six CH groups for $D_{6h}$ and $D_{3h}$ geometries. The results of the $\Delta E_{\pi}$ components, confirms the suggestion of Shaik and others [108] (see Chapter 12) that the $\pi$ bonding in benzene favors a $D_{3h}$ structure, whereas the $\sigma$ bonding favors the $D_{6h}$ optimized structure. Table 6.5 also shows that the $\Delta E_{\pi}$ component of $D_{6h}$ benzene dissociated into three $C_2H_2$ fragments is larger than that of the 1,3,5,7-*all-cis*-octatetraene dissociated into two $C_2H_3$ and two $C_2H_2$ fragments. The difference between $\Delta E_{\pi}$ values of the two compounds gives an estimation

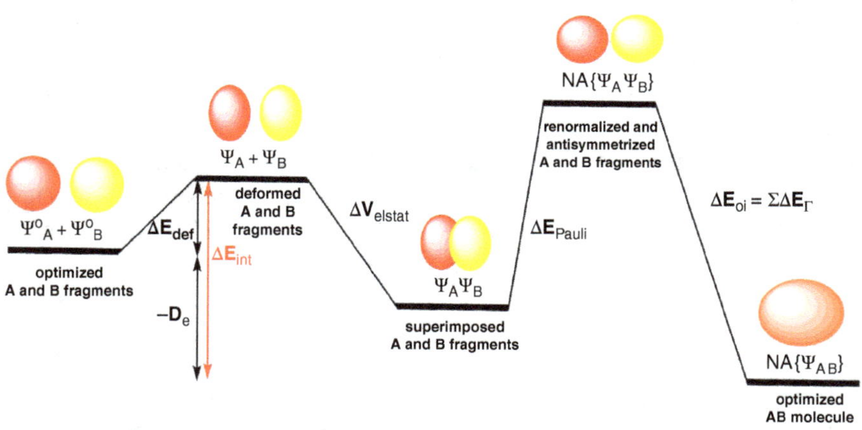


**Figure 6.5** Schematic representation of the different steps in the Ziegler-Rauk energy decomposition analysis for the formation of a molecule AB from two fragments A and B.

**Table 6.5** EDA (kcal/mol) for benzene and for the acyclic reference compound 1,3,5,7-*all*-*cis*-octatetraene at the BP86/TZ2P level of theory [103].[a]

| Molecule | $\Delta E_{int}$ | $\Delta E_{Pauli}$ | $\Delta V_{elstat}$ | $\Delta E_{oi}$ | $\Delta E_{\sigma}$ | $\Delta E_{\pi}$ |
|---|---|---|---|---|---|---|
| $D_{6h}$  | −996.7 | 1573.7 | −1072.1 | −1498.3 | −1280.6 | −217.7 |
| $D_{3h}$ 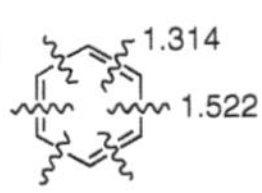  | −997.8 | 1531.8 | −1036.8 | −1472.8 | −1250.6 | −222.2 |
| $D_{6h}$ 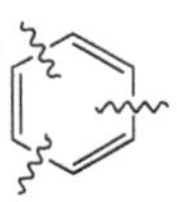 | −513.3 | 1180.0 | −540.4 | −1152.8 | −1045.2 | −107.7 |
| $C_{2h}$ 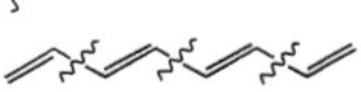 | −489.5 | 1064.9 | −512.5 | −1041.9 | −976.7 | −65.2 |

a) CH fragments in their quartet state, $C_2H_2$ fragments in their open singlet state, $C_2H_3$ fragments in doublet state.

for the ASE of 42.5 kcal/mol. The EDAs applied to $C_5H_5E$ (E=CH, N–Bi) and $(HB{=}NH)_3$ calculated in this manner predict strength of ASE in the order N > CH > P > As > Bi » borazine.

For nonplanar species, the determination of ASE is more complicated due to the strain energy of some of the bonds. A key problem in nonplanar polycyclic conjugated systems is to find a proper reference structure that fully balances its strain. Eqs. (6.15)–(6.17) show three equations that can be used to determine the ASE of corannulene [109]. The first two provide ASEs of 37.7 and 42.5 kcal/mol at the B3LYP/6-311G(d,p) level of theory. The two reactions fulfill the homodesmotic criteria. The difference of 4.8 kcal/mol between the two reactions can be ascribed to different number of *syn* and *anti* butadiene subunits and different number of H···H repulsive interactions. In this sense, reaction (6.17) has balanced *syn*/*anti* units and H···H interactions, so the estimated ASE using reaction (6.17) of 46.7 kcal/mol is more precise. It is possible to devise at least 18 homodesmotic reactions similar to (6.17) with fully balanced *syn*/*anti* butadiene topology and other structural effects. The obtained values are extremely consistent, with the mean value of ASE of coronene of 44.5 kcal/mol and estimated standard deviation of 1.2 kcal/mol [109]. Similar equations have been designed to estimate the ASE of fullerenes (results show that the cyclic π electron delocalization does not stabilize $C_{60}$ [110]), highly bent pyrenophanes [111], and nonplanar linear acenes [112]. The latter show very good dependence of aromatic stabilization energies on the extent of bent(twist), showing that even extremely distorted systems retain much of aromaticity.

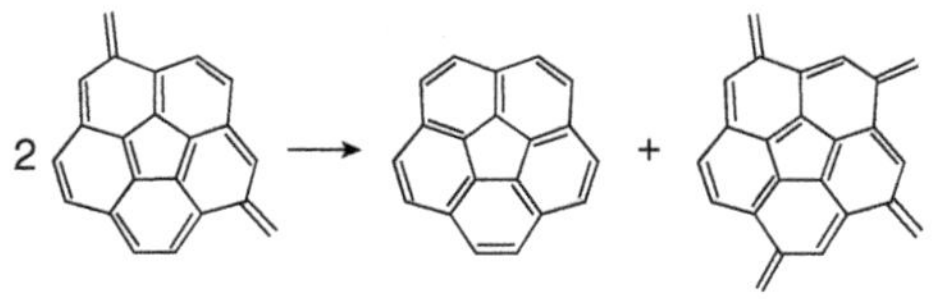

(6.15)

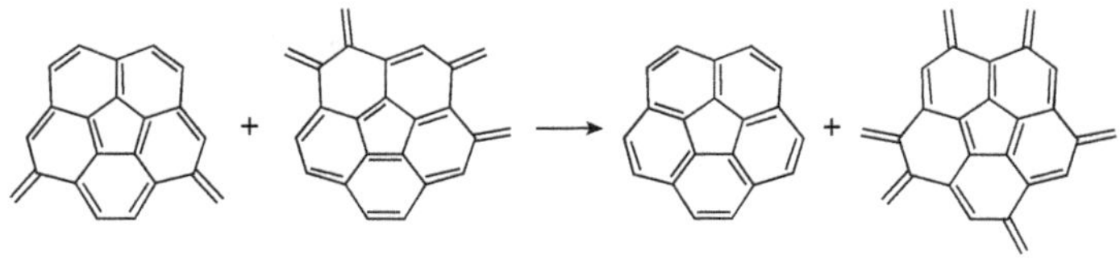

(6.16)

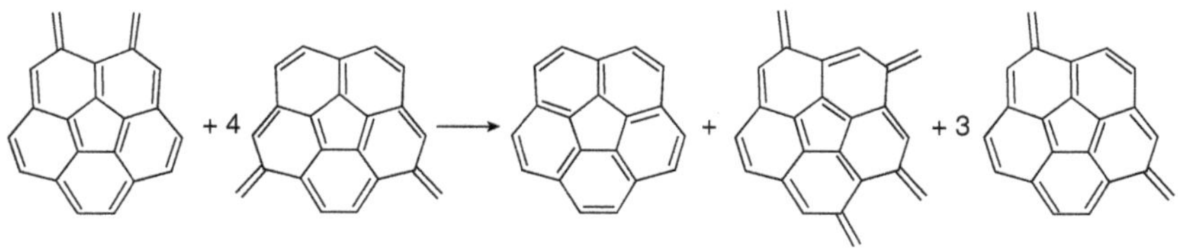

(6.17)

## References


1 Cox, J.D. and Pilcher, G. (1970). *Thermochemistry of Organic and Organometallic Compounds*. New York: Academic Press.
2 Pedley, J.B. (1986). *Thermochemical Data of Organic Compounds*. London: Springer.
3 Piela, L. (2006). *Ideas of Quantum Chemistry*. Oxford: Elsevier Science.
4 Pauling, L. and Sherman, J. (1933). *J. Chem. Phys.* 1: 606–617.
5 Kistiakowsky, G.B., Ruhoff, J.R., Smith, H.A., and Vaughan, W.E. (1936). *J. Am. Chem. Soc.* 58: 146–153.
6 Cyrański, M.K. (2005). *Chem. Rev.* 105: 3773–3811.
7 Minkin, V.I. (1999). *Pure Appl. Chem.* 71: 1919–1981.
8 Minkin, V.I., Glukhovtsev, M.N., and Simkin, B.Y. (1994). *Aromaticity and Antiaromaticity: Electronic and Structural Aspects*. New York: Wiley.
9 March, J. (1993). *Advanced Organic Chemistry: Reactions, Mechanisms, and Structure*, 4e. New York: Wiley.
10 Pauling, L. and Wheland, G.W. (1933). *J. Chem. Phys.* 1: 362–374.
11 Norbeck, J.M. and Gallup, G.A. (1974). *J. Am. Chem. Soc.* 96: 3386–3393.
12 Balaban, A.T., Oniciu, D.C., and Katritzky, A.R. (2004). *Chem. Rev.* 104: 2777–2812.

**13** Bergmann, E.D. and Pullman, B. (ed.) (1971). *Aromaticity, Pseudo-aromaticity, Anti-aromaticity, Proceedings of an International Symposium held in Jerusalem 31 March – 3 April 1970*. Jerusalem: Israel Academy of Science and Humanities.

**14** Krygowski, T.M., Cyrański, M.K., Czarnocki, Z. et al. (2000). *Tetrahedron* 56: 1783–1796.

**15** Lewis, D. and Peters, D. (1975). *Facts and Theories of Aromaticity*. London: Macmillan Press.

**16** Katritzky, A.R., Jug, K., and Oniciu, D.C. (2001). *Chem. Rev.* 101: 1421–1449.

**17** Cohen, N. and Benson, S.W. (1993). *Chem. Rev.* 93: 2419–2438.

**18** Krygowski, T.M., Ciesielski, A., Bird, C.W., and Kotschy, A. (1995). *J. Chem. Inf. Comput. Sci.* 35: 203–210.

**19** Johnston, H.S. and Parr, C. (1963). *J. Am. Chem. Soc.* 85: 2544–2551.

**20** Pauling, L. (1947). *J. Am. Chem. Soc.* 69: 542–553.

**21** Exner, K. and Schleyer, P.v.R. (2001). *J. Phys. Chem. A* 105: 3407–3416.

**22** Grimme, S. (1996). *J. Am. Chem. Soc.* 118: 1529–1534.

**23** Howard, S.T., Cyranski, M.K., and Stolarczyk, L.Z. (2001). *Chem. Commun.* 197–198.

**24** Karolak-Wojciechowska, J. (1985). *Phosphorus Sulfur Rel. Elem.* 25: 229–235.

**25** Krygowski, T.M. (1988). *Structure and Reactivity* (ed. J.F. Liebman and A. Greenberg), 231–254. Weinheim: VCH.

**26** Krygowski, T.M. (1991). *Progress in Physical Organic Chemistry*, vol. 17 (ed. R.W. Taft), 239–291. New York: Wiley.

**27** Krygowski, T.M., Anulewicz, R., and Kruszewski, J. (1983). *Acta Crystallogr., Sect. B: Struct. Sci.* B39: 732–739.

**28** Krygowski, T.M., Anulewicz, R., and Wisiorowski, M. (1995). *Pol. J. Chem.* 69: 1579–1584.

**29** Grabowski, S.J. and Krygowski, T.M. (2002). *Advances in Quantitative Structure – Property Relationships*, vol. 3 (ed. M. Charton and B.I. Charton), 27–66. Amsterdam: JAI Press.

**30** Krygowski, T.M., Szatyłowicz, H., Stasyuk, O.A. et al. (2014). *Chem. Rev.* 114: 6383–6422.

**31** Madura, I.D., Krygowski, T.M., and Cyrański, M.K. (1998). *Tetrahedron* 54: 14913–14918.

**32** Zborowski, K.K. and Proniewicz, L.M. (2009). *Pol. J. Chem.* 83: 477–484.

**33** Andrzejak, M., Kubisiak, P., and Zborowski, K.K. (2013). *Struct. Chem.* 24: 1171–1184.

**34** Hess, B.A. and Schaad, L.J. (1971). *J. Am. Chem. Soc.* 93: 305–310.

**35** Hess, B.A. and Schaad, L.J. (1971). *J. Org. Chem.* 36: 3418–3423.

**36** Randić, M. (1977). *Tetrahedron* 33: 1905–1920.

37 Ciesielski, A., Krygowski, T.M., Cyrański, M.K., and Balaban, A.T. (2011). *Phys. Chem. Chem. Phys.* 13: 3737–3747.

38 Bird, C.W. (1985). *Tetrahedron* 41: 1409–1414.

39 Bird, C.W. (1986). *Tetrahedron* 42: 89–92.

40 Bird, C.W. (1997). *Tetrahedron* 53: 13111–13118.

41 Personal Communication from C. Reichardt to T. M. Krygowski (Marburg) 1981.

42 Roberts, J.D., Streitwieser, A., and Regan, C.M. (1952). *J. Am. Chem. Soc.* 74: 4579–4582.

43 Streitwieser, A. Jr., (1961). *A Molecular Orbital Theory for Organic Chemists.* New York: Wiley.

44 Schaad, L.J. and Hess, B.A. (2001). *Chem. Rev.* 101: 1465–1476.

45 Hess, B.A. and Schaad, L.J. (1971). *J. Am. Chem. Soc.* 93: 2413–2416.

46 Schaad, L.J. and Hess, B.A. (1972). *J. Am. Chem. Soc.* 94: 3068–3074.

47 Hess, B.A. and Schaad, L.J. (1973). *J. Am. Chem. Soc.* 95: 3907–3912.

48 Hess, B.A., Schaad, L.J., and Agranat, I. (1978). *J. Am. Chem. Soc.* 100: 5268–5271.

49 Hückel, E. (1931). *Z. Phys.* 70: 104–186.

50 Hückel, E. (1932). *Z. Phys.* 76: 628–648.

51 Dewar, M.J.S. and Gleicher, G.J. (1965). *J. Am. Chem. Soc.* 87: 692–696.

52 Dewar, M.J.S., Harget, A.J., and Trinajstic, N. (1969). *J. Am. Chem. Soc.* 91: 6321–6325.

53 Dewar, M.J.S. and de Llano, C. (1969). *J. Am. Chem. Soc.* 91: 789–795.

54 Pariser, R. and Parr, R.G. (1953). *J. Chem. Phys.* 21: 466–471.

55 Pariser, R. and Parr, R.G. (1953). *J. Chem. Phys.* 21: 767–776.

56 Hehre, W.J., Ditchfiel, R., Radom, L., and Pople, J.A. (1970). *J. Am. Chem. Soc.* 92: 4796–4801.

57 Pople, J.A., Radom, L., and Hehre, W.J. (1971). *J. Am. Chem. Soc.* 93: 289–300.

58 Hehre, W.J., Radom, L., Schleyer, P.v.R., and Pople, J.A. (1986). *Ab initio Molecular Orbital Theory*. New York: Wiley.

59 George, P., Trachtman, M., Bock, C.W., and Brett, A.M. (1976). *J. Chem. Soc., Perkin Trans. 2* 1222–1227.

60 George, P., Trachtman, M., Brett, A.M., and Bock, C.W. (1977). *J. Chem. Soc., Perkin Trans. 2* 1036–1047.

61 Cyrański, M.K., Schleyer, P.v.R., Krygowski, T.M. et al. (2003). *Tetrahedron* 59: 1657–1665.

62 Cyrański, M.K., Krygowski, T.M., Katritzky, A.R., and Schleyer, P.v.R. (2002). *J. Org. Chem.* 67: 1333–1338.

63 George, P., Bock, C.W., and Trachtman, M. (1984). *J. Chem. Educ.* 61: 225–227.

**64** Afeely, H.Y., Liebman, J.F., and Stein, S.E. (2021). *Neutral Thermochemical Data. In NIST Chemistry WebBook, NIST Standard Reference Database No. 69*. Gaithersburg MD: National Institute of Standards and Technology.

**65** George, P. (1975). *Chem. Rev.* 75: 85–111.

**66** Pedley, J.B., Naylor, R.D., and Kirby, S.P. (1986). *Thermodynamical Data of Organic Compounds*. London: Chapman & Hall.

**67** Schleyer, P.v.R. and Pühlhofer, F. (2002). *Org. Lett.* 4: 2873–2876.

**68** Bally, T., Hasselmann, D., and Loosen, K. (1985). *Helv. Chim. Acta* 68: 345–354.

**69** Bartmess, J.E. (1982). *J. Am. Chem. Soc.* 104: 335–337.

**70** Wannere, C.S., Moran, D., Allinger, N.L. et al. (2003). *Org. Lett.* 5: 2983–2986.

**71** Suresh, C.H. and Koga, N. (2002). *J. Org. Chem.* 67: 1965–1968.

**72** Schleyer, P.v.R., Jiao, H.J., van Eikema Hommes, N.J.R. et al. (1997). *J. Am. Chem. Soc.* 119: 12669–12670.

**73** Wu, J.I., Dobrowolski, M.A., Cyrański, M.K. et al. (2009). *Mol. Phys.* 107: 1177–1186.

**74** Wodrich, M.D., Wannere, C.S., Mo, Y. et al. (2007). *Chem. Eur. J.* 13: 7731–7744.

**75** Ciesielski, A., Stępień, D.K., Dobrowolski, M.A. et al. (2012). *Chem. Commun.* 48: 10129–10131.

**76** Fishtik, H. and Datta, R. (2003). *J. Phys. Chem. A* 107: 10471–10476.

**77** Fishtik, I. (2011). *J. Phys. Chem. A* 115: 6657–6663.

**78** Fishtik, I. and Grimme, S. (2012). *Phys. Chem. Chem. Phys.* 14: 15888–15896.

**79** Mo, Y. (2003). *J. Chem. Phys.* 119: 1300–1306.

**80** Mo, Y., Hiberty, P.C., and Schleyer, P.v.R. (2010). *Theor. Chem. Acc.* 127: 27–38.

**81** Zhu, J., An, K., and Schleyer, P.v.R. (2013). *Org. Lett.* 15: 2442–2445.

**82** Trinajstić, N. (1983). *Chemical Graph Theory*. Boca Raton: CRC Press.

**83** Graovać, A., Gutman, I., and Trinajstić, N. (1977). *Lecture Notes in Chemistry*. Berlin: Springer-Verlag.

**84** Cyvin, S.J., Brunvoll, J., and Cyvin, B.N. (1991). *Theory of Coronoid Hydrocarbons*. Berlin: Springer-Verlag.

**85** Gutman, I., Milun, M., and Trinajstić, N. (1977). *J. Am. Chem. Soc.* 99: 1692–1704.

**86** Trinajstić, N. (1977). *Int. J. Quantum Chem.* 12: 469–477.

**87** Aihara, J. (1976). *J. Am. Chem. Soc.* 98: 6840–6844.

**88** Aihara, J. (1977). *J. Am. Chem. Soc.* 99: 2048–2053.

**89** Aihara, J. (1978). *J. Am. Chem. Soc.* 100: 3339–3342.

**90** Aihara, J.-i. (2016). *Bull. Chem. Soc. Jpn.* 89: 1425–1454.

**91** Trinajstić, N. (1977). *Croat. Chem. Acta* 49: 593–633.

**92** Dias, J.R. (1992). *J. Chem. Educ.* 69: 695–700.

93 Gutman, I. (1978). *Bull. Chem. Soc. Chim. Beograd* 43: 191–197.

94 Randić, M. (2003). *Chem. Rev.* 103: 3449–3605.

95 Randic, M. (1977). *J. Am. Chem. Soc.* 99: 444–450.

96 Randić, M. (1976). *Chem. Phys. Lett.* 38: 68–70.

97 Hückel, E. (1937). *Z. Elektrochem.* 43: 752–788, 827–849.

98 Clar, E. (1972). *The Aromatic Sextet*. New York: Wiley.

99 Solà, M. (2013). *Front. Chem.* 1: 22.

100 Ciesielski, A., Krygowski, T.M., Cyrański, M.K. et al. (2009). *J. Chem. Inf. Model.* 49: 369–376.

101 Poater, J., Visser, R., Solà, M., and Bickelhaupt, F.M. (2007). *J. Org. Chem.* 72: 1134–1142.

102 Ciesielski, A., Krygowski, T.M., and Cyrański, M.K. (2008). *J. Chem. Inf. Model.* 48: 1358–1366.

103 Fernández, I. and Frenking, G. (2007). *Faraday Discuss.* 135: 403–421.

104 Hopffgarten, M. and Frenking, G. (2012). *WIREs Comput. Mol. Sci.* 2: 43–62.

105 te Velde, G., Bickelhaupt, F.M., Baerends, E.J. et al. (2001). *J. Comput. Chem.* 22: 931–967.

106 Morokuma, K. (1977). *Acc. Chem. Res.* 10: 294–300.

107 Ziegler, T. and Rauk, A. (1977). *Theor. Chim. Acta* 46: 1–10.

108 Shaik, S.S., Shurki, A., Danovich, D., and Hiberty, P.C. (2001). *Chem. Rev.* 101: 1501–1539.

109 Dobrowolski, M.A., Ciesielski, A., and Cyrański, M.K. (2011). *Phys. Chem. Chem. Phys.* 13: 20557–20563.

110 Cyrański, M.K., Howard, S.T., and Chodkiewicz, M.L. (2004). *Chem. Commun.* 2458–2459.

111 Dobrowolski, M.A., Cyrański, M.K., Merner, B.L. et al. (2008). *J. Org. Chem.* 73: 8001–8009.

112 Dobrowolski, M.A., Cyrański, M.K., and Wróbel, Z. (2016). *Phys. Chem. Chem. Phys.* 18: 11813–11820.

# 7

# Descriptors of Aromaticity: Magnetic Criteria

*"In the world of human thought generally and in physical science in particular, the most important and fruitful concepts are those to which it is impossible to attach a well-defined meaning"*

Hans Kramers, Dutch physicist

## 7.1 Introduction

According to the law of induction reported in 1831 by Michael Faraday (the same scientist who discovered benzene in 1825), when a metallic loop is subjected to an external magnetic field, an electric current is induced. This electric current, in turn, generates an induced magnetic field that opposes the external magnetic field. Aromatic molecules contain delocalized electrons in a closed cyclic circuit and, therefore, in the presence of an external magnetic field, these molecules show a ring current and the corresponding induced magnetic field (see Figure 7.1) that is proportional to the external magnetic field ($\vec{B}_o$) as shown:

$$\vec{B}_{ind} = -\vec{\vec{\sigma}}\vec{B}_o \tag{7.1}$$

where $\vec{\vec{\sigma}}$ is the magnetic shielding tensor defined by:

$$\sigma_{\alpha\beta} = \frac{\partial^2 E}{\partial \mu_\alpha \partial B_\beta} \tag{7.2}$$

with E being the electronic energy of the molecule; **B**, the external magnetic field; and **μ**, the magnetic moment (in a loop μ = I.A, intensity per area). This shielding tensor of nine components is symmetric ($\sigma_{\alpha\beta} = \sigma_{\beta\alpha}$, with α and β = x, y, z) and different at each point in space. When calculated at the center of the aromatic ring, we get an idea of the intensity of the ring currents around the ring.

*Aromaticity and Antiaromaticity: Concepts and Applications*, First Edition.
Miquel Solà, Alexander I. Boldyrev, Michał K. Cyrański, Tadeusz M. Krygowski, and Gabriel Merino.

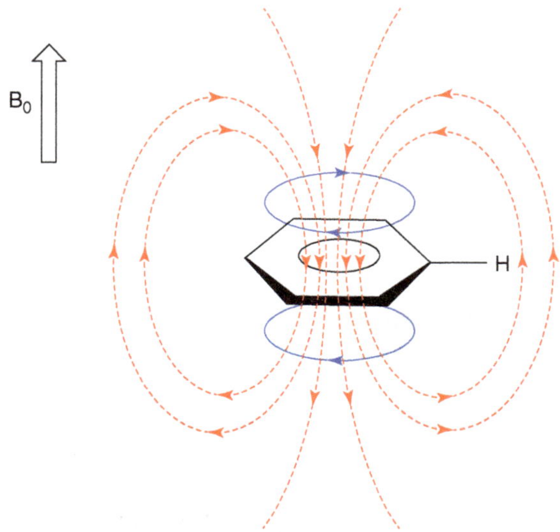


**Figure 7.1** Ring current (in blue) and induced magnetic field (in red) in benzene in the presence of an external magnetic field.

The ring current model (RCM) was first formulated by Pauling, Lonsdale, and London [1–3] in an attempt to explain the large diamagnetic susceptibility anisotropy of aromatic molecules and it is still currently used to rationalize the magnetic properties of molecules [4]. Later on, the RCM theory was extended by Pople to the case of nonuniform magnetic fields [5]. Pople was the first to interpret the anomalous $^1$H nuclear magnetic resonance (NMR) chemical shift of the hydrogen atoms of benzene with the RCM [6]. As seen in Figure 7.1, the H atoms are deshielded by the induced magnetic field and, therefore, they appear downfield. In 1958, Johnson and Bovey [7] made a further refinement, assuming that the π electrons precess in two circular paths separated by 1.28 Å, a value that was empirically adjusted. All these theories were based on a Hückel Hamiltonian concept. Modern calculations start with the quantum mechanical treatment of McWeeny [8] and employ wave functions derived from more sophisticated *ab initio* or density functional theory.

In the presence of an external magnetic field, the total wave function $\Psi(\mathbf{r})$ is composed of a real, $\Psi_R$, and an imaginary, $\Psi_I$, components. The formal expression of the current density $J(\mathbf{r})$ calculated as the first-order response to the applied magnetic field is given by [9]:

$$J(\vec{r}) = \frac{i}{2}(\psi_R \nabla \psi_I - \psi_I \nabla \psi_R + 2i\psi_R A(\vec{r})\psi_R) \tag{7.3}$$

where $\mathbf{A(r)} = \frac{1}{2}\mathbf{B_0}\text{x}(\mathbf{r}\text{-}\mathbf{R_0})$ is a vector potential for the external magnetic field. The first two terms in Eq. (7.3) are the paramagnetic contribution to the ring current and the last term is the diamagnetic one. The calculated ring currents is origin dependent. Several attempts have been performed to circumvent the gauge origin

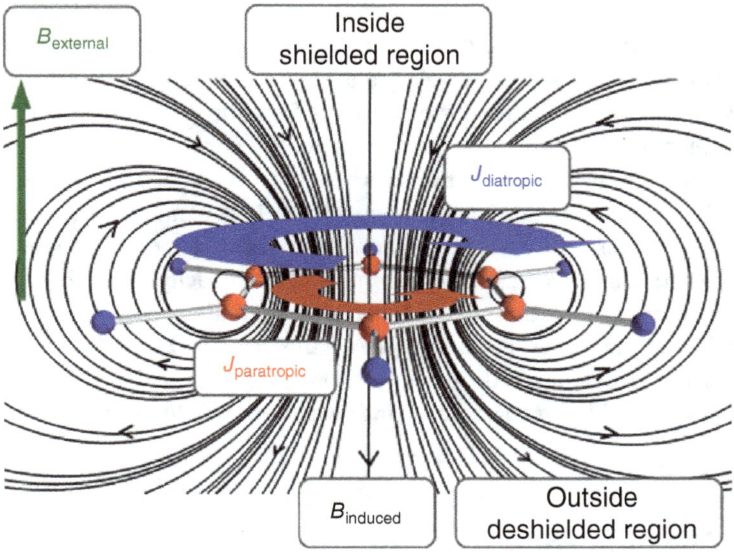


**Figure 7.2** The diatropic (clockwise, blue) and paratropic (anticlockwise, red) directions of the induced current flow relative to the external magnetic field (green) using benzene as an example. Source: Reprinted with permission of Wiley from Ref. [9]. Copyright © 2016, Wiley.

problem and generate gauge-independent current densities, probably the most successful being those that used a separate gauge origin for each point in space [10, 11]. The gauge-including magnetically induced currents (GIMIC) method has been used frequently in the last years [12]. In benzene, the GIMIC plot of the ring currents shows a strong diatropic ring current outside the ring and a less intense paratropic one inside the ring (see Figure 7.2).

Although the analogy between the classical metallic loop and the π-electrons in benzene is frequently done, there are significant differences between the classical mechanics (metallic loop) and quantum mechanics (molecule) phenomena [13]. In classical mechanics, if we pull the magnet back out, the direction of the induced electric current will be reversed as compared to that found when we push the magnet towards the metallic loop. In molecules, the direction of the induced electric current depends only on their molecular and electronic structure. Thus, for the classical monocyclic π-conjugated (anti)aromatic species, application of an external magnetic field yields a diatropic ring current for the monocyclic systems containing a number of $4n+2$ π-electrons, whereas in those with $4n$ π-electrons, it generates a paratropic ring current. An explanation based on the symmetry of the highest occupied molecular orbital-lowest unoccupied molecular orbital (HOMO–LUMO) frontier orbitals was provided by Steiner and

Fowler [14, 15]. Because the intensity of the current is inversely proportional to the energy of the transition, the smaller HOMO–LUMO gap in monocyclic π-conjugated aromatic species with $4n$ π-electrons lead to stronger paratropic ring currents as compared to the diatropic ring currents found in $4n+2$ π-electrons species. The different direction of the ring currents (by convention clockwise diatropic vs. anticlockwise paratropic) provides a method to distinguish between aromatic $4n+2$ and antiaromatic $4n$ π-electrons species. In general, it is assumed that the ring current can be used as a "measure" of electron delocalization and that the stronger the ring currents, the higher the (anti)aromaticity of the ring.

Diatropic ring currents form an induced magnetic field that opposes the external magnetic field in the center of the ring (see Figures 7.1 and 7.2). In the next sections, we discuss several magnetic indices of aromaticity that are based on the rings' currents generated in (anti)aromatic rings when exposed to external magnetic fields [16].

## 7.2 NMR Chemical Shifts

Let us start by emphasizing that neither ring currents nor aromaticity are physical observables. One of the clearest indirect experimental measures of the existence of ring currents are the NMR chemical shifts. Ring currents in aromatic molecules are relevant to NMR spectroscopy because the induced magnetic field dramatically influences the $^1$H NMR chemical shifts of the H's attached to (anti)aromatic rings [17]. As said before, in the benzene ring, protons experience a deshielding effect because the induced magnetic field has the direction same as the external field in the position of H atoms and the chemical shift of these H atoms is about 7.3 ppm. A remarkable difference of the chemical shift for protons in the benzene ring is obvious when compared to 5.6 ppm of a vinyl proton in cyclohexene. It is worth noting that this model was challenged by Wannere and Schleyer [18], but the response by Zanasi and coworkers [19] to this challenge advocated that the model is still valid. This deshielding effect is observed for the protons located outside the aromatic ring. On the other hand, the protons located inside the aromatic rings are shielded. For instance, in the [18] annulenes (see Figure 7.3), the 12 protons outside of the ring show normal chemical shift at ca. 9 ppm, whereas the six protons inside the ring appear at about −3 ppm [20]. As said before, the direction of the diatropic ring currents in the aromatic rings is opposite to that found in antiaromatic compounds with paratropic ring currents. Consequently, the protons that are shielded in aromatic species are deshielded in antiaromatic compounds. The degree of (de)shielding of the $^1$H NMR can be used as a magnetic descriptor of aromaticity. However, if one wants to establish the relative aromaticity of two compounds by comparing the chemical shifts of their internal protons, the systems

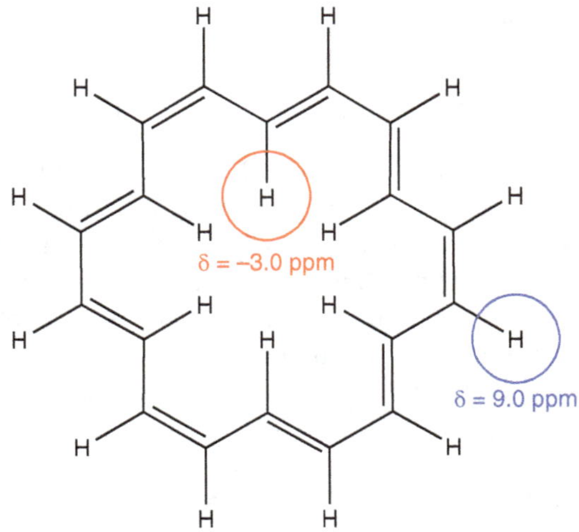


**Figure 7.3** The chemical shifts of the external H atoms (in blue) and the internal hydrogen atoms (in red) in [18] annulene $C_{18}H_{18}$.

must have similar geometries, ring areas, and number of π-electrons. In addition, care must be exercised that the protons are similarly situated with respect to the center of the rings [17]. Otherwise, comparisons are not reliable.

Besides the characteristic chemical shifts of proton NMR, the exaltation of magnetic susceptibility (Λ) has been also used as a magnetic criterion for the quantification of aromaticity. Like NMR chemical shifts, the exaltation of magnetic susceptibility is an experimentally measurable quantity that result from ring currents. The exaltation of magnetic susceptibility is defined as the difference between the true diamagnetic susceptibility, $\chi_M$, and the one calculated by an additive scheme using atom and double bond increments, $\chi_{M'}$. The exaltations are negative (diamagnetic) for aromatic compounds and positive (paramagnetic) for antiaromatic compounds. For instance, at the B3LYP/6-31G* level of theory Λ is −62.9 ppm cgs for benzene and 28.7 ppm cgs for cyclobutadiene. It can also be applied to nonclassical aromatic compounds. As an example, the exaltation of magnetic susceptibility of the closo borane $B_6H_6^{2-}$ is −49.6 ppm cgs [21].

## 7.3 Nucleus Independent Chemical Shifts

The nucleus independent chemical shift (NICS) was proposed in 1996 by Schleyer and coworkers [22, 23] (although Bühl and van Wüllen [24] calculated essentially the same thing for the same purpose a year earlier without naming it). NICS is probably the most widely used index of aromaticity because of its simplicity and availability. It is defined as a negative value of the absolute isotropic shielding computed at the center of a ring or at some other interesting point of the system. It is obtained by averaging the diagonal elements of the magnetic shielding tensor (see Eq. [7.1]) over all directions, as shown:

$$NICS = -\frac{1}{3}(\sigma_{xx} + \sigma_{yy} + \sigma_{zz}) = -\sigma_{av} \tag{7.4}$$

Rings with large negative NICS values in the center are indicative of aromatic character. The more negative the NICS values, the more aromatic the rings are. Several improvements have been introduced since its original definition. In fact, NICS's validity to indicate diamagnetic ring currents is limited by the potential spurious contributions from the components of in-plane tensor that, at least in classical organic aromatic compounds, are not related to aromaticity. For instance, the NICS at the ring center of benzene, which is −8.9 ppm, is the result of a negative NICS(π) of −20.7 ppm and a positive NICS(σ) of 13.8 ppm [23]. The effect of σ-electrons on the total NICS value is partially eliminated by using NICS(1), that is considered to reflect more exactly the π-electron effects [25, 26] or with the corresponding out-of-plane tensor component ($NICS(1)_{zz}$). If the goal is to quantify π-electron effects only, it is even better to use the π contribution to the out-of-plane magnetic shielding tensor component computed at the center of ring or at 1 Å above ($NICS(0)_{\pi zz}$ and $NICS(1)_{\pi zz}$, respectively). $NICS(1)_{zz}$ and $NICS(0)_{\pi zz}$ have been reported to be the best descriptors of aromaticity among different NICS-related definitions in organic molecules [27, 28]. It is possible to obtain individual canonical molecular orbital (CMO) contributions to NICS [23]. When applying the CMO-NICS decomposition to annulenes, it is found that the lowest energy π orbital gives the largest contribution to NICS (−15.2 ppm in the case of benzene) [29]. $NICS_{\pi}$ values can be calculated through the decomposition of NICS indices into molecular orbital components [30].

In polycyclic conjugated hydrocarbons with several fused rings that may contain more than one ring current circuit, the NICS value at a certain point may be the result of different induced magnetic fields. This may lead to wrong assignments of the (anti)aromaticity of the systems studied. In these cases, Gershoni-Poranne and Stanger [31] devised a NICS-based method, the so-called NICS-*XY*-scan, for the determination of local and global ring currents in polycyclic conjugated hydrocarbons. The method involves placing the NICS probes along the X axis, and if needed, along the Y axis, at a constant Z height (the authors suggest at 1.7 Å) above the system under study. Following the change in the NICS values along these axes allows the identification of global and local induced currents. Figure 7.4, for instance, shows the presence of a unique ring current in benzene and anthracene, whereas two local ring currents in the two six-membered rings are found in biphenylene.

NICS is not exempt from criticism [4, 32–34]. One problem is related to the dependence of NICS values on the size of ring [35]. In addition, the coupling of magnetic fields from different regions of the molecule can have a large influence

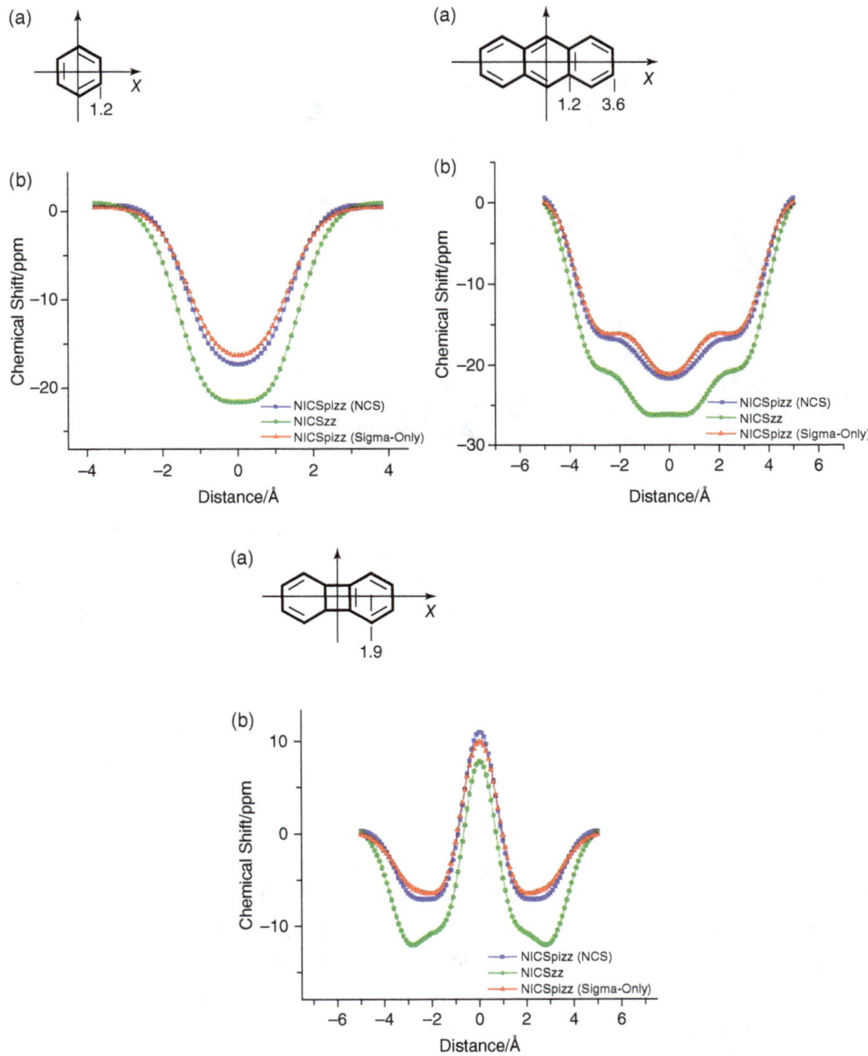


**Figure 7.4** NICS-X-scan of benzene (right), anthracene (middle), and biphenylene. (a) Scheme of axes; (b) NICS-X-scan. All scans performed at a height of 1.7 Å. Blue: $NICS_{\pi zz}$; red: $NICS_{\pi zz}$($\sigma$-only); green: $NICS_{zz}$. Source: Reprinted with permission of Wiley from Ref. [31]. Copyright © 2014, Wiley.

on the NICS value of a given ring [34, 36]. The HF trimer is a case where NICS calculations dramatically fail to predict the correct aromatic character [37]. In general, it is recommended [38, 39] to plot NICS or its variants as a function of the distance from the molecular plane, as NICS values obtained at the center of the ring or at 1 Å above the ring plane can provide contradictory results. It is worth

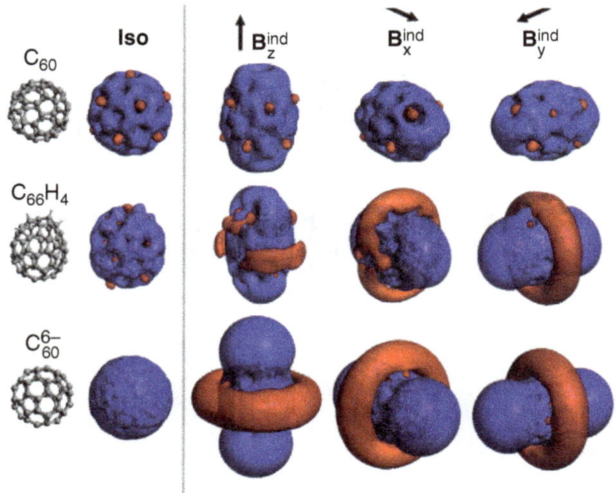


**Figure 7.5** Three-dimensional representations of the induced magnetic field ($B^{ind}$) accounting for the orientational-averaged (Iso), and specific orientations of the applied field ($B^{ind}{}_i$; $i$ = x, y, z) for $C_{60}$, $C_{66}H_4$, and $C_{60}{}^{6-}$. Isosurface set at ±2.0 ppm, Blue: shielding; red, deshielding. Source: Reprinted with permission of Wiley from Ref. [42]. Copyright © 2021, Wiley.

noting that a database of NICS and $NICS_{zz}$ values of 660 neutral benzenoid and cyclopenta-fused polycyclic aromatic hydrocarbons is available [40].

In 2001, Klod and Kleinpeter [41], presented for the first time an iso-chemical-shielding surface (ICSS) by using a three-dimensional grid of NICS probes. The plots show the effect of the shielding and deshielding due to ring currents in the different molecular regions (see Figure 7.5 for the ICSS in several fullerenes). Aromatic molecules, such as benzene or $C_{60}{}^{6-}$, show the typical shielding cone of aromatic species [43]. Interestingly, while for planar aromatics, the cone is reserved only for an out-of-plane applied field, for spherical aromatic compounds, the three-dimensional cage allows the formation of a shielding cone according to the given orientation of the external field.

Let us finally mention that if the external magnetic field ($\vec{B}_0$) perpendicular to the ring is assumed to be 1 T, profiles of $\vec{B}_{ind}$ or $\vec{B}^{z}_{ind}$ *are* equivalent to those of the NICS or $NICS_{zz}$ index, as can be seen from the application of Eq. (7.1) [44, 45]. In this sense, three-dimensional plots of the induced magnetic field are equivalent to ICSSs.

## 7.4 Magnetically Induced Current Densities

The magnetically induced current density shows how electron charges move in molecular systems when exposed to external magnetic fields. The basic quantity obtained from quantum chemical calculations is the tensor of the current density susceptibility. The nine components of the tensor at each point in space correspond to the current flow in the three Cartesian coordinates when external magnetic fields are applied in the three Cartesian directions. It provides the change in the magnitude of the current density as a function of the strength of the external field and its units are $AT^{-1}$. To obtain more detailed information about current strength and current pathways in macrocyclic species, we can determine by numerical integration the current strength for each bond. As a reference, the ring current strength in benzene is 11.8 nA/T (a combination of diatropic 16.7 nA/T and paratropic −4.9 nA/T) and −19.9 nA/T for cyclobutadiene [12].

In planar molecules, the external field is oriented perpendicularly to the molecular plane. The result obtained in the plot depends on the distance of the plotting plane from the molecule. Visualization of the current density is usually done in a plane placed at 1 bohr above the molecular plane. As an example, Figure 7.6 shows different possibilities to visualize the ring currents in naphthalene.

In nonplanar three-dimensional molecules, the representation is somewhat more difficult. Figure 7.7 shows the current-density vector field for the $[B_{12}H_{12}]^{2-}$

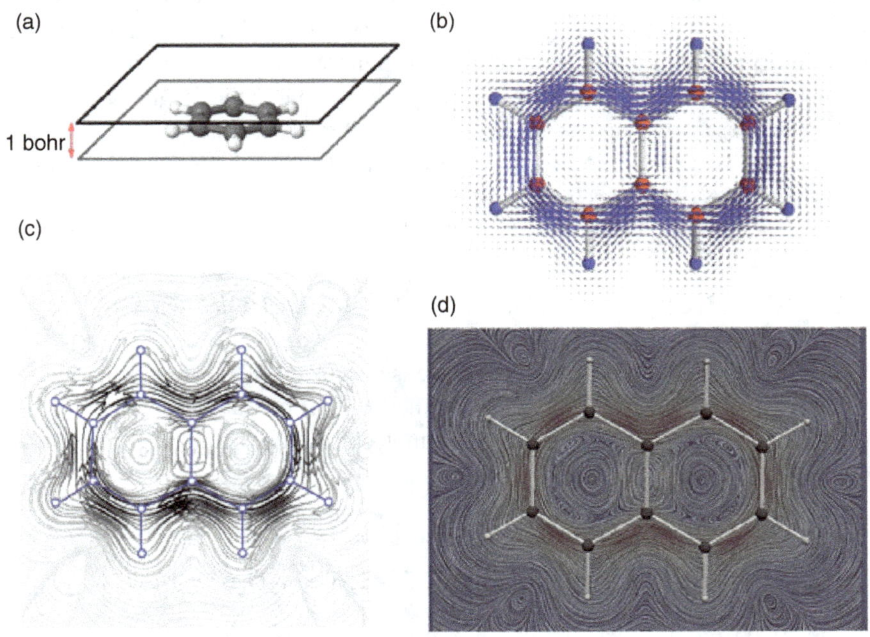


**Figure 7.6** (a) Sketch with the placement of the visualization plane parallel to the molecular plane. (b) Vector plot visualization of the current density of naphthalene in a plane placed 1 bohr above the molecular plane. Clockwise circulations correspond to diatropic currents. (c) Streamline graphic of the current density in the same plane. (d) Line integral convolution plot of the current density in the same plane. Source: Reprinted with permission of Wiley from Ref. [9]. Copyright © 2016, Wiley.

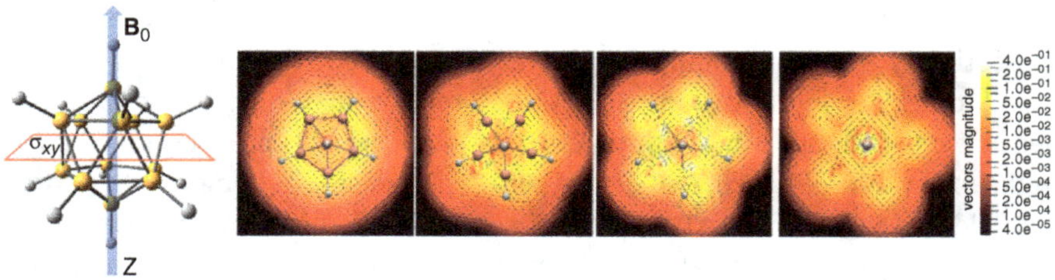


**Figure 7.7** GIMIC representation of the current-density vector field for the $[B_{12}H_{12}]^{2-}$ system. Top view of the currents in the perpendicular plane with respect to the magnetic field vector $B_0$ located at 0, 1, 2 and 3 bohrs (from left to right, respectively) above the central plane. Units are nA $T^{-1}$. Source: Adapted from Ref. [46].

system in perpendicular planes (with respect to the magnetic external field $B_0$) located at 0, 1, 2 and 3 bohrs above the central plane of the molecule. As can be seen, the highest ring current strengths are seen in the central plane of the icosahedron. Like in benzene, we can observe stronger diatropic ring currents in the external polyhedron and weaker paratropic currents inside the polyhedron.

Although it was found that there is a good correlation between NICS indices and π-electron bond current strenghts in monocycles [47, 48], it is generally believed that ring-current strength susceptibilities are a more reliable approach than NICS values to determine the degree of aromaticity according to the magnetic criterion [49]. Indeed, a given ring current leads to a specific NICS value in a certain point of space. However, it is not possible to reconstruct the ring current from a limited discrete set of NICS values [50].

## 7.5 Anisotropy of the Induced Current Density Tensor

The anisotropy of the induced current density tensor (ACID) was proposed by Herges and Geuenich [51] as a scalar function that can be used to determine the degree of electron delocalization of molecules [52]. The ACID quantity is defined as:

$$\Delta J^2(\vec{r}) = \frac{1}{3}\left[\left(J_x^x(\vec{r}) - J_y^y(\vec{r})\right)^2 + \left(J_y^y(\vec{r}) - J_z^z(\vec{r})\right)^2 + \left(J_z^z(\vec{r}) - J_x^x(\vec{r})\right)^2\right] + \frac{1}{2}\left[\left(J_x^y(\vec{r}) - J_x^y(\vec{r})\right)^2 + \left(J_x^z(\vec{r}) - J_z^x(\vec{r})\right)^2 + \left(J_y^z(\vec{r}) - J_z^y(\vec{r})\right)^2\right] \quad (7.5)$$

The ACID method was applied, for instance, to determine the aromaticity of the triplet state of biphenylene [53]. According to ACID, it was found that, among the three options shown in Figure 7.8, the best representation of the biphenylene in its lowest-lying triplet state is given by resonant structure $\mathbf{T_1}$-**I**. The ACID plot in yellow contains the depicted ring currents that indicate a

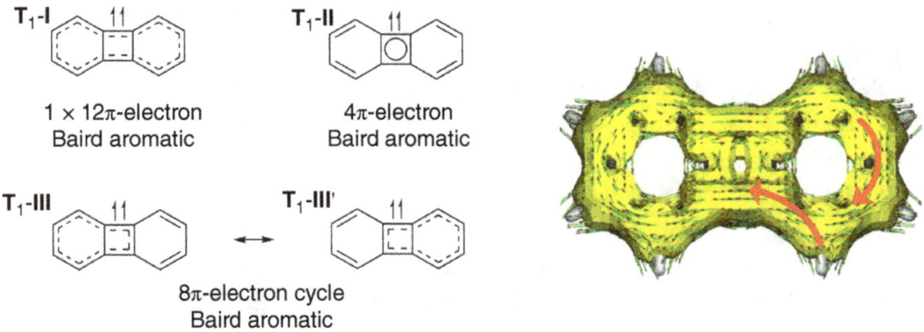


**Figure 7.8** On the left, a schematic representation of the different possible resonant forms in the triplet state of biphenylene (**T$_1$-I**, **T$_1$-II**, and **T$_1$-III**) and on the right the ACID picture. Source: Adapted from Ref. [53].

diatropic circulation (as expected in a Baird-aromatic species, see Chapter 14) around the whole perimeter, as expected in a 12π-electron Baird aromatic species.

In summary, magnetic indicators of aromaticity are based on the rings currents generated on the (anti)aromatic rings when exposed to external magnetic fields. It is assumed that the stronger the ring currents, the higher the electron delocalization, and, therefore, the larger the aromaticity. Although these indices are very useful, they have also shown some pitfalls. In particular, there are several examples of aromatic compounds according to energetic descriptors of aromaticity that hold paramagnetic ring currents [54], showing that paratropicity and aromaticity are not always exclusive. Among these examples, we can mention the $D_{2h}$ benzene radical anion which has a strong paratropic ring current (NICS = 136 ppm) but it retains about one third of the aromatic stabilization energy of benzene [54]. In this sense, to determine the aromatic character of a given ring is advisable to use indices based on several physical properties, the magnetic ones among them.

## References

**1** Pauling, L. (1936). *J. Chem. Phys.* 4: 673–677.

**2** Lonsdale, K.Y. and Bragg, W.H. (1937). *Proc. R. Soc. Lond., A Math. Phys. Sci.* 159: 149–161.

**3** London, F. (1937). *J. Chem. Phys.* 5: 837–838.

**4** Lazzeretti, P. (2000). *Prog. Nucl. Magn. Reson. Spectrosc.* 36: 1–88.

**5** Pople, J.A. (1958). *Mol. Phys.* 1: 175–180.

**6** Pople, J.A. (1956). *J. Chem. Phys.* 24: 1111–1111.

**7** Johnson, C.E. and Bovey, F.A. (1958). *J. Chem. Phys.* 29: 1012–1014.

**8** McWeeny, R. (1958). *Mol. Phys.* 1: 311–321.

9 Sundholm, D., Fliegl, H., and Berger, R.J.F. (2016). *WIREs Comput. Mol. Sci.* 6: 639–678.
10 Keith, T.A. and Bader, R.F.W. (1993). *Chem. Phys. Lett.* 210: 223–231.
11 Coriani, S., Lazzeretti, P., Malagoli, M., and Zanasi, R. (1994). *Theor. Chem. Acc.* 89: 181–192.
12 Fliegl, H., Taubert, S., Lehtonen, O., and Sundholm, D. (2011). *Phys. Chem. Chem. Phys.* 13: 20500–20518.
13 Gershoni-Poranne, R. and Stanger, A. (2015). *Chem. Soc. Rev.* 44: 6597–6615.
14 Steiner, E. and Fowler, P.W. (2001). *J. Phys. Chem. A* 105: 9553–9562.
15 Steiner, E. and Fowler, P.W. (2001). *Chem. Commun.* 2220–2221.
16 Gomes, J.A.N.F. and Mallion, R.B. (2001). *Chem. Rev.* 101: 1349–1383.
17 Mitchell, R.H. (2001). *Chem. Rev.* 101: 1301–1315.
18 Wannere, C.S. and Schleyer, P.v.R. (2003). *Org. Lett.* 5: 605–608.
19 Viglione, R.G., Zanasi, R., and Lazzeretti, P. (2004). *Org. Lett.* 6: 2265–2267.
20 Wannere, C.S., Sattelmeyer, K.W., Schaefer, H.F. III, and Schleyer, P.v.R. (2004). *Angew. Chem. Int. Ed.* 43: 4200–4206.
21 Schleyer, P.v.R. and Jiao, H. (1996). *Pure Appl. Chem.* 68: 209–218.
22 Schleyer, P.v.R., Maerker, C., Dransfeld, A. et al. (1996). *J. Am. Chem. Soc.* 118: 6317–6318.
23 Chen, Z., Wannere, C.S., Corminboeuf, C. et al. (2005). *Chem. Rev.* 105: 3842–3888.
24 Bühl, M. and van Wüllen, C. (1995). *Chem. Phys. Lett.* 247: 63–68.
25 Corminboeuf, C., Heine, T., Seifert, G. et al. (2004). *Phys. Chem. Chem. Phys.* 6: 273–276.
26 Schleyer, P.v.R., Manoharan, M., Jiao, H.J., and Stahl, F. (2001). *Org. Lett.* 3: 3643–3646.
27 Fallah-Bagher-Shaidaei, H., Wannere, C.S., Corminboeuf, C. et al. (2006). *Org. Lett.* 8: 863–866.
28 Feixas, F., Matito, E., Poater, J., and Solà, M. (2008). *J. Comput. Chem.* 29: 1543–1554.
29 Heine, T., Schleyer, P.v.R., Corminboeuf, C. et al. (2003). *J. Phys. Chem. A* 107: 6470–6475.
30 Corminboeuf, C., Heine, T., and Weber, J. (2003). *Phys. Chem. Chem. Phys.* 5: 246–251.
31 Gershoni-Poranne, R. and Stanger, A. (2014). *Chem. Eur. J.* 20: 5673–5688.
32 Lazzeretti, P. (2004). *Phys. Chem. Chem. Phys.* 6: 217–223.
33 Aihara, J. (2002). *Chem. Phys. Lett.* 365: 34–39.
34 Poater, J., Solà, M., Viglione, R.G., and Zanasi, R. (2004). *J. Org. Chem.* 69: 7537–7542.
35 Katritzky, A.R., Jug, K., and Oniciu, D.C. (2001). *Chem. Rev.* 101: 1421–1449.

36 Poater, J., Bofill, J.M., Alemany, P., and Solà, M. (2006). *J. Org. Chem.* 71: 1700–1702.

37 Islas, R., Martínez-Guajardo, G., Jiménez-Halla, J.O.C. et al. (2010). *Theory Comput.* 6: 1131–1135.

38 Stanger, A. (2006). *J. Org. Chem.* 71: 883–893.

39 Jiménez-Halla, J.O.C., Matito, E., Robles, J., and Solà, M. (2006). *J. Organomet. Chem.* 691: 4359–4366.

40 Alvarez-Ramírez, F. and Ruiz-Morales, Y. (2020). *J. Chem. Inf. Model.* 60: 611–620.

41 Klod, S. and Kleinpeter, E. (2001). *J. Chem. Soc., Perkin Trans. 2* 1893–1898.

42 Chen, D., Szczepanik, D.W., Zhu, J. et al. (2021). *Chem. Eur. J.* 27: 802–808.

43 Muñoz-Castro, A. (2017). *Phys. Chem. Chem. Phys.* 19: 12633–12636.

44 Merino, G., Vela, A., and Heine, T. (2005). *Chem. Rev.* 105: 3812–3841.

45 Islas, R., Heine, T., and Merino, G. (2012). *Acc. Chem. Res.* 45: 215–228.

46 Poater, J., Viñas, C., Bennour, I. et al. (2020). *J. Am. Chem. Soc.* 142: 9396–9407.

47 Báez-Grez, R., Ruiz, L., Pino-Rios, R., and Tiznado, W. (2018). *RSC Adv.* 8: 13446–13453.

48 Radenković, S. and Đorđević, S. (2021). *Phys. Chem. Chem. Phys.* 23: 11240–11250.

49 Kumar, C., Fliegl, H., and Sundholm, D. (2017). *J. Phys. Chem. A* 121: 7282–7289.

50 Van Damme, S., Acke, G., Havenith, R.W.A., and Bultinck, P. (2016). *Phys. Chem. Chem. Phys.* 18: 11746–11755.

51 Herges, R. and Geuenich, D. (2001). *J. Phys. Chem. A* 105: 3214–3220.

52 Geuenich, D., Hess, K., Köhler, F., and Herges, R. (2005). *Chem. Rev.* 105: 3758–3772.

53 Ayub, R., Bakouri, O.E., Jorner, K. et al. (2017). *J. Org. Chem.* 82: 6327–6340.

54 Mauksch, M. and Tsogoeva, S.B. (2021). *Chem. Eur. J.* 27: 14660–14671.

# 8

# Descriptors of Aromaticity: Electronic Criteria

*"The more accurate the calculations become, the more the concepts tend to vanish into thin air."*

R. S. Mulliken (*J. Chem. Phys.* **1965**, 43, S2)

## 8.1 Introduction

The concepts of electron localization and delocalization are often used in the discussion of many topics of general modern chemistry [1]. Thus, delocalization of π-electrons is usually invoked to rationalize the stability, molecular structure, chemical reactivity, and magnetic properties of π-conjugated organic molecules [2, 3]. Classical organic and inorganic aromatic species are always characterized by the presence of highly delocalized electrons in their electronic structure [4]. Indeed, Chen, Schleyer, and coworkers [5] defined *aromaticity* as "a manifestation of electron delocalization in closed circuits, either in two or three dimensions." Arguments based on electronic delocalization are also put forward to discuss the origin of phenomena like conjugation, hyperconjugation, or the chemical bonding in metals. In addition, electron delocalization plays a role in the analysis of properties like electron fluctuation or electron correlation effects that are important in the development of new density functionals. Electronic delocalization reveals itself in several chemical phenomena that can be experimentally interrogated to have a measure of the degree of electron delocalization in our system, the most widely used being those based on the magnetic properties of molecules [6]. In descriptive chemistry, electron localization is quantified to know where local groups of electrons like core or valence electrons, electron pairs, bonding pairs, or unpaired electrons are located or to separate them into subsystems of, for instance, π- and σ-electrons. Electron pair localization is important in chemistry to explain

*Aromaticity and Antiaromaticity: Concepts and Applications*, First Edition.
Miquel Solà, Alexander I. Boldyrev, Michał K. Cyrański, Tadeusz M. Krygowski, and Gabriel Merino.

the nature of the chemical bond and to establish a link between classical concepts derived from the Lewis theory [7] and the more rigorous but abstract information derived from the wave function in quantum chemistry calculations.

The wave picture of the electron implies an always-present electron delocalization. This electronic delocalization is not an observable, and therefore there is not an experimental property that allows measuring it directly. From a theoretical point of view, the fact that electron localization/delocalization is not an observable implies that there is not a unique and generally accepted measure of this property. A number of ways of putting this concept on a theoretically sound quantum mechanical basis have been developed, many of them based in one way or another on the electron pair density. Tools such as the electron localization function (ELF) [8–10], which provides a picture of the regions where the electron localization is high, or the quantum theory of atoms in molecules (QTAIM) [11–13], which provides with a wealth of resources to tackle the electron structure in organic and inorganic species, are nowadays routinely used. In the following sections, we discuss first a series of density functions relevant to the description of electron delocalization; second, we show the basis of the ELF and the QTAIM indices used to quantify electron delocalization; third, we list a number of electronic descriptors of aromaticity that employ electron delocalization descriptors.

## 8.2 Density Functions

Let us consider a system of N electrons described by a wave function $\Psi(\vec{x}_1, \vec{x}_2, \ldots, \vec{x}_N)$, where x contains the position ($\vec{r}$) and spin (s) coordinates of each electron. The product $\Psi(\vec{x}_1, \vec{x}_2, \ldots, \vec{x}_N)\Psi^*(\vec{x}_1, \vec{x}_2, \ldots, \vec{x}_N)d\vec{x}_1 d\vec{x}_2 \ldots d\vec{x}_N$ gives us the probability of finding electron 1 between $\vec{x}_1$ and $\vec{x}_1 + d\vec{x}_1$, electron 2 between $\vec{x}_2$ and $\vec{x}_2 + d\vec{x}_2, \ldots,$ and electron N between $\vec{x}_N$ and $\vec{x}_N + d\vec{x}_N$. The probability of finding electron 1 between $\vec{x}_1$ and $\vec{x}_1 + d\vec{x}_1$, independently of where the other electrons are found, is given by:

$$d\vec{x}_1 \int \Psi(\vec{x}_1, \vec{x}_2, \cdots \vec{x}_N)\Psi^*(\vec{x}_1, \vec{x}_2, \cdots \vec{x}_N)d\vec{x}_2 \cdots d\vec{x}_N \tag{8.1}$$

and because the electrons are indistinguishable, the probability of finding an electron between $\vec{x}_1$ and $\vec{x}_1 + d\vec{x}_1$, independently of where the others are found reads:

$$\rho(\vec{x}_1)d\vec{x}_1 = N\, d\vec{x}_1 \int \Psi(\vec{x}_1, \vec{x}_2, \cdots \vec{x}_N)\Psi^*(\vec{x}_1, \vec{x}_2, \cdots \vec{x}_N)d\vec{x}_2 \cdots d\vec{x}_N \tag{8.2}$$

$\rho(\vec{x})$ is the so-called *density function*. The *electron density*, $\rho(\vec{r})$, is obtained integrating with respect to the spin coordinate:

$$\rho(\vec{r}_1) = \int \rho(\vec{x}_1)ds_1 = N \int \Psi(\vec{x}_1, \vec{x}_2, \cdots \vec{x}_N)\Psi^*(\vec{x}_1, \vec{x}_2, \cdots \vec{x}_N)ds_1 d\vec{x}_2 \cdots d\vec{x}_N \tag{8.3}$$

And because $\Psi$ is normalized:

$$\int \rho(\vec{r})d\vec{r} = N \tag{8.4}$$

In an N electron system, $N_\alpha$ electrons have spin $\alpha$ and the remaining $N_\beta$ electrons have spin $\beta$. It is possible to define the electron densities corresponding to the $\alpha$ and $\beta$ electrons separately, $\rho^\alpha(\vec{r})$ and $\rho^\beta(\vec{r})$. For the particular case of $\rho^\alpha(\vec{r})$ we have that:

$$\rho^\alpha(\vec{r}_1) = N_\alpha \int \Psi(\vec{x}_1,\vec{x}_2,\cdots\vec{x}_N)\Psi^*(\vec{x}_1,\vec{x}_2,\cdots\vec{x}_N)ds_1 d\vec{x}_2\cdots d\vec{x}_N \tag{8.5}$$

delivers the probability of finding an electron $\alpha$ between $\vec{r}_1$ and $\vec{r}_1 + d\vec{r}_1$.

On the other hand, the integral:

$$d\vec{x}_1\, d\vec{x}_2 \int \Psi(\vec{x}_1,\vec{x}_2,\cdots\vec{x}_N)\Psi^*(\vec{x}_1,\vec{x}_2,\cdots\vec{x}_N)d\vec{x}_3\cdots d\vec{x}_N \tag{8.6}$$

gives the probability of finding electron 1 between $\vec{x}_1$ and $\vec{x}_1 + d\vec{x}_1$ and electron 2 between $\vec{x}_2$ and $\vec{x}_2 + d\vec{x}_2$, independently of where the others are found. The *second order density*, $\gamma_2(\vec{x}_1,\vec{x}_2)$, is defined as [14, 15]:

$$\gamma_2(\vec{x}_1,\vec{x}_2) = N(N-1) \int \Psi(\vec{x}_1,\vec{x}_2,\cdots\vec{x}_N)\Psi^*(\vec{x}_1,\vec{x}_2,\cdots\vec{x}_N)d\vec{x}_3\cdots d\vec{x}_N \tag{8.7}$$

where the product N(N−1) are the possible electron pairs that can be formed. Because electrons are indistinguishable, $\gamma_2(\vec{x}_1,\vec{x}_2)d\vec{x}_1 d\vec{x}_2$ represents the probability of finding an electron between $\vec{x}_1$ and $\vec{x}_1 + d\vec{x}_1$ and another one between $\vec{x}_2$ and $\vec{x}_2 + d\vec{x}_2$. Integrating with respect to the spin variables, we obtain the *two-electron density, spinless pair density* or *pair function*:

$$\gamma_2(\vec{r}_1,\vec{r}_2) = \int \gamma_2(\vec{x}_1,\vec{x}_2)ds_1 ds_2 \tag{8.8}$$

which gives us the probability of finding any two electrons, one between $\vec{r}_1$ and $\vec{r}_1 + d\vec{r}_1$ and the other between $\vec{r}_2$ and $\vec{r}_2 + d\vec{r}_2$, under any spin combination ($\alpha\alpha$,$\alpha\beta$,$\beta\alpha$,$\beta\beta$). Thus, one can separate the pair function in the following four contributions:

$$\gamma_2(\vec{r}_1,\vec{r}_2) = \gamma_2^{\alpha\alpha}(\vec{r}_1,\vec{r}_2) + \gamma_2^{\alpha\beta}(\vec{r}_1,\vec{r}_2) + \gamma_2^{\beta\alpha}(\vec{r}_1,\vec{r}_2) + \gamma_2^{\beta\beta}(\vec{r}_1,\vec{r}_2) \tag{8.9}$$

where

$$\gamma_2^{\alpha\alpha}(\vec{r}_1,\vec{r}_2) = N_\alpha(N_\alpha - 1) \int \Psi(\vec{r}_1 s_1,\vec{r}_2 s_2,\vec{x}_3\cdots\vec{x}_N) \times \Psi^*(\vec{r}_1 s_1,\vec{r}_2 s_2,\vec{x}_3\cdots\vec{x}_N)ds_1 ds_2 d\vec{x}_3\cdots d\vec{x}_N \tag{8.10}$$

is the probability density of finding two $\alpha$ electrons in positions defined by vectors $\vec{r}_1$ and $\vec{r}_1 + d\vec{r}_1$ and $\vec{r}_2$ and $\vec{r}_2 + d\vec{r}_2$. Likewise, $\gamma_2{}^{\alpha\beta}(\vec{r}_1, \vec{r}_2)$ is given by:

$$\gamma_2^{\alpha\beta}(\vec{r}_1,\vec{r}_2) = N_\alpha N_\beta \int \Psi(\vec{r}_1 s_1,\vec{r}_2 s_2,\vec{x}_3\cdots\vec{x}_N) \times \Psi^*(\vec{r}_1 s_1,\vec{r}_2 s_2,\vec{x}_3\cdots\vec{x}_N)ds_1 ds_2 d\vec{x}_3\cdots d\vec{x}_N \tag{8.11}$$

The spinless pair density can be split into an uncorrelated pair density part and a part that gathers all exchange and correlation effects,

$$\gamma_2(\vec{r}_1,\vec{r}_2) = \rho(\vec{r}_1)\rho(\vec{r}_2) + \gamma_{xc}(\vec{r}_1,\vec{r}_2) \tag{8.12}$$

The uncorrelated part of the pair density, given by the product $\rho(\vec{r}_1)\rho(\vec{r}_2)$, provides the probability density of finding simultaneously two *independent* electrons at positions $\vec{r}_1$ and $\vec{r}_2$. The difference between $\gamma_2(\vec{r}_1,\vec{r}_2)$ and $\rho(\vec{r}_1)\rho(\vec{r}_2)$ is the *exchange-correlation density*, $\gamma_{xc}(\vec{r}_1,\vec{r}_2)$, which is a measure of the degree to which density is excluded at $\vec{r}_2$ because of the presence of an electron at $\vec{r}_1$. Given the fact that the one-electron density and the pair density are normalized to $N$ electrons and $N(N-1)$ electron pairs, respectively, the double integration of the exchange-correlation density, as defined in Eq. (8.12), yields $-N$ electrons. By dividing $\gamma_2(\vec{r}_1,\vec{r}_2)$ by $\rho(\vec{r}_1)$, one gets the so-called *conditional pair density*, $P(\vec{r}_1,\vec{r}_2)$, which gives the probability of finding an electron at position $\vec{r}_2$ when one electron is known to be at reference position $\vec{r}_1$:

$$P(\vec{r}_1,\vec{r}_2) = \frac{\gamma(\vec{r}_1,\vec{r}_2)}{\rho(\vec{r}_1)} \tag{8.13}$$

Since $\gamma(\vec{r}_1,\vec{r}_2)$ can be further partitioned in same spin and unlike spin electron contributions (Eq. [8.9]), using Eq. (8.12) it is possible to write the $P^{\sigma\sigma}(\vec{r}_1,\vec{r}_2)$ and $P^{\sigma\sigma'}(\vec{r}_1,\vec{r}_2)$ contributions to the conditional pair density as:

$$P^{\sigma\sigma}(\vec{r}_1,\vec{r}_2) = \rho^{\sigma}(\vec{r}_2) + \rho_{xc}^{\sigma\sigma}(\vec{r}_1,\vec{r}_2) \tag{8.14}$$

$$P^{\sigma\sigma'}(\vec{r}_1,\vec{r}_2) = \rho^{\sigma'}(\vec{r}_2) + \rho_{xc}^{\sigma\sigma'}(\vec{r}_1,\vec{r}_2). \tag{8.15}$$

In Eqs. (8.14) and (8.15), the $\rho_{xc}^{\sigma\sigma}(\vec{r}_1,\vec{r}_2)$ and $\rho_{xc}^{\sigma\sigma'}(\vec{r}_1,\vec{r}_2)$ ($\sigma = \alpha, \beta$) terms are the so-called *Fermi and Coulomb holes*, respectively. The Fermi hole is a nonpositive quantity determining the decrease in the probability of finding another electron with the same spin relative to a fixed position of the electron of reference located at $\vec{r}_1$. Its integral over $\vec{r}_2$ equals $-1$ (the so-called sum rule), corresponding to the removal of a same-spin electron from the σ-spin density, $\rho^{\sigma}(\vec{r}_2)$. The sum of the Fermi and Coulomb holes gives the exchange-correlation hole, $\rho_{xc}(\vec{r}_1,\vec{r}_2)$. Figure 8.1 provides a schematic picture of the Fermi hole, the Coulomb hole, and the sum of both for a hydrogen molecule with internuclear distances of 1.4 and 5.0 bohrs obtained from a correlated wave function [16]. The Fermi hole displays a very similar shape to that of the orbital of the reference electron, in this case the $1\sigma_g$ bonding orbital of $H_2$. This hole comprises an electron that prevents another electron of the same spin from occupying the same orbital. For this reason, the shape of the Fermi hole does not vary very much with the change in the position of the reference electron, that is, it has a static nature. The electron density moved away from around the nucleus containing the reference electron only corresponds to half an electron (the other half correspond to the other H nucleus).

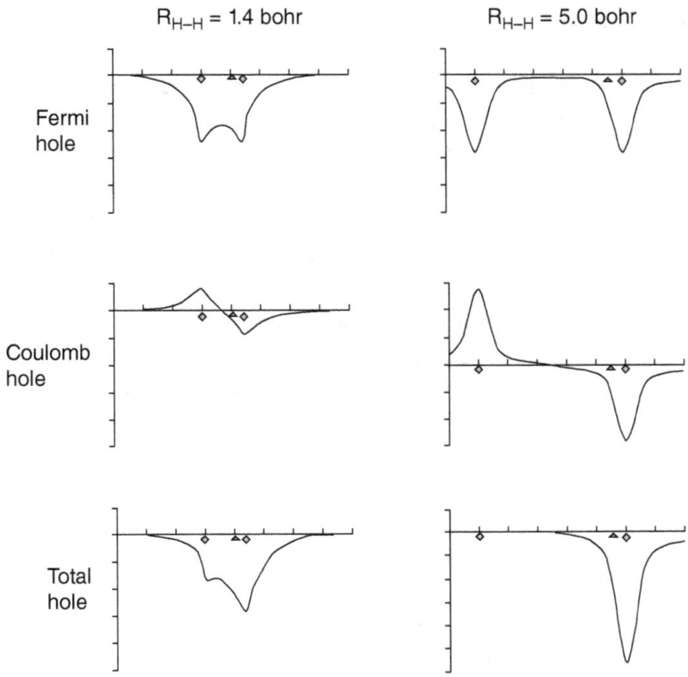


**Figure 8.1** Fermi hole, Coulomb hole, and total exchange-correlation hole for the hydrogen molecule at a H–H bond distances of 1.4 and 5.0 bohrs. Rhombuses indicate the position of the protons and triangles show the position of the reference electron. Source: Adapted from Ref. [16].

On the other hand, the Coulomb hole has a dynamic nature because its shape depends on the position of the reference electron. It gives zero by integration. The Coulomb hole reduces the probability of finding an electron of a different spin from that of the reference electron in the region near the reference electron, but on the other hand, the probability of finding it increases in regions further away from the reference electron. The total hole, considering that it is the sum of the two previous holes, gives an electron by integration. While the Fermi hole and the total exchange-correlation hole have similar shapes at equilibrium distances, at greater distances the difference between the two holes is substantial. The Hartree-Fock (HF) method includes only the Fermi hole and, as a consequence, it is understood that it gives an acceptable result for the equilibrium distance but that it is clearly inadequate to study bond dissociation.

The relationship of the exchange-correlation hole and the exchange-correlation density is given by:

$$\gamma_{xc}(\vec{r}_1, \vec{r}_2) = \rho(\vec{r}_1)\rho_{xc}(\vec{r}_1, \vec{r}_2) \tag{8.16}$$

## 8.3 Measures of Electron Delocalization

There are many ways to determine electron localization/delocalization. Some of them use the components of the wave function like, for instance, all methods that localize molecular orbitals. Others are obtained from the electron density and its derivatives. In this group, one can refer to the Laplacian of the electron density ($\nabla^2\rho$) [11–13], the ellipticity along the bond path [17], or the noncovalent interaction index (NCI) [18]. However, the most abundant are those that are derived from the pair density or higher-order densities. The reason is that the location of electron pairs involves two spatial coordinates and, therefore, methods based on functions of two (or more) positions such as the two-electron density or pair density are more suitable to analyze electron localization and delocalization. Among the methods based on the pair density, probably the most widely used are the electron localization function (ELF) [8–10] and the electron sharing indices (ESI) [19–21]. These two methods are briefly discussed in the following subsections.

### 8.3.1 The Electron Sharing Indices (ESI)

Bader and Stephens [19, 22] showed that the extent of the localization or delocalization of an electron is determined by the corresponding localization of its exchange-correlation hole. At the HF level the Coulomb hole is zero in whole space and the total hole is directly the Fermi hole. As can be seen in Figure 8.1, in the $H_2$ molecule, the total exchange-correlation hole at 1.4 bohr is more delocalized in the whole molecule than at 5.0 bohr. In this latter case, the exchange-correlation hole is almost totally localized in the H atom of the right, indicating that for this bond distance the electron is more localized in one of the H atoms. On the other hand, for the equilibrium distance the electrons are more delocalized. The same is found if one plots the exchange-correlation density given by Eq. (8.16). Based on this property of the exchange-correlation density, Bader et al. defined the localization indices (LIs) and delocalization indices (DIs) from the double integration of the exchange-correlation density over two atomic domains [19]:

$$\lambda(\mathrm{A}) = -\int_A\int_A \gamma_{XC}(\vec{r}_1,\vec{r}_2)\mathrm{d}\vec{r}_1\mathrm{d}\vec{r}_2 \tag{8.17}$$

$$\delta(\mathrm{A},\mathrm{B}) = -2\int_A\int_B \gamma_{XC}(\vec{r}_1,\vec{r}_2)\mathrm{d}\vec{r}_1\mathrm{d}\vec{r}_2 \tag{8.18}$$

The DI or two-center electron sharing index (2c-ESI), $\delta(A,B)$, gives a quantitative measure of the number of electron pairs delocalized or shared between atomic basins $A$ and $B$, whereas the LI, $\lambda(A)$, is a measure of the average number of electrons localized on atom $A$ [20]. The DI is considered by some authors a measure of covalent bond order and the difference between the formal bond multiplicity and the DI can be taken as a measure of bond polarity [21].

Equations (8.17) and (8.18) require integration over atomic domains. In QTAIM, these domains are the *atomic basins*, defined as the regions in real space bound by zero-flux surfaces in $\rho(\vec{r})$ or by infinity [11–13, 23]. The atomic contributions to any molecular property can be calculated by integration of the proper operator through the atomic basins. For instance, the electron population of an atom A is defined as follows:

$$N(A) = \int_A \rho(\vec{r})d\vec{r} \tag{8.19}$$

where the subscript $A$ indicates that the integration has to be carried out only through the space corresponding to the atomic basin of atom $A$. Summation of all the atomic populations in a molecule yields the total number of electrons, $N$. In addition, the following sum rule can be demonstrated:

$$N(A) = \lambda(A) + \frac{1}{2}\sum_{B\neq A}\delta(A,B) \tag{8.20}$$

Equation (8.20) shows that the total number of electrons belonging to a given atomic domain can be exactly partitioned into its localized, λ(A), and delocalized, $\frac{1}{2}\sum_{B\neq A}\delta(A,B)$, components. The percentage of electron localization in a given basin can be calculated as the ratio $\%\lambda(A) = 100 \cdot \frac{\lambda(A)}{N(A)}$.

The calculation of LIs and DIs can also be done using other atomic partitions like the Mulliken-like partitioning in the Hilbert space spanned by the basis functions, the fuzzy-atom approach [24], or the topological fuzzy Voronoi cell (TFVC) [25]. The 2c-ESIs of nonpolarized covalent bonds give qualitatively the same results regardless the atomic partition used [26, 27].

Equations (8.17) and (8.18) can be used at any level of theory, provided that the first- and second-order density functions are known. For monodeterminantal wave functions (as those corresponding to the HF method and also to the wave function generated from Kohn-Sham orbitals in the density functional theory (DFT)), the exchange-correlation density is given by:

$$\gamma_{xc}(\vec{r}_1,\vec{r}_2) = -\sum_{i=1}^{N}\sum_{j=1}^{N}\phi_i^*(\vec{r}_1)\phi_i(\vec{r}_2)\phi_j(\vec{r}_1)\phi_j^*(\vec{r}_2) \tag{8.21}$$

for a system of N electrons and the expressions of the LIs and Dis are:

$$\lambda(A) = \sum_{\mathrm{i,j}}(\mathrm{S_{ij}}(A))^2 \tag{8.22}$$

$$\delta(A,B) = 2\sum_{\mathrm{i,j}}\mathrm{S_{ij}}(A)\mathrm{S_{ij}}(B) \tag{8.23}$$

where $\mathrm{S_{ij}}(A)$ stands for the atomic overlap matrix (AOM) of atom $A$ (overlap between molecular orbitals integrated within the basin of atom $A$) and the summations run over all the pairs of occupied Molecular Spin-Orbitals (MSOs) in the molecule. The quantity $V(A) = \sum_{B\neq A A}\delta(A,B) = 2[N(A) - \lambda(A)]$ is taken as a measure of number of electrons on $A$ that are delocalized over other atoms. At the HF and DFT levels, DIs in Eq. (8.23) reduce to Wiberg–Mayer bond orders if integrations over atomic basins are replaced by a Hilbert space partitioning of the corresponding integrals, that is:

$$\delta^{WM}(A,B) \equiv B_{AB}^{WM} = 2\sum_{\mu\in A}\sum_{\nu\in B}(PS)_{\mu\nu}(PS)_{\nu\mu} \tag{8.24}$$

The definition of the ESI can be generalized to analyze multicenter delocalization or sharing of electrons. It is possible to define a multicenter DI or *Mc*-ESI [28] between the $M$ centers $A_1$ to $A_M$ that for a closed-shell monodeterminantal wave functions reads:

$$I_{A_1,\cdots,A_M} = 2^M \sum_{i_1\cdots i_M} S_{i_1 i_2}(A_1)S_{i_2 i_3}(A_2)\cdots S_{i_M i_1}(A_M) \tag{8.25}$$

where the sum runs over all doubly occupied MOs. This formulation assumes that the electron sharing occurs only between neighboring atoms. Bultinck and coworkers [29] suggested another definition that takes into account all possible permutations of the $M$ atoms, the multicenter electron delocalization index (MCI):

$$MCI(A_1,\cdots,A_M) = \frac{1}{2M}\sum_{P(A_1,\cdots,A_M)} I_{A_1,\cdots,A_M} \tag{8.26}$$

where $P$ represents all possible permutations between centers $A_1$ to $A_M$. For planar species, MCI like $I_{A_1,\cdots,A_M}$ can be exactly split into σ- and π-contributions, namely, $\mathrm{MCI}_\sigma$ and $\mathrm{MCI}_\pi$, respectively, because $S_{ij}(A_k) = 0$ for $i\in\sigma$ and $j\in\pi$ orbital symmetries.

### 8.3.2 The Electron Localization Function (ELF)

The ELF has been proved so far a valuable tool to determine the location of electrons pairs. Because of that, the ELF has been widely used to understand the nature of the chemical bonding and to discuss the mechanism of chemical reactions. The definition of the ELF by Becke and Edgecombe [8] is based on the spherically averaged same-spin conditional pair density (Eq. [8.14], see Figure 8.2) [2, 8, 9, 30, 31].

The spherically averaged same-spin conditional pair density is given by:

$$\begin{aligned}\langle e^{\vec{s}\cdot\vec{\nabla}} P^{\sigma\sigma}(\vec{r},\vec{r}+\vec{s})\rangle &= \frac{1}{4\pi}\left(\int_0^{\pi}\int_0^{2\pi} e^{s\nabla\theta\cos\theta}\sin\theta d\theta d\phi\right) P^{\sigma\sigma}(\vec{r},\vec{r}+\vec{s})|_{\vec{s}=\vec{0}} \\ &= \frac{\sinh(s\nabla)}{s\nabla} P^{\sigma\sigma}(\vec{r},\vec{r}+\vec{s})|_{\vec{s}=\vec{0}}\end{aligned} \tag{8.27}$$

that after using the Taylor expansion of *sinh* around the reference electron (**s=0**) generates an expression whose leading terms reads:

$$\begin{aligned}\langle e^{\vec{s}\cdot\vec{\nabla}} P^{\sigma\sigma}(\vec{r},\vec{r}+\vec{s})\rangle &= \left(1+\frac{1}{6}s^2\nabla_s^2+\ldots\right) P^{\sigma\sigma}(\vec{r},\vec{r}+\vec{s})|_{\vec{s}=0} \\ &\approx \frac{1}{6}s^2\,\nabla_s^2 P^{\sigma\sigma}(\vec{r},\vec{r}+\vec{s})\Big|_{\vec{s}=\vec{0}}\end{aligned} \tag{8.28}$$

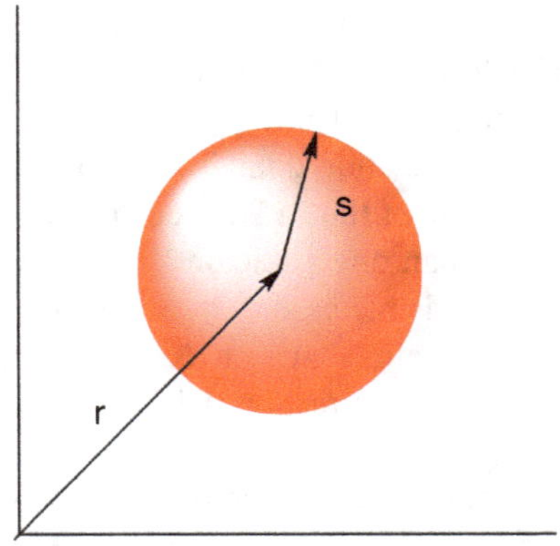


**Figure 8.2** The basis of the definition of the electron localization function is the spherically averaged same-spin conditional pair probability.

In the central expression of Eq. (8.28), the first leading term is zero because the Pauli principle forces the same spin pair density at the coalescence point to be zero, $\gamma^{\sigma\sigma}(\vec{r},\vec{r}) = 0$. Higher than second order terms are considered negligible. Thus, the Taylor expansion of the spherically averaged same-spin conditional pair probability is found to be proportional to:

$$D_\sigma = \frac{1}{2}\nabla_s^2 P^{\sigma\sigma}(\vec{r},\vec{r}+\vec{s})\Big|_{\vec{s}=\vec{0}} = \frac{1}{2}\sum_{j=1}^{N_\sigma}|\nabla\phi_j(\vec{r})|^2 - \frac{1}{8}\frac{|\nabla\rho^\sigma(\vec{r})|^2}{\rho^\sigma(\vec{r})} \tag{8.29}$$

where the r.h.s. of Eq. (8.29) holds only for monodeterminantal wave functions. $D_\sigma$ can be understood as a measure of the excess kinetic energy due to the Pauli exclusion principle. Becke and Edgecombe [8] used the relative ratio of $D_\sigma$ with respect to the same quantity for the homogenous electron gas assuming a monodeterminantal situation [32]:

$$D_\sigma^0 = \frac{3}{5}(6\pi^2)^{2/3}(\rho^\sigma)^{5/3} = \frac{3}{10}(3\pi^2)^{2/3}\rho^{5/3} = c_F\rho^{5/3} \tag{8.30}$$

where $c_F = \frac{3}{10}(3\pi^2)^{2/3}$ is the Fermi constant. Hence, for monodeterminantal wave functions the relative ratio reads:

$$D(\vec{r}) = \frac{D_\sigma}{D_\sigma^0} = \frac{\frac{1}{2}\sum_{j=1}^{N_\sigma}|\nabla\phi_j(\vec{r})|^2 - \frac{1}{8}\frac{|\nabla\rho^\sigma(\vec{r})|^2}{\rho^\sigma(\vec{r})}}{3/5(6\pi^2)^{2/3}[\rho^\sigma(\vec{r})]^{5/3}} \tag{8.31}$$

The probability of finding an electron with spin $\sigma$ when there is already another electron with the same spin nearby is lower when the former is localized. Therefore, the ratio in Eq. (8.31) accounts for electron localization; the higher it is, the lower the localization. The ELF was defined as a function of the aforementioned ratio in order to range the electron localization measure in the interval $[0,1]$ with a Lorentzian form such as [8]:

$$ELF \equiv \eta = \frac{1}{1+D(\vec{r})^2} = \frac{1}{1+\left(\frac{D_\sigma}{D_\sigma^0}\right)^2} \tag{8.32}$$

ELF = 1 corresponds to a completely localized situation, 0 corresponds to a delocalized system, and ELF = 0.5 is the value one should obtain for the homogenous electron gas.

A localization domain is a volume bounded by a given isosurface, defined by the value of the ELF. A localization domain is said to be reducible if it contains more than one attractor. When the ELF value defining a reducible localization domain is increased, this latter is split for a given ELF value, giving rise to two new domains sharing a (3,−1) critical point. The ELF value where a domain splits is called a bifurcation point. It is then possible to draw the so-called localization reduction diagram, whose pattern indicates in which part of the molecule the delocalization occurs. Figure 8.3 contains the localization domains and bifurcation diagram of the benzene molecule. On the left side of Figure 8.3 the localization domains of benzene bounded by the $\text{ELF}(\vec{r}) = 0.65$ isosurface characterized by the aromatic domain in green are shown while on the right side the bifurcation diagram is presented. In this latter, the successive bifurcations occur for $\text{ELF}(r) = 0.1$ (core-valence), 0.61 (V(C,H) from aromatic domain) and 0.65 (aromatic domain).

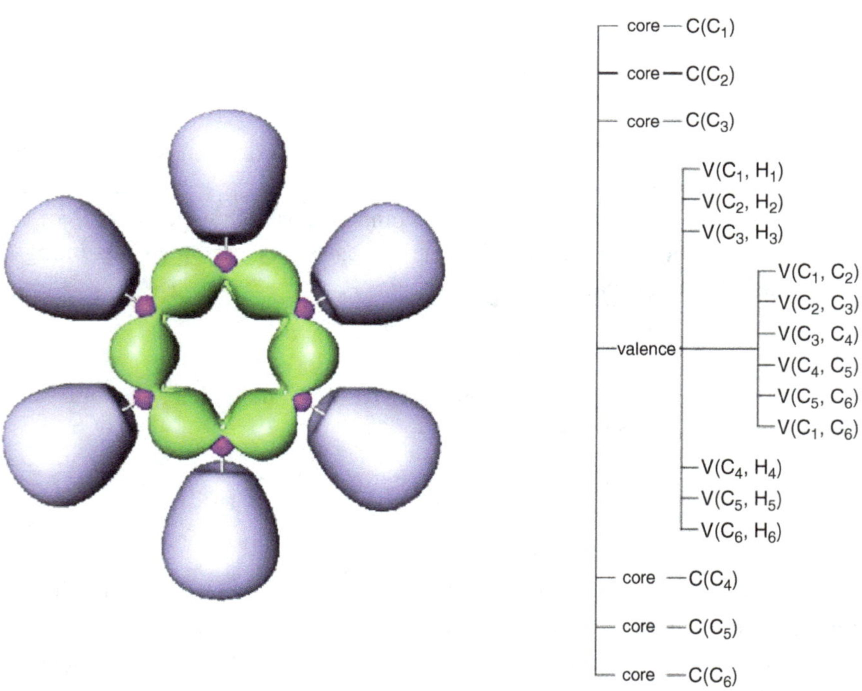


**Figure 8.3** Localization domains and bifurcation diagram of the benzene molecule. Core C(C) domains in magenta, valence V(C,C) domains in green and valence V(C,H) domains in lila. Source: Reprinted with permission from Figure 2 of Ref. [2]. Copyright © 2005, American Chemical Society.

## 8.4 Electronic Descriptors of Aromaticity

Cyclic delocalization of mobile electrons in two or three dimensions is one of the key aspects that characterize aromatic compounds. However, it has not been until the last two decades that electron delocalization descriptors have been widely employed to quantify aromaticity.

The origin of these new descriptors of aromaticity can be traced back to 1974 when Fringelli et al. [33] proposed to use the sum of the bond order differences ($\sum \Delta N$) in a ring as a descriptor of aromaticity. The bond orders were calculated by applying the Gordy equation for the bond order:

$$N = aR^{-2} + b \tag{8.33}$$

where a and b are constants that depend on the type of bond considered. Strictly speaking, this Fringelli index is a geometric descriptor of aromaticity because it makes use of only geometric parameters. However, since it is based on the evaluation of the bond order, it could be also considered a precursor of the current electronic based indices. In 1983, Jug [34] suggested that aromaticity could be estimated by the minimal bond order in a given ring structure. Two years later, Bird [35] defined a new descriptor of aromaticity based on the statistical deviation of the bond orders in the ring as follows:

$$V_n = \frac{100}{N_{av}} \left[ \frac{1}{n} \sum_{i=1}^{n} (N_{av} - N_i)^2 \right]^{\frac{1}{2}} \tag{8.34}$$

where $n$ is the number of bonds in the ring, $N_i$ is the Gordy bond order of each bond in the ring, and $N_{av}$ is the mean ring Gordy bond order. The variance $V_n$ is then shifted and scaled to provide an index in the range [0,100] given by the following equation:

$$I_n = 100 \left( 1 - \frac{V_n}{V_n^K} \right) \tag{8.35}$$

in which ${V_n}^K$ refers to the variance in the corresponding localized Kekulé structure with alternating single and double bonds (${V_n}^K = 33.3$ for a six-membered ring (6-MR) and 35 for a 5-MR). Also in 1985, Pozharskii [36] proposed the average fluctuation of the Gordy bond orders in a given cycle as a descriptor of aromaticity:

$$\Delta \overline{N} = \frac{1}{n} \sum_{i,j=1}^{n} (N_i - N_j) \tag{8.36}$$

and Kotelevski and Prezhdo [37] in 2001 divided the Pozharskii index by the mean ring Gordy bond order to obtain their own index of aromaticity:

$$V(\%) = 100 \frac{\Delta \overline{N}}{N_{av}} \tag{8.37}$$

The *para*-delocalization index (PDI) [38] was proposed in 2003 and represents the first index that can be considered strictly electronic because it only uses information derived from the wave function (pair density). It is obtained using the delocalization index (DI) [20] as defined in the framework of the QTAIM. The PDI is an average of the three DIs of *para*-related carbon atoms in a given 6-MR. PDI can only be used to quantify the aromaticity of 6-MRs. For 5-MRs, it was suggested to use the ΔDI descriptor, defined as the difference between DIs of formally single and double bonds in the 5-MR [38]. In much the same spirit of PDI, the *para* linear response (PLR) indicator of aromaticity was defined by Sablon et al. [39] as the linear response kernel [40] integrated over the atoms in *para* positions of a 6-MR. Also in 2003, Matta [41] introduced the θ aromaticity index that has a similar expression to that of Bird (Eq. [8.34]) but using DIs instead of Gordy bond orders:

$$\vartheta = 1 - \frac{c}{n}\left[\sum_{i=1}^{n}(\delta_o - \delta_i)^2\right]^{\frac{1}{2}} \tag{8.38}$$

In this expression, $c$ is a constant such that $\theta = 0$ for cyclohexane, $n = 6$ for benzenoids, $\delta_o$ is a reference value corresponding to the total electron delocalization of a carbon atom of benzene with all other C atoms in that molecule, and $\delta_i$ is the actual total electron delocalization of a carbon atom with the other carbon atoms forming a ring in a given benzenoid.

The aromatic fluctuation index (FLU) [42] is another electronic-based indicator of aromaticity proposed in 2005 that takes into account the uniformity of the electron delocalization along the molecular ring and its bonding difference with respect to some aromatic reference, and it is given by:

$$FLU(\mathcal{A}) = \frac{1}{N}\sum_{i=1}^{N}\left[\left(\frac{V(A_i)}{V(A_{i-1})}\right)^{\alpha}\left(\frac{\delta(A_i, A_{i-1}) - \delta_{ref}(A_i, A_{i-1})}{\delta_{ref}(A_i, A_{i-1})}\right)\right]^2 \tag{8.39}$$

where the ring considered is formed by atoms in the string $\{\boldsymbol{A}\} = \{A_1, A_2, \dots A_N\}$, $A_0 \equiv A_N$, and the atomic delocalization is defined as:

$$V(A) = \sum_{A \neq B} \delta(A, B) \tag{8.40}$$

Finally, α is a simple function to make sure that the first term in Eq. (8.39) is always greater or equal to 1,

$$\alpha = \begin{cases} 1 & V(A_i) > V(A_{i-1}) \\ -1 & V(A_i) \leq V(A_{i-1}) \end{cases} \tag{8.41}$$

The $\delta_{ref}$(C,C) and $\delta_{ref}$(C, N) reference values are taken from benzene and pyridine in their ground state. FLU is close to 0 in aromatic species and increases as the molecule departs from the aromatic reference. $FLU^{1/2}$ is sometimes preferred over FLU because $FLU^{1/2}$ values are scattered over a wider range and, therefore, the trends derived are clearer [43].

In the same year, Bultinck et al. [29] defined the so-called BOIA (bond order index of aromaticity) index. BOIA has the same structure as the HOMA index, but used bond orders or DIs between each pair of adjacent atoms in the considered ring ($\delta_i$) instead of bond lengths. As a reference, the bond order in benzene ($\delta_{C6H6}$) is used. So, the expression of the BOIA index is:

$$BOIA = 1 - \frac{1}{6}\sum_{i=1}^{n}(\delta_{C6H6} - \delta_i)^2 \tag{8.42}$$

As said before, the definition of DI can be generalized to study multicenter bonding by defining a multicenter DI between the $N$ centers $A_1$ to $A_M$. Giambiagi et al. [44], defined the $I_{ring}$ index by applying Eq. (8.25) to the $M$ atoms that form a given ring. Extension of this $I_{\text{ring}}$ index by Bultinck et al. [29] resulted in the so-called MCI index given by Eq. (8.26).

There is a normalized version of the $I_{ring}$ index, the so-called $I_{NG}$ [45], which is expected to be less dependent on the ring size than its unnormalized homologues, and it is written for aromatic species as:

$$I_{NB}(\mathcal{A}) = \frac{\pi^2}{4}\frac{G(N_\pi)}{NN_\pi}I_{ring}(\mathcal{A})^{1/N} \tag{8.43}$$

where N is the total number of C atoms in the ring and $N_\pi$ the total number of π electrons and $G(N_\pi)$ equals 1 and −3 for aromatic and antiaromatic systems, respectively. There is also a normalized version of the MCI index for aromatic rings, the so-called $I_{NB}$ [45], given by:

$$I_{NB}(\mathcal{A}) = C\frac{G(N_\pi)}{NN_\pi}MCI(\mathcal{A})^{1/N} \tag{8.44}$$

where $C \approx 1.5155$. These indices have the serious disadvantage of relying on $G(N_\pi)$, which might not be known *a priori*.

The value of ELF at which two basins merge is called a bifurcation value (BV). The ring-closure bifurcation value (RCBV) is the ELF value at which all basins in the ring merge to form one continuous cyclic basin. The π-component of the electron localization function ($ELF_\pi$) can be computed from Eqs. (8.31) and (8.32) using only the occupied π molecular orbitals and the π density [46, 47]. To be aromatic, molecules should obey at least these two properties: (i) The maximal difference between the bifurcation values, $\Delta BV(ELF_\pi)$ should be small, and (ii) the $RCBV(ELF_\pi)$ should be above a certain threshold (0.64–0.70).

The electron density of delocalized bonds (EDDBs) method decomposes the one-electron density (ED) in several "layers" representing different levels of electron delocalization [48], namely, the electron density localized on atoms (EDLAs) representing inner shells, lone pairs, and so on; the electron density of localized bonds (EDLBs) representing typical (2-center 2-electron) Lewis-like

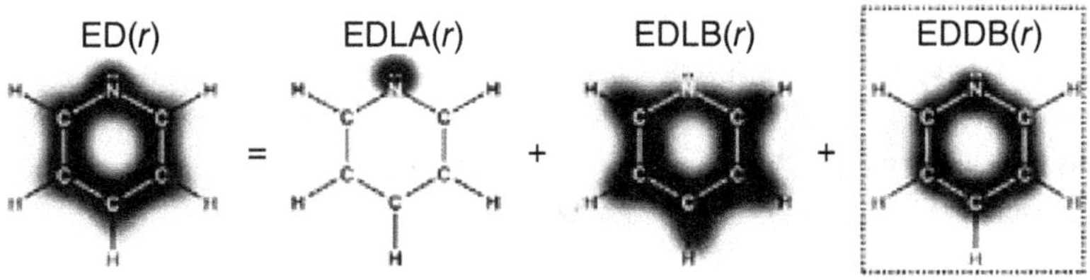


**Figure 8.4** Different levels of electron delocalization from the one-electron density decomposition scheme by Eq. (8.45) for pyridine (only valence electrons). Source: Reprinted with permission from Figure 1a of Ref. [49]. Copyright © 2017, Royal Society of Chemistry.

bonds; and EDDB, which represents electron density that cannot be assigned to atoms or bonds due to its (multicenter) delocalized nature (see Figure 8.4).

$$\mathrm{ED}(\vec{r}) = \mathrm{EDLA}(\vec{r}) + \mathrm{EDLB}(\vec{r}) + \mathrm{EDDB}(\vec{r}) \tag{8.45}$$

In the basis of natural atomic orbitals (NAOs) [50], $\{\chi_{\mu,\nu}(\vec{r})\}$, and in a single-determinant molecular wave function, the spinless *global* electron density of delocalized bonds function, $\mathrm{EDDB}_\mathrm{G}(\vec{r})$, is defined as [51, 52]:

$$\mathrm{EDDB}_\mathrm{G}(\vec{r}) = \sum_{\mu,\nu} \chi^{\dagger}_{\mu}(\vec{r}) D^{\Omega_\mathrm{G}}_{\mu,\nu} \chi_\nu(\vec{r}), \tag{8.46}$$

where

$$\boldsymbol{D}^{\Omega_\mathrm{G}} = 2\sum_{\sigma=\alpha,\beta} \mathbf{P}^{\sigma}\left[\sum_{a,b}^{\Omega_\mathrm{G}} C^{\sigma}_{a,b}\varepsilon^{\Omega_\mathrm{G},\sigma}_{a,b}\left(\lambda^{\sigma}_{a,b}\right)^{2} C^{\sigma\dagger}_{a,b}\right]\mathbf{P}^{\sigma}. \tag{8.47}$$

In Eqs. (8.46) and (8.47), $\mathrm{P}^\sigma$ ($\sigma = \alpha,\beta$) is the standard σ spin-resolved one-electron density matrix; $C^{\sigma}_{a,b}$, collects linear-combination coefficients of the orthogonalized two-center bond orbitals (2cBO) [53, 54]; $\lambda_{\alpha\beta}$, diagonal matrix collecting the corresponding 2cBO squared occupation numbers; $\varepsilon^{\Omega_\mathrm{G},\sigma}_{a,b}$, diagonal matrix of the bond-orbital delocalization factors ($\varepsilon^{\Omega_\mathrm{G},\sigma}_{a,b}$ is close to 1 for delocalized bonds and approaches 0 for the localized ones); and $\Omega = \{(\mathrm{X}_a, \mathrm{X}_b)\}$ denotes a set of all the atomic pairs in a molecular ring [49]. The definition of $\varepsilon^{\Omega_\mathrm{G},\sigma}_{a,b}$ involves a series of projections of 2cBO onto their three-center counterparts, followed by the projection onto the occupied molecular orbitals (MOs) [51, 55]. For planar aromatic systems, the π-ED and π-EDDB components are obtained from the subset of occupied π-MOs. Finally, the $\mathrm{EDDB^k}$ index is the EDDB without taking into account the cross-ring electron delocalization. $\mathrm{EDDB^k}$ and EDDB have the same relationship as $\mathrm{I_{ring}}$ and MCI, respectively.

Finally, Krygowski and coworkers [56, 57] suggested employing the density at the ring critical point of an aromatic ring and other QTAIM properties like the density of total electron energy, H, to quantify aromaticity. In addition, some authors

[58, 59] proposed the use of the hardness as a descriptor of aromaticity. The hardness, η, is a measure of the resistance of a chemical species to change its electronic configuration and, therefore, of its stability. It is defined as the second-order partial derivative of the total electronic energy with respect to the total number of electrons. Using a finite difference approximation and the Koopmans' theorem [60], we obtain the following working definition of the hardness:

$$\eta = \varepsilon_{LUMO} - \varepsilon_{HOMO} \tag{8.48}$$

where $\varepsilon_{LUMO}$ and $\varepsilon_{HOMO}$ are the energies of the lowest unoccupied molecular orbital (LUMO) and the highest occupied molecular orbital (HOMO), respectively. One can expect that the higher the hardness, the larger the aromatic character of a molecule.

Table 8.1 collects the electronic indices computed for a series of aromatic molecules that are depicted in Figure 8.5. The larger the $I_6/I_5$, PDI, MCI, $I_{NB}$, $I_{ring}$, $I_{NG}$, and EDDB$^k$ and the lower the ΔDI and FLU$^{1/2}$, the higher the aromaticity. For the series of 6-MRs, the expected order of aromaticity is:

**Table 8.1** $I_6/I_5$, PDI, ΔDI, FLU$^{1/2}$, MCI, $I_{NB}$, $I_{ring}$, $I_{NG}$, results for species depicted in Figure 8.4. PDI, ΔDI, MCI, and $I_{ring}$ are given in electrons.

| Compound | $I_6/I_5$ | PDI | ΔDI | FLU$^{1/2}$ | MCI | $I_{NB}$ | $I_{ring}$ | $I_{NG}$ | EDDB$^k$ |
|---|---|---|---|---|---|---|---|---|---|
| Benzene | 100 | 0.103 | N/A | 0.000 | 0.072 | 0.0411 | 0.0478 | 0.0413 | 5.525 |
| Pyridine | 86.2 | 0.103 | N/A | 0.022 | 0.068 | 0.0407 | 0.0456 | 0.0410 | 5.571 |
| Pyridazine | — | 0.105 | N/A | 0.041 | 0.069 | 0.0408 | 0.0469 | 0.0412 | 5.027 |
| Pyrimidine | — | 0.101 | N/A | 0.032 | 0.065 | 0.0404 | 0.0440 | 0.0407 | 5.589 |
| Pyrazine | — | 0.108 | N/A | 0.042 | 0.065 | 0.0404 | 0.0435 | 0.0406 | 5.629 |
| Triazine | — | 0.096 | N/A | 0.043 | 0.063 | 0.0402 | 0.0427 | 0.0405 | 5.788 |
| Cyclopentadienyl anion | — | N/A | 0.000 | 0.021 | 0.068 | 0.0467 | 0.0475 | 0.0447 | 5.125 |
| Pyrrole | 73.0 | N/A | 0.214 | 0.071 | 0.043 | 0.0426 | 0.0306 | 0.0410 | 3.063 |
| Furan | 32.2 | N/A | 0.376 | 0.091 | 0.028 | 0.0392 | 0.0224 | 0.0385 | 1.773 |
| Cyclopentadiene | — | N/A | 0.647 | 0.237 | 0.003 | 0.0251 | 0.0079 | 0.0312 | 0.407 |
| Borole | — | N/A | 0.855 | N/A[b] | −0.003 | −0.0379 | 0.0043 | 0.0416 | 0.131 |
| Cyclopentadienly cation | — | N/A | 0.704 | 0.218 | −0.038 | −0.0625 | 0.0001 | 0.0211 | 0.839 |
| Phenanthrene, ring A | — | 0.081 | N/A | 0.073 | 0.046 | 0.0382 | 0.0323 | 0.0386 | 4.220 |
| Phenanthrene, ring B | — | 0.046 | N/A | 0.143 | 0.018 | 0.0326 | 0.0139 | 0.0336 | 1.830 |

a) PDI, FLU$^{1/2}$, MCI, $I_{NB}$, $I_{ring}$, and $I_{NG}$ from Ref. [43], ΔDI from Ref. [38], $I_6/I_5$ values from Ref. [61] and EDDB$^k$ values from Ref. [49].

b) There is no reference parameter for the B—C bond.

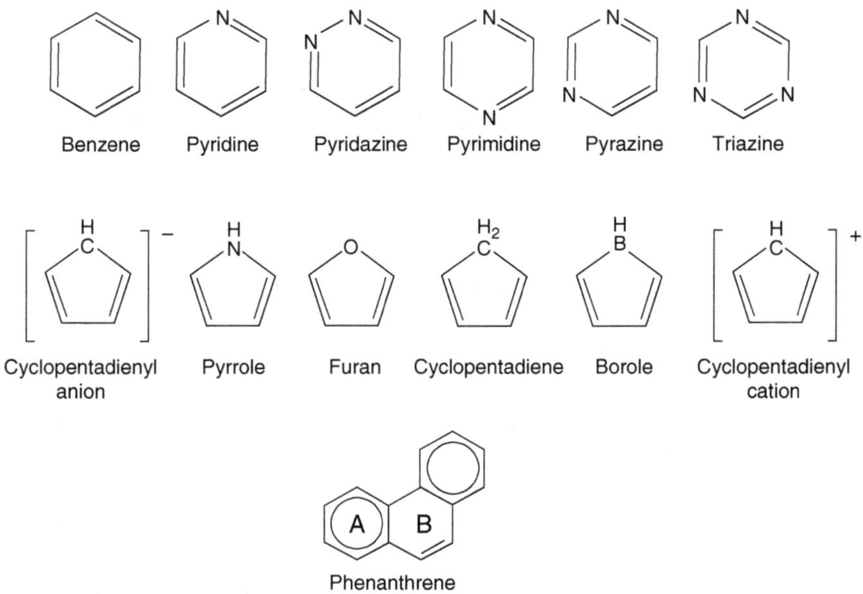


**Figure 8.5** Schematic molecular structure of molecules listed in Table 8.1.

benzene > pyridine > pyridazine ≈ pyrimidine ≈ pyrazine > triazine. This is the order given by all indices with some exceptions. PDI is unsuccessful by considering pyrazine and pyridazine more aromatic than benzene itself and MCI, $I_{ring}$, together with their normalized versions fail only in the ordering of pyridazine that is considered more aromatic than pyridine. $EDDB^k$ in general fails but $EDDB^k(\pi)$ gives the expected results except for pyridazine. For the 5-MRs $C_4H_4X$, the expected order of aromaticity for this set of molecules is $CH^- > NH > O > CH_2 > BH > CH^+$. MCI, $I_{NB}$, and $I_{ring}$ succeed in given the expected trend. On the other hand, ΔDI, $FLU^{1/2}$, and $I_{NG}$ fail in the ordering of cyclopentadienyl anion, whereas $EDDB^k$ fails for the cyclopentadiene cation. Finally, for phenanthrene, all methods agree in assigning a larger aromaticity to the outer ring as expected from the Clar's π-sextet rule discussed in Chapter 2.

## References

1 Matito, E. and Solà, M. (2009). *Coord. Chem. Rev.* 253: 647–665.
2 Poater, J., Duran, M., Solà, M., and Silvi, B. (2005). *Chem. Rev.* 105: 3911–3947.
3 Merino, G., Vela, A., and Heine, T. (2005). *Chem. Rev.* 105: 3812–3841.

4 Feixas, F., Matito, E., Poater, J., and Solà, M. (2015). *Chem. Soc. Rev.* 44: 6434–6451.

5 Chen, Z., Wannere, C.S., Corminboeuf, C. et al. (2005). *Chem. Rev.* 105: 3842–3888.

6 Mitchell, R.H. (2001). *Chem. Rev.* 101: 1301–1315.

7 Lewis, G.N. (1916). *J. Am. Chem. Soc.* 38: 762–785.

8 Becke, A.D. and Edgecombe, K.E. (1990). *J. Chem. Phys.* 92: 5397–5403.

9 Silvi, B. and Savin, A. (1994). *Nature* 371: 683–686.

10 Feixas, F., Matito, E., Duran, M. et al. (2010). *J. Chem. Theory Comput.* 6: 2736–2742.

11 Bader, R.F.W. (1985). *Acc. Chem. Res.* 18: 9–15.

12 Bader, R.F.W. (1990). *Atoms in Molecules: A Quantum Theory*. Oxford: Clarendon.

13 Bader, R.F.W. (1991). *Chem. Rev.* 91: 893–928.

14 Davidson, E.R. (1976). *Reduced Density Matrices in Quantum Chemistry*. New York: Academic Press.

15 Cioslowski, J. (2000). *Many-Electron Densities and Reduced Density Matrices*. New York: Kluwer Academic/Plenum Publishers.

16 Baerends, E.J. and Gritsenko, O.V. (1997). *J. Phys. Chem. A* 101: 5383–5403.

17 Scherer, W., Sirsch, P., Shorokhov, D. et al. (2003). *Chem. Eur. J.* 9: 6057–6070.

18 Johnson, E.R., Keinan, S., Mori-Sánchez, P. et al. (2010). *J. Am. Chem. Soc.* 132: 6498–6506.

19 Bader, R.F.W. and Stephens, M.E. (1975). *J. Am. Chem. Soc.* 97: 7391–7399.

20 Fradera, X., Austen, M.A., and Bader, R.F.W. (1999). *J. Phys. Chem. A* 103: 304–314.

21 Fradera, X., Poater, J., Simon, S. et al. (2002). *Theor. Chem. Acc.* 108: 214–224.

22 Bader, R.F.W. and Stephens, M.E. (1974). *Chem. Phys. Lett.* 26: 445–449.

23 Matta, C.F. and Boyd, R.J. (ed.) (2007). *The Quantum Theory of Atoms in Molecules. From Solid State to DNA and Drug Design*. Weinheim: Wiley-VCH Verlag GmbH & Co. KGaA.

24 Mayer, I. and Salvador, P. (2004). *Chem. Phys. Lett.* 383: 368–375.

25 Salvador, P. and Ramos-Cordoba, E. (2013). *J. Chem. Phys.* 139: 071103.

26 Matito, E., Poater, J., Solà, M. et al. (2005). *J. Phys. Chem. A* 109: 9904–9910.

27 Matito, E., Salvador, P., Duran, M., and Solà, M. (2006). *J. Phys. Chem. A* 110: 5108–5113.

28 Giambiagi, M., de Giambiagi, M.S., and Mundim, K.C. (1990). *Struct. Chem.* 1: 423–427.

29 Bultinck, P., Ponec, R., and Van Damme, S. (2005). *J. Phys. Org. Chem.* 18: 706–718.

**30** Savin, A., Nesper, R., Wengert, S., and Fassler, T.F. (1997). *Angew. Chem. Int. Ed. Engl.* 36: 1809–1832.

**31** Savin, A., Becke, A.D., Flad, J. et al. (1991). *Angew. Chem. Int. Ed. Engl.* 30: 409–412.

**32** Becke, A.D. (1985). *Int. J. Quantum Chem.* 27: 585–594.

**33** Fringuelli, F., Marino, G., Taticchi, A., and Grandolini, G. (1974). *J. Chem. Soc., Perkin Trans. 2* 332–337.

**34** Jug, K. (1983). *J. Org. Chem.* 48: 1344–1348.

**35** Bird, C.W. (1985). *Tetrahedron* 41: 1409–1414.

**36** Pozharskii, A.F. (1985). *Chem. Heterocycl. Compd.* 21: 717–749.

**37** Kotelevskii, S.I. and Prezhdo, O.V. (2001). *Tetrahedron* 57: 5715–5729.

**38** Poater, J., Fradera, X., Duran, M., and Solà, M. (2003). *Chem. Eur. J.* 9: 400–406.

**39** Sablon, N., De Proft, F., Solà, M., and Geerlings, P. (2012). *Phys. Chem. Chem. Phys.* 14: 3960–3967.

**40** Geerlings, P., Fias, S., Boisdenghien, Z., and De Proft, F. (2014). *Chem. Soc. Rev.* 43: 4989–5008.

**41** Matta, C.F. and Hernández-Trujillo, J. (2003). *J. Phys. Chem. A* 107: 7496–7504.

**42** Matito, E., Duran, M., and Solà, M. (2005). *J. Chem. Phys.* 122: 014109.

**43** Feixas, F., Matito, E., Poater, J., and Solà, M. (2008). *J. Comput. Chem.* 29: 1543–1554.

**44** Giambiagi, M., de Giambiagi, M.S., dos Santos, C.D., and de Figueiredo, A.P. (2000). *Phys. Chem. Chem. Phys.* 2: 3381–3392.

**45** Cioslowski, J., Matito, E., and Solà, M. (2007). *J. Phys. Chem. A* 111: 6521–6525.

**46** Santos, J.C., Andres, J., Aizman, A., and Fuentealba, P. (2005). *J. Chem. Theor. Comput.* 1: 83–86.

**47** Santos, J.C., Tiznado, W., Contreras, R., and Fuentealba, P. (2004). *J. Chem. Phys.* 120: 1670–1673.

**48** Szczepanik, D.W. (2016). *Comput. Theor. Chem.* 1080: 33–37.

**49** Szczepanik, D.W., Andrzejak, M., Dominikowska, J. et al. (2017). *Phys. Chem. Chem. Phys.* 19: 28970–28981.

**50** Reed, A.E., Weinstock, R.B., and Weinhold, F. (1985). *J. Chem. Phys.* 83: 735–746.

**51** Szczepanik, D.W., Żak, E., Dyduch, K., and Mrozek, J. (2014). *Chem. Phys. Lett.* 593: 154–159.

**52** Szczepanik, D.W., Andrzejak, M., Dyduch, K. et al. (2014). *Phys. Chem. Chem. Phys.* 16: 20514–20523.

**53** Szczepanik, D. and Mrozek, J. (2012). *Chem. Phys. Lett.* 521: 157–160.

**54** Szczepanik, D. and Mrozek, J. (2013). *J. Math. Chem.* 51: 1619–1633.

**55** Szczepanik, D.W. and Mrozek, J. (2013). *Comput. Theor. Chem.* 1023: 83–87.

**56** Howard, S.T. and Krygowski, T.M. (1997). *Can. J. Chem.* 75: 1174–1181.
**57** Palusiak, M. and Krygowski, T.M. (2007). *Chem. Eur. J.* 13: 7996–8006.
**58** Zhou, Z. and Parr, R.G. (1989). *J. Am. Chem. Soc.* 111: 7371–7379.
**59** De Proft, F. and Geerlings, P. (2001). *Chem. Rev.* 101: 1451–1464.
**60** Koopmans, T. (1933). *Physica (Utrecht)* 1: 104–137.
**61** Krygowski, T.M., Szatyłowicz, H., Stasyuk, O.A. et al. (2014). *Chem. Rev.* 114: 6383–6422.

# 9

# Heteroaromaticity

*"Every set of theoretical approximations, every level of theory, generates its own chemistry. This is what John Pople taught us."*

*R. Hoffman, P. v. R. Schleyer, and H. F Schaefer III (Angew. Chem. Int. Ed. **2008**, 47, 7164)*

## 9.1 Introduction

Heteroaromatic compounds contain one or a few various heteroatoms within an aromatic cyclic π-electron system [1]. According to Balaban et al. [2], half of the chemical compounds known to date contain at least one heteroaromatic ring. Many of these compounds play an important role in the chemistry, biochemistry, and pharmaceutical industries [3]. The progress in the synthesis of novel compounds, investigations of their physicochemical properties, and new applications is extremely rapid. This can be easily exemplified by articles published, for example, in special issues of *Chemical Communications* [4], and more recently *ChemPlusChem* [5, 6]. Moreover, one of the volumes of the Springer series *Topics in Heterocyclic Chemistry* has been devoted to "Aromaticity in Heterocyclic Compounds" [7]. Last but not least, it is worth mentioning that the International Union of Pure and Applied Chemistry (IUPAC) International Symposia on Novel Aromatic (ISNA) Compounds have been organized since 1970 initially every three, and now every two years (see for instance [8]). Due to almost unlimited combinations of carbon, hydrogen, and heteroatoms, an inexhaustible number of novel compounds can be devised following any physicochemical property aimed to be reached, but also the researcher's fantasy.

The first discovered heteroaromatic compound with a six-membered ring (6-MR) was pyridine ($C_5H_5N$), which results from the substitution of a CH group

*Aromaticity and Antiaromaticity: Concepts and Applications*, First Edition.
Miquel Solà, Alexander I. Boldyrev, Michał K. Cyrański, Tadeusz M. Krygowski, and Gabriel Merino.

in benzene by an isoelectronic N atom. It was obtained in 1849 by Thomas Anderson [9, 10] from the distillation of bone oil and other animal matter. Some years before, however, Friedlieb F. Runge [11] already detected pyrrole ($C_4H_5N$) from the distillation of coal. Many other organic heteroaromatic species were discovered or synthesized in the second half of the 19th century.

Borazine, $B_3N_3H_6$, was synthesized by Stock and Pohland [12] in 1926 from the reaction of diborane and ammonia. It is considered the first inorganic benzene and is the result of triple C=C substitution by isoelectronic B=N units in benzene. In 1979, Thorn and Hoffmann [13] predicted that metallabenzenes might be synthesized as stable molecules. These heteroaromatic compounds results from the substitution of CH groups by transition metals with their ancillary ligands [14]. Three years later an osmabenzene was synthesized by Roper and coworkers [15]. Other metalla-aromatic species such as metallafurans, metallapyridines, metallapyridiniums, metallapyridynes, and metallabenzynes have been also prepared and characterized [16–21]. Paneque et al. [22] extended the range of metalla-aromatics from monocyclic to polycyclic systems, completing the preparation of metallanaphthalene. In 2014, efforts to expand the aromatic family led to the discovery of the first metallapentalene (osmapentalene) [23]. The first metallaanthracene was prepared by Wright and Frogley in 2017 [24]. Several metalla-heteroaromatic compounds like aza-metallapentalenes have been also synthesized [25, 26].

In 2001, Boldyrev, Wang, and coworkers [27, 28] synthesized a series of bimetallic clusters containing an aromatic four-membered ring of $Al_4^{2-}$ face capped by a $M^+$ cation (M = Li, Na, Cu) with a laser vaporization technique. They characterized these species from the comparison between the experimental and theoretical photoelectron spectra. $Al_4^{2-}$ was the first discovered all-metal aromatic species. The first detected all-metal heteroaromatic cluster was $XAl_3^-$ (X = Si, Ge, Sn, and Pb) [29]. In the last two decades has brought great advancement in the field of all-metal heteroaromatic species. We can mention here the $M_3O_9^-$ and $M_3O_9^{2-}$ (M = W, Mo), which are the first all-metal heteroaromatic compounds showing *d*-orbital aromaticity [30] or $Ta_3O_3^-$, the first metallic cluster with δ-aromaticity [31] or, more recently, the observation of Möbius heteroaromatic planar metallaborocycles $ReB_4^-$ and $ReB_4$ [32].

All the preceding described types of heteroaromatic rings are summarized in Figure 9.1. Some of them are discussed in the following sections.

The largest group of heterocycle molecules is that generated by substitution of C atoms by heteroatoms in classic organic aromatic molecules. Compared to carbocyclic species, the extent of cyclic π-electron delocalization is modified by several factors such as the nature of heteroatom(s) in the ring(s), their number and position(s), but also topology in ring system(s).

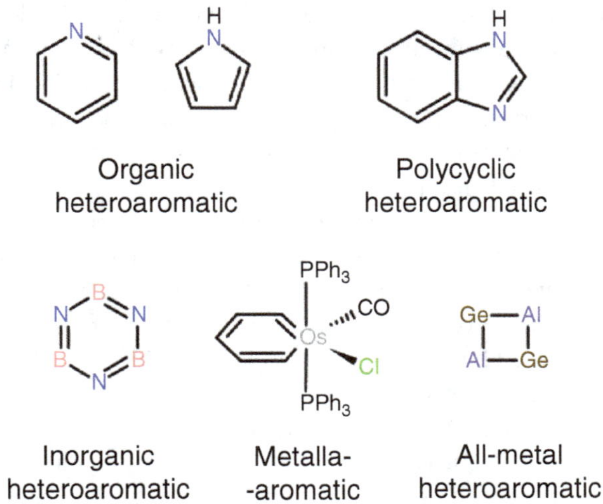


**Figure 9.1** Representation of several heteroaromatic compounds.

Considering the nature of heteroatoms, they differ from carbon atom by two important features: their electronegativity and valencies. Heteroatoms may bring equal, smaller or greater number of electrons to the system, while replacing a carbocyclic fragment, and in some cases they may also introduce other types of orbitals than those of $\sigma$ and $\pi$ type [33]. Taking main-group heteroatoms of the second and further rows, the aromatic character of the heterosubstituted benzenes is generally diminished compared to the parent hydrocarbons [33]. On the other hand, the heterosubstitution in antiaromatic molecules with the number of $\pi$-electrons remaining unchanged may remove the degeneracy of incompletely filled $\pi$-levels [34]. The presence of heteroatoms in rings may further modify the properties of $\pi$-electron systems with two additional, albeit extremely important factors, which are not present in carbocyclic compounds: tautomerism (involving also zwitterionic, mesoionic forms) [35–43] and weak nonbonding interactions with hydrogen bonds as most distinctive examples [44, 45]. Furthermore, these effects may not only significantly influence the structure/properties but also they may operate simultaneously. In this context, the role of solvents, the intermolecular environment [46–48], substituent effects [49, 50], but also other factors like temperature [51, 52] or pressure [53] is much important than in the case of purely carbocyclic systems [54].

Apart from the nature of heteroatoms and their spatial proximity, the difference in electronegativity between hetero and/or carbon atoms is another factor influencing the effectiveness of electron delocalization. A smaller difference favors more efficient delocalization [33]. The relative stability of ring systems can be rationalized by topological charge stabilization [55]. The maximal stabilization is then attained by insertion of heteroatoms with large electronegativities (as compared with carbons) in conjugated hydrocarbons in sites which are characterized by the greatest densities in the isoelectronic conjugated hydrocarbons [55]. Contrary to many carbocyclic species, the presence of one or more electronegative heteroatoms may also change the symmetry of aromatic rings, what additionally modifies their physicochemical behavior [56–63].

Planarity of heterocyclics is another extremely important issue that has to be mentioned. Although for years it has been considered as a paradigm of aromaticity, due to electronic structure of heteroatoms, the nonplanar heterocyclics have probably never been considered as an oxymoron [64], unlike to curved [65–68] or twisted hydrocarbons [69, 70]. The rigidity of heterocyclic systems is reduced and, in most extreme cases, a twist towards Möbius-like topology causes reversed π-electron (de)localization pattern [71].

Last but not least, chemical reactivity is another characteristic of heteroaromatics that should be mentioned. Although it is determined by the interplay of thermodynamic and kinetic factors, generally it is similar to that of saturated hydrocarbons. They have a propensity towards substitution reaction, rather than to addition, what reflects the strong tendency towards restoration of the initial electronic structure [2, 33, 72]. Although pyridine-like heteroatoms decrease the availability of π-electrons and the tendency toward electrophilic substitution, allowing for addition and/or nucleophilic substitution in "π-deficient heteroatoms" [73], pyrrole-like heteroatoms in "π-excessive heterocyles" induce the tendency toward electrophilic substitution [2, 74].

These introductory words shortly outline the great diversity of heterocyclic compounds and complexity of problems, also factors influencing their chemical behavior. It is rather obvious that the description of their aromatic properties in a frame of some simple and general regularities is an extremely difficult task. Therefore, in this chapter we shall attempt to trace some characteristic trends and present arguably most interesting cases, rather than to provide full systematic analysis. As mentioned, the amount of literature is vast and an arbitrary choice had been made.

## 9.2 Six-Membered Organic and Inorganic Heterocycles

Undoubtedly benzene is the archetypical aromatic compound [75], and many descriptors of aromaticity take values estimated for benzene as a reference. Important to note is that descriptors of aromaticity do not necessarily speak with the same voice [76–79] and hence in most cases several descriptors should be applied to end up with a more reliable conclusion [80, 81]. The replacement of the —CH= group in an alternant hydrocarbon with a heteroatom gives rise to the alternation of the π-electron density at the ring atoms. As a result, the aromaticity should be somewhat reduced.

We start by analyzing the aromaticity of a series of azines that are generated by isoelectronic substitution of CH groups by N atoms [60, 82–84]. These nitrogen atom(s) contain a lone electron pair lying in the plane of a molecule bearing its basicity. Additionally, nitrogen atom contains one $2p_z$ electron that is part of a whole π-electron system, resembling the π-electron structure of benzene. Starting from benzene, we could anticipate a decrease of the aromaticity of these species as the number of replacements increase, that is, the order given in Figure 9.2.

This is exactly the trend obtained using the resonance energy (RE) criterion except for triazine [85]. For the set of molecules with the same number of nitrogen atoms the most aromatic are those with the largest number of N—N bonds [86]. Among the different tests suggested in Ref. [78] to assess the performance of aromaticity indices, reproducing the order given in Figure 9.2 was the most difficult one for the aromaticity indices.

As can be seen in Table 9.1, the Bird index $I_6$ fails to provide a good ordering for almost all azines. The aromatic fluctuation index (FLU) fails only on finding pyrimidine more aromatic than pyridazine. The *para*-delocalization index (PDI)

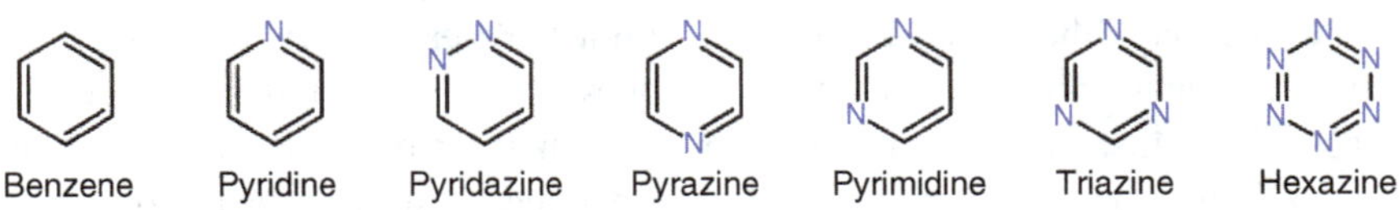


**Figure 9.2** The series of azines discussed.

**Table 9.1** $I_6$, PDI, $FLU^{1/2}$, HOMA, NICS(0), NICS(1), $NICS(1)_{zz}$, and resonance energies (RE, kcal mol$^{-1}$) for azines. NICS values in ppm. Results of PDI, $FLU^{1/2}$, HOMA, NICS(0), NICS(1), $NICS(1)_{zz}$ obtained at the B3LYP/6-311++G(d,p) level of theory.

| | PDI | $FLU^{1/2}$ | $I_6$ a) | HOMA | NICS (0) | NICS (1) | $NICS(1)_{zz}$ | RE b) |
|---|---|---|---|---|---|---|---|---|
| Benzene | 0.103 | 0.000 | 100 | 0.989 | −8.05 | −10.22 | −29.25 | 48.7 |
| Pyridine | 0.103 | 0.022 | 85.7 | 0.995 | −6.82 | −10.12 | −28.59 | 44.1 |
| Pyridazine | 0.105 | 0.041 | 78.9 | 0.981 | −5.35 | −10.52 | −28.28 | 43.9 |
| Pyrimidine | 0.101 | 0.032 | 84.3 | 0.999 | −5.53 | −10.00 | −27.36 | 40.1 |
| Pyrazine | 0.108 | 0.042 | 88.8 | 0.996 | −5.36 | −10.24 | −28.37 | 38.8 |
| Triazine | 0.096 | 0.043 | 100 | 1.000 | −4.07 | −9.66 | −25.54 | 44.9 |
| Hexazine | 0.131 | 0.081 | 100 | 0.984 | −7.29 | −12.85 | −33.41 | – |

a) From Ref. [74].
b) From Ref. [85] in kcal/mol.
Source: Data from Ref. [78].

fails to reproduce the expected aromaticity trend in the heteroaromatic series. It considers pyrazine, pyridazine, and hexazine more aromatic than benzene itself. Since PDI estimates the electron delocalization across the ring, the lone pairs of nitrogen, which do not account for the aromaticity of the ring but are part of nitrogen's basin, may increase the delocalization of the C–N pair and give artificially larger PDI values. The harmonic oscillator model of aromaticity (HOMA) index, which uses only geometrical parameters, fails completely on giving a reasonable trend for the series. The nucleus-independent chemical shift (NICS) indices also fail on reproducing the aromaticity of these heteroaromatic species. NICS(1) and NICS$(1)_{zz}$ break down indicating hexazine and triazine to be the most and the least aromatic species of the series, and they also fail to give the correct trend for the species with one or two nitrogen atoms. Similar to FLU, the multicenter indices (MCI, $I_{ring}$, $I_{NG}$, $I_{NB}$) fail only on finding pyridazine less aromatic than pyrimidine. These results were obtained at the B3LYP/6-311++G(d,p) level of theory, however, if the CCSD/6-311++G(d,p) method is used the expected trend is perfectly reproduced by the multicenter indices, although some compounds in the series are found equally aromatic [78]. The multicenter indices perform better than other aromaticity indices but they are quite dependent on the level of theory when differences are small as those reported in this heteroaromatic series.

The presence of the nitrogen atom(s) in the 6-MR increases dramatically its electron withdrawing abilities. This, however, does not affect too much aromaticity of azines. For instance, most indices show a small variation of aromaticity when moving from benzene to pyridine. Maps of the induced current densities by a perpendicular magnetic field (computed at HF/cc-pVDZ//MP2/cc-pVTZ) [87], (see Figure 9.3) support the previously mentioned similarity of aromaticity of pyridine and benzene.

A systematic study on aromaticity of heterocyclic analogues of benzene ($C_5H_5X$ with X = SiH, GeH, N, P, As, $O^+$, $S^+$, $Se^+$) [87] and the substituent effects on their aromaticity have been undertaken with application of a wide set of theoretical tools [50]. Due to different electronegativities and valencies of heteroatoms, the resonance structures providing maximal contribution (greater than 3%) to the total structure of heterocycles differ as shown in Scheme 9.1.

Obviously, so great differences in π-electron structures are a reason of substantial differences in substituent effects on their aromatic character, but do not differentiate their aromaticities unless substituted systems are not taken into account. The aromatic stabilization energy (ASE) value for benzene is 37.36 kcal/mol whereas these values for heteroanalogues are in the range between 29.05 ($C_5H_5As$) and 32.65 kcal/mol for $C_5H_5N$. Differences in ρ(RCP) (RCP = ring critical point in the Atoms-in-Molecules (AIM) theory [88–90]) are small, in the range 0.027 ($C_5H_5N$) and 0.017 ($C_5H_5GeH$) with benzene ρ(RCP) = 0.025 a.u. Finally, the NICS$(1)_{zz}$ values are also not too much differentiated: for benzene

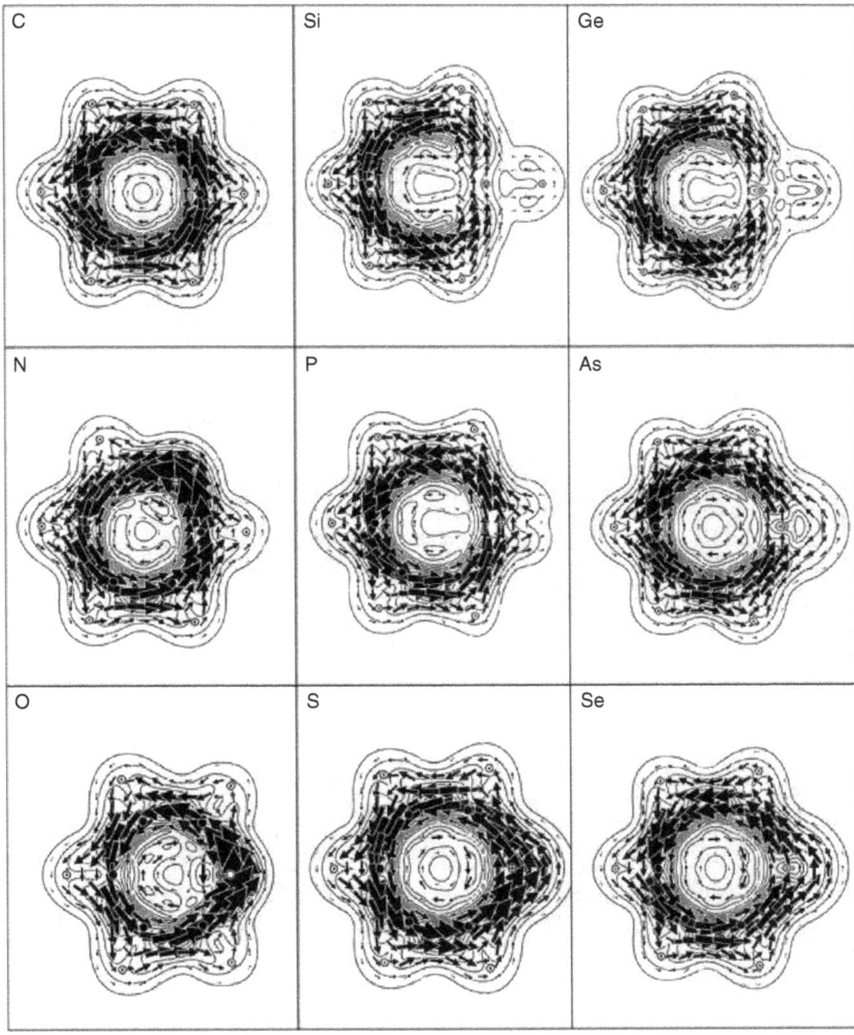


**Figure 9.3** Map of the HF/cc-pVDZ//MP2/cc-pVTZ induced current densities by a perpendicular magnetic field of $C_5H_5X$ (X = CH, SiH, GeH, N, P, As, $O^+$, $S^+$, $Se^+$). Contributions of all electrons are shown, plotted at a height of $1a_0$ above the molecular plane with a diatropic ring current represented through counterclockwise circulation. Positions of nuclei are marked with Dalton symbols except for fourth period elements where a generic symbol is used. Source: Reprinted with permission from Ref. [87]. Copyright © 2011, Royal Society of Chemistry.

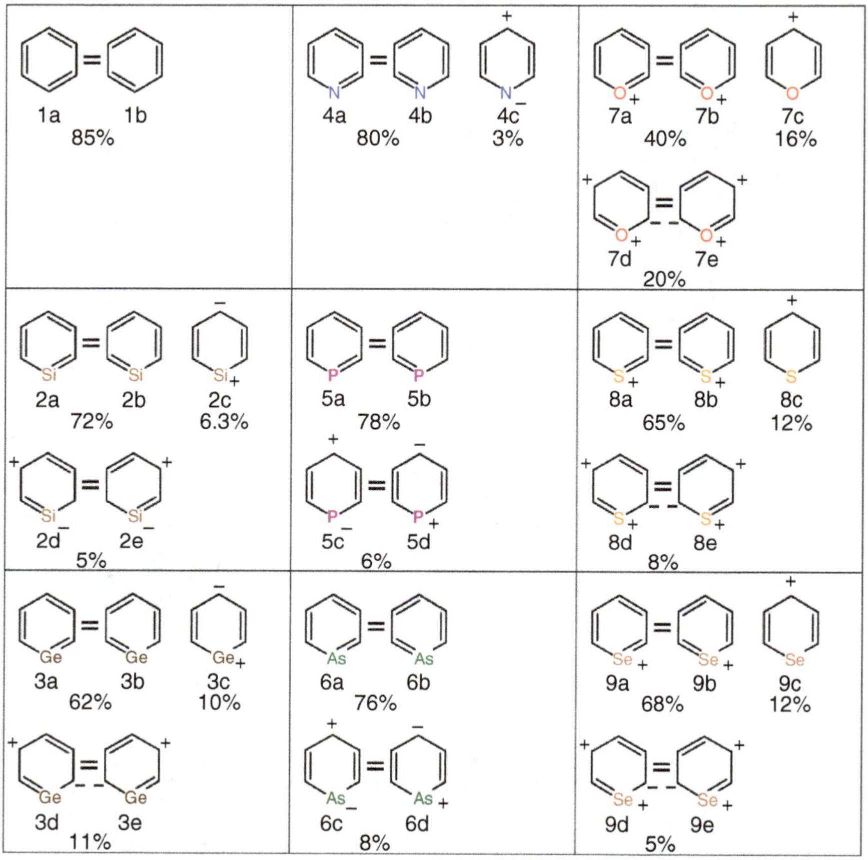

**Scheme 9.1** Resonance structures providing maximal contributions to the total structure of heterocycles. Only structures with contributions higher than 3 percent are listed [87].

it is −30.4 ppm, and for others are in the range between −30.2 ($C_5H_5N$) and −25.2 ppm ($C_5H_5GeH$).

The idea of the paper by Shishkin and coworkers [50] was to observe dependence in changes of aromaticity characteristics ($NICS(1)_{zz}$, ASE, and Bird's $I_6$) on substituents strongly differing in their electron properties: electron donating $NH_2$ and electron withdrawing $NO_2$. In both cases, aromaticity of $C_6H_5X$ (X = N, P, As, SiH, GeH, $O^+$, $S^+$, and $Se^+$) systems estimated by the three previously mentioned characteristics is always lower than that of the mother π-electron system, benzene. In the case of amino substituted systems, the decrease of aromaticity in comparison to the unsubstituted ones is greater than for $NO_2$ substituted systems. Usually the greatest increase of aromaticity is associated with substitution in position 3. In dependence on position, the substitution $NH_2$ differentiates aromaticity stronger than $NO_2$. As expected, the charged systems ($X = O^+, S^+, Se^+$) are much stronger affected by substitution by $NH_2$ than by $NO_2$. A similar study devoted to topological characteristics of the RCPs and aromaticity of groups IIIA to VIA hetero-benzenes [91] revealed that descriptors such as NICS(0), NICS(0)$_{zz}$, NICS(1), NICS(1)$_{zz}$, Isomerization Stabilization Energy (ISE), magnetic susceptibility for zz direction, and RCP properties as, electron density, Laplacian of the density, and densities of kinetic and potential energies, correlate relatively well in groups of similar structure. However even for the four kinds of NICS's, sometimes the correlation is weak. Notably, always the AIM derived data estimated in RCP correlate exceptionally well.

An important group of nitrogen containing six-membered monocycles are borapyridines and phosphapyridines. B and P atoms do not differ in their electronegativity: in Pauling scale they are 2.0 and 2.1 for B and P, respectively. However, boron atoms are electron deficient whereas phosphorus is electron excessive. This difference is well observed in changes of aromatic character for the series. Complete studies of their aromatic character based on NICS(0) and NICS(1), for all combinations of B and P atoms on the ring, sequentially up to five of them were carried out at BLYP/6-311+G** level of computation [92]. Since NICS(1) was recommended as being a better descriptor of π-electron delocalization than NICS(0) [93], NICS(1) data are subject of these analyses. The range of changes in NICS(1) values for borapyridines is very large and cover values from +0.01 for 2,3,4,5-borapyridine down to −18.43 ppm for 2,3,5,6-borapyridines. The NICS(1) values depend not only on the number of B atoms, but also depend strongly on their position in the ring. Contrary to this picture, the range of NICS(1) changes for phosphopyridines is much smaller, and amounts between −6.45 for 2,3,5,6-phosphapyridine and −9.51 ppm for 4-phosphapyridine. These may be compared with the range NICS(1) for azapyridines, where the least and most aromatic are 3,5-diazapyridine with NICS(1) = −9.65, and 2,3-diazapyridine with NICS(1) = −10.80 ppm, respectively. Protonation of azapyridines only slightly modifies their aromaticity.

Borazine is an important symmetric combination of BH and NH units in a 6-MR known as an inorganic benzene. Its aromaticity is questionable. Schleyer et al. [93] claim that due to the highly polar BN bonds, a little evidence for ring current is observed [94] and hence borazine is weakly aromatic, if at all. Comparison of current density maps of π-currents for benzene, triazine, borazine, and boroxine illustrates the differences very nicely and shows a very weak π-current in borazine (see Figure 9.4). Similar results are obtained for a number of inorganic benzene valence isomers [95] and for a recently synthesized 1,4-diaza-2,3,5,6-tetraborinine derivative [96].

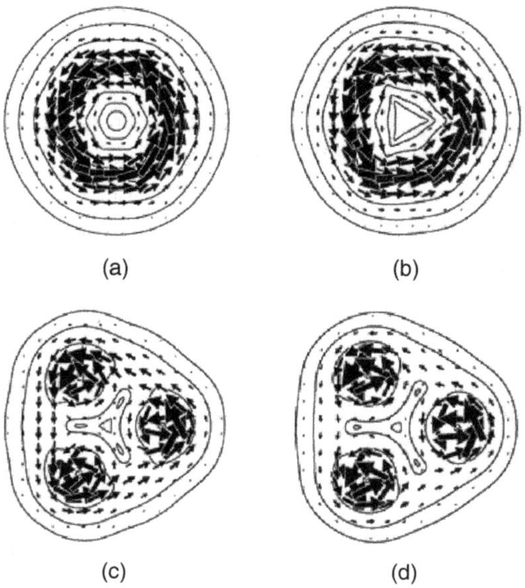

Figure 9.4 (a) π current density maps in benzene ($C_6H_6$), (b) s-triazine ($C_3N_3H_3$), (c) borazine ($B_3N_3H_6$), and (d) boroxine ($B_3O_3H_3$). The maps show the current density field j in a plane 1 bohr above the molecular plane. Source: Reprinted with permission from Ref. [94]. Copyright © 1997, American Chemical Society.

Detailed study of borazine at B3LYP 6-311++G(d,p) level of computation [97] revealed that to understand aromatic character of borazine, it is necessary to separate σ and π contribution and then $B^{ind}$ and ELF analyses (abbreviation of Electron Localization Function, for details see Chapter 8) show that the π system of borazine is (partly) aromatic despite that accumulation of electron density at the nitrogen atoms precludes an effective π-electron delocalization. This observation is in line with substituent effects on cyclic π-electron delocalization in symmetric B- and N-trisubstituted borazine derivatives (computations carried out at HF/6-311+G** and B3LYP/6-311+G** levels of theory) [98] where mutual correlation between NICS's, extra cyclic resonance energy (ECRE), and PDI were well fulfilled whereas geometry based HOMA exhibited no correlation with these indices. Calculated protonation and methylation of borazine with the B3LYP/6-311+G** method [99] indicated that stability of the obtained σ-complex is a good indicator of aromaticity. Comparing to that for benzene, it could be estimated as only one third of that for benzene aromaticity.

Application of aromaticity indices PDI, average two-center index (ATI), and FLU to study substituent effect on local aromaticity in mono- and di-substituted heterocyclic analogs of naphthalene [100] revealed that substituents decrease aromaticity of the system in study irrespective whether the substituent is electron attracting or donating.

Recently, there have appeared studies with application of the σ electron donor-acceptor (sEDA) and π electron donor-acceptor (pEDA) descriptors [102] on the heteroatom incorporation in σ- and π-electron systems [101]. sEDA is defined as the difference between the sum of the populations of the s, $p_x$, and $p_y$ atomic orbitals, contributing to the σ valence orbitals and that of the reference molecule (benzene). pEDA is the same, but considering only the population of the $p_z$ orbitals. In the studied systems, diheterocyclohexadiene (DHCHD) and monoheterocyclopentadiene (MHCPD), see Scheme 9.2 [101], 31 heteroatoms or heteroatom containing groups were subject of computational study by means of the B3LYP/aug-cc-pVDZ method. It was found that aromaticity index $NICS(1)_{zz}$ plotted against pEDA follows a curvilinear relation for MHCPD derivatives ($R = 0.990$) and a linear relationship for DHCHD ($R = 0.969$) (see Figure 9.5). When HOMA for MHCPD derivatives were plotted against pEDA and $NICS(1)_{zz}$, linear relationships with $R = 0.950$ and $R = 0.959$, respectively, were found as shown in Figure 9.6 [101].

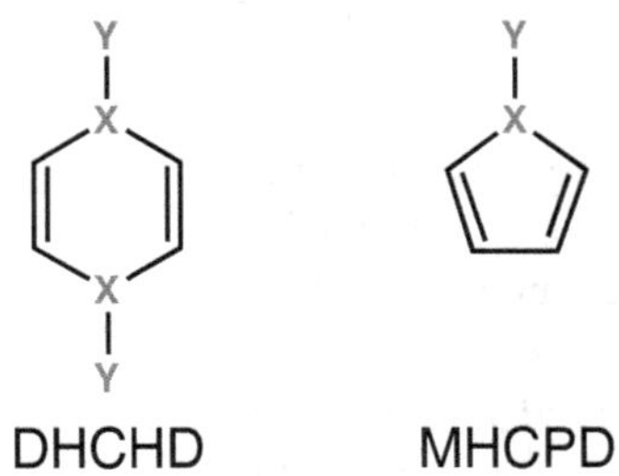


**Scheme 9.2** Diheterocyclohexadiene (DHCHD) and monoheterocyclopentadiene (MHCPD) [101].

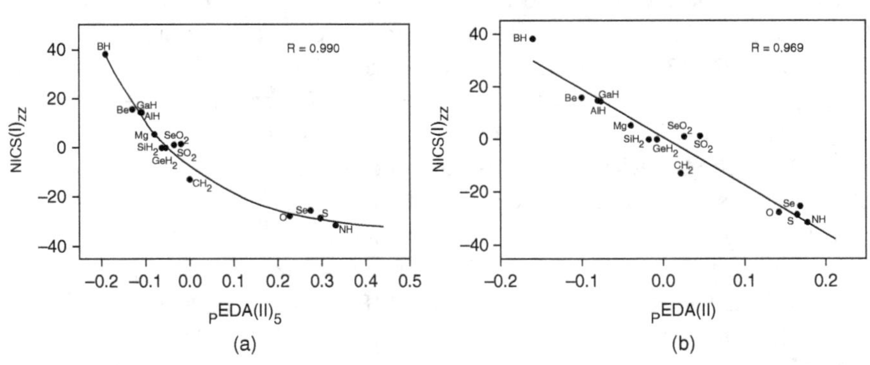


**Figure 9.5** Exponential decay between the $pEDA(II)_5$ and the $NICS(1)_{ZZ}$ aromaticity index of the MHCPD molecules (a) and linear correlation between pEDA(II) descriptor constructed on the basis of DHCHD molecules and the $NICS(1)_{ZZ}$ aromaticity index in MHCPD molecules calculated at the B3LYP/aug-cc-pVDZ level (b). Source: Reprinted with permission from Ref. [101]. Copyright © 2012, American Chemical Society.

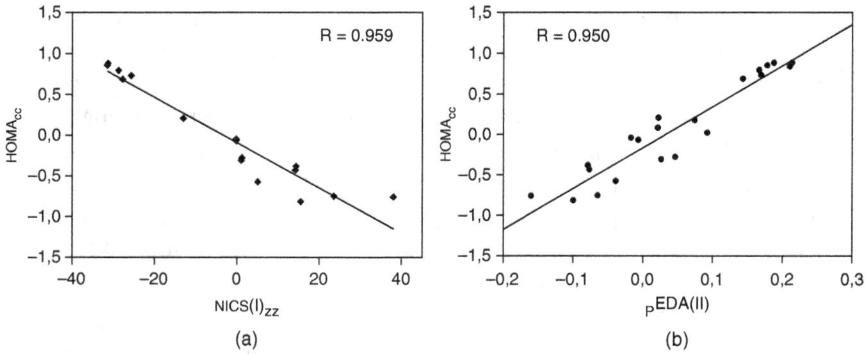


**Figure 9.6** (a) Linear correlation between the NICS(1)$_{zz}$ aromaticity index and the HOMA$_{CC}$ index truncated to CC bonds in the MHCPD system and (b) the linear correlation between the pEDA(II) descriptor and the HOMA$_{CC}$ index calculated at the B3LYP/aug-cc-pVDZ level. Source: Reprinted with permission from Ref. [101]. Copyright © 2012, American Chemical Society.

One particular example of 6-MR heterocycles corresponds to those containing metals in the ring, which are formally obtained by substitution of a $CH^+$ group in benzene by an isolobal metal group, the so-called metallabenzenes [103–106]. The first stable metallabenzene was synthesized by Roper and coworkers [15] in 1982 and was an osmabenzene obtained by a cyclization reaction involving a thiocarbonyl ligand and two ethyne molecules. Interestingly, the possible existence of metallabenzenes was predicted by Hoffmann and Thorn in 1979, three years before [13]. There is a controversy on whether metallabenzenes should be considered species with 6π-electrons like in benzene [107–109] or 10π-electrons [105, 110–112] with standard Hückel aromaticity or even 8π-electrons in a Möbius topology [103, 113]. The different electron counting is due to the debatable participation of the metal *d* orbitals in π-bonding. Depending on the degree of C $2p_z$ and metal *d* orbital mixing, one can consider that metallabenzenes are 6, 8, or 10π-electron species. Indeed, π-electron counting in these species is not as obvious as it is in classical organic molecules.

As an example, let us discuss the relative energy and aromaticity of the *ortho*, *meta*, and *para* $MClY(XC_4H_4)(PH_3)_2$ heterometallabenzenes (M = Ir, Rh; X = N, P; Y = Cl, see Figure 9.7) [112]. For all these 18-electron complexes, the ground state is the singlet. Table 9.2 gathers the B3LYP/cc-pVDZ-PP//BP86/TZ2P relative energies of the 12 *ortho*, *meta*, and *para* heterometallabenzenes studied, together with their values for two indicators of aromaticity, namely, NICS(1)$_{zz}$ and MCI. The results show that the *meta* isomers are the most stable for the **IrN** and **RhN** species by about 10 kcal/mol with respect to the *ortho* and *para* isomers,

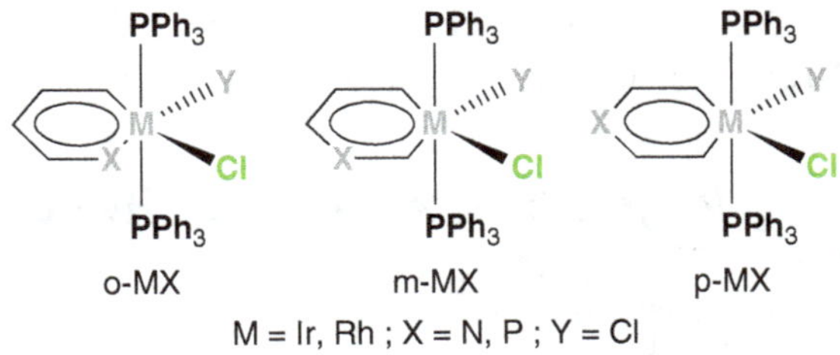


**Figure 9.7** Ortho, meta, and para isomers of $MClY(XC_4H_4)(PH_3)_2$ complexes (M = Ir, Rh; X = N, P; Y = Cl).

**Table 9.2** B3LYP/cc-pVDZ-PP//BP86/TZ2P relatives energies (units are kcal/mol), MCI values (in electrons), and $NICS(1)_{zz}$ (in ppm) values of the $MClY(XC_4H_4)(PH_3)_2$ complexes (M = Ir, Rh, X = N, P; Y = Cl).

| Compound | $\Delta E$ | $NICS(1)_{zz}$ | MCI |
|---|---|---|---|
| *o*-IrN | 10.34 | −5.7 | 0.027 |
| *m*-IrN | 0.00 | −8.8 | 0.024 |
| *p*-IrN | 9.60 | −12.1 | 0.029 |
| *o*-IrP | 0.00 | −5.9 | 0.024 |
| *m*-IrP | 20.35 | −6.3 | 0.023 |
| *p*-IrP | 18.30 | −5.5 | 0.022 |
| *o*-RhN | 9.59 | −10.9 | 0.023 |
| *m*-RhN | 0.00 | −10.2 | 0.023 |
| *p*-RhN | 11.26 | −13.0 | 0.028 |
| *o*-RhP | 0.00 | −7.3 | 0.023 |
| *m*-RhP | 22.59 | −7.9 | 0.022 |
| *p*-RhP | 20.06 | −6.9 | 0.021 |

Source: Data from Ref. [112].

while the *ortho* species are the lowest-lying in energy for all metallaphosphinines studied by c. 20 kcal/mol. The aromaticity of the different isomers is supported by the negative values of the out-of-plane component of the NICS indicator of aromaticity (**o-IrN**: $NICS(1)_{zz}$ = −5.7 ppm; **m-IrN**: $NICS(1)_{zz}$ = −8.8 ppm; **p-IrN**: $NICS(1)_{zz}$ = −12.1 ppm), and the positive value of the electronic delocalization multicenter index (**o-IrN**: MCI = 0.027; **m-IrN**: MCI = 0.024; **p-IrN**: MCI = 0.029). Compared to benzene with $NICS(1)_{zz}$ = −28.9 ppm and MCI = 0.056, it is found that aromaticity in this series of ruthenabenzenes is about half as strong as for benzene [105]. The smaller impact of aromaticity in these systems is illustrated by the fact that the most stable isomer is not always the most aromatic according to $NICS(1)_{zz}$ and MCI descriptors of aromaticity.

## 9.3 Polycyclic Heteroaromatic Hydrocarbons

A special class of heterocyclic compounds are those built up of a few rings. A general question may arise: How aromaticity of this system differs from that of the mother one? Aromaticity analyzed for 24 aza analogues of 7 polycyclic benzenoid hydrocarbons (naphthalene, anthracene, phenanthrene, chrysene, perylene, triphenylene, and helicene) by use of HOMA estimated from experimental geometry [114] revealed that the presence of nitrogen atom in the ring increases its aromaticity, and also increases the aromatic character of the neighboring ring.

The pEDA/sEDA modeling was applied for description of the effect of the incorporation of the ring junction heteroatom [115]. Effect of incorporation of the ring junction heteroatoms in derivatives of 3a,8a-diheteropyrene and cyclopenta[fg] acenaphthylene, see Scheme 9.3, cannot be related in any clear way with changes in aromaticity of the systems studied.

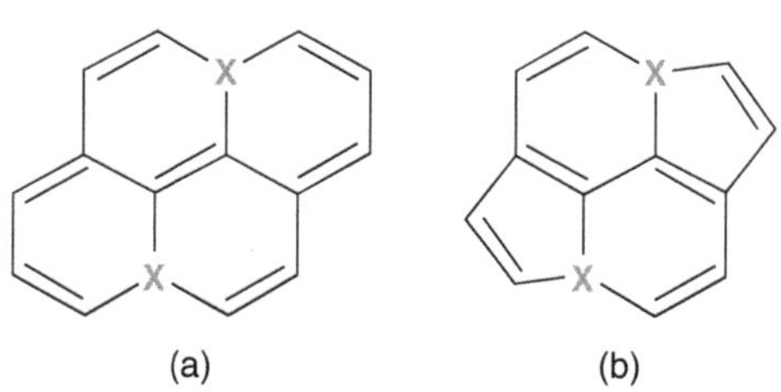


**Scheme 9.3** (a) 3a,8a-diheteropyrene, (b) cyclopenta[fg]acenaphthylene.

Application of HOMA, PDI, and FLU indices obtained using B3LYP with a 6-31+G(d,p) basis set to study aromaticity of benzene, pyrene, chrysene, triphenylene, and tetracene in which CC bonds are replaced by BN revealed that their aromaticity strongly depends on the position of location of the replaced bond and also that local aromaticity depends on the position of the BN bond [116]. Comparison of geometry patterns of benzenoid hydrocarbons and their heteroatom analogues with XY = BB, BC, BN, CN, NN computed at the B3LYP/6-311++G** level of theory revealed [117] that peripheral bond lengths of the systems studied correlate well with CC bonds lengths with the worst $R = 0.965$ (for BB bond lengths, see Figure 9.8) [117]. However, the internal bond lengths not only are uncorrelated, but also they are usually longer than the peripheral ones. This is a reason why the HOMA values for all bonds of a system in question are always lower than that for peripheral ones, as shown in Table 9.3 [117].

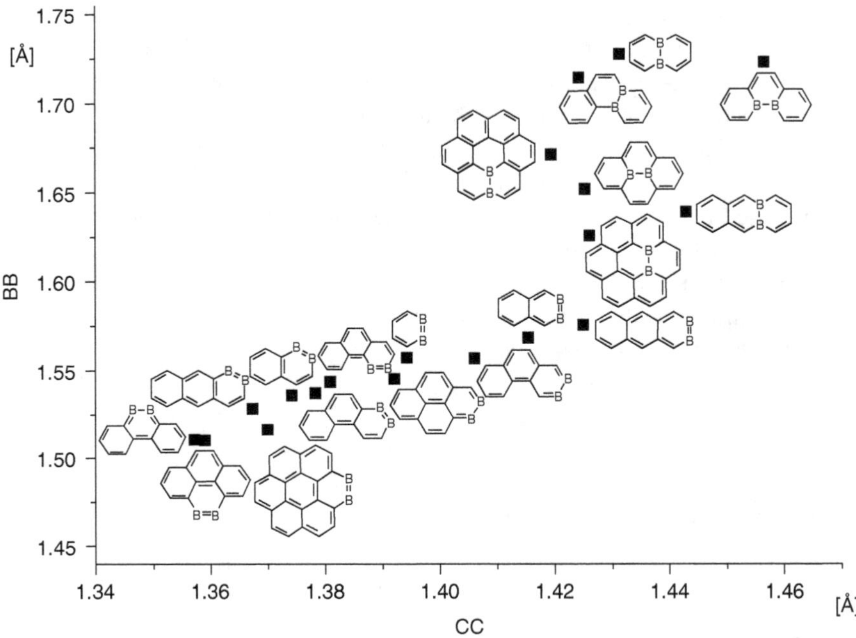


**Figure 9.8** Dependence of BB bond lengths on the equivalent CC bond lengths in hydrocarbon molecules. Source: Reprinted with permission from Ref. [117]. Copyright © 2014, Elsevier.

**Table 9.3** HOMA values for bonds: peripheral (per), inside the ring (in) and for the entire molecule (total) [117].

| XY bond | $HOMA_{per}$ | $HOMA_{in}$ | $HOMA_{total}$ |
|---|---|---|---|
| CC | 0.82 | 0.54 | 0.79 |
| BB | 0.54 | −0.50 | 0.34 |
| NN | 0.74 | 0.29 | 0.62 |
| BC | 0.70 | 0.12 | 0.58 |
| BN | 0.68 | 0.34 | 0.61 |
| CN | 0.79 | 0.53 | 0.71 |

## 9.4 Five-Membered Organic Heterocycles

As in the case of six-membered heterocyclic systems, the most relevant five-membered organic heterocycles are nitrogen containing π-electron systems – pyrrole and azole. They contain two kinds of nitrogen atoms. One is bound to the hydrogen atom and forms an NH group with a $2p_z$ electron pair and the other type of nitrogen atom is the same as that in pyridine. Similarly, as in the case of azines, presence of electronegative heteroatom(s) as nitrogen in both forms, increase electron accepting ability of the heterocyclic system.

Presence of nitrogen atoms of two kinds as mentioned earlier bears possibilities for tautomerism, illustrated in Scheme 9.4 by an example of tautomerism of C-substituted 1,2,3-triazoles that is strongly dependent on the kind of substituent revealing that, for all substituents applied, the N2-H tautomer is the most stable (computation at B3PW91/6-311++G**) [118]. A very interesting observation is that two dramatically different substituents as $NO_2$ and $NH_2$ groups affect the 1,2,3-triazole ring in a similar way.

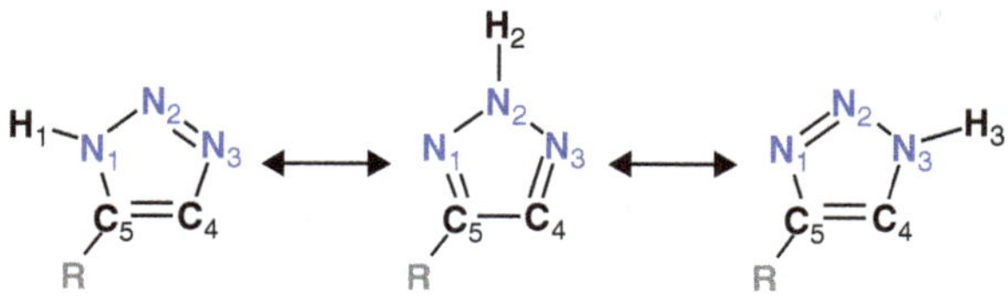


**Scheme 9.4** Tautomerism of substituted 1,2,3-triazoles [118].

One of the most studied series of five-membered organic heterocycles is the one constituted by $C_4H_4X$, X = $CH^-$, NH, O, $CH_2$, BH, $CH^+$ (see Scheme 9.5) [78]. According to the electron counting based on molecular orbital occupation, heterocycles with X = $CH^-$, NH, and O have six $\pi$-electrons and, thus, these three rings should be considered aromatic. Moreover, since the nitrogen atom is less electronegative than the oxygen atom, it is expected a lower reduction of the electron delocalization for the NH group than for the O with respect to the totally delocalized cyclopentadienyl anion ($CH^-$). Therefore, pyrrole is more aromatic than furan. On the other hand, the system with X = $CH_2$ is predicted to be nonaromatic, whereas heterocycles with X = BH and $CH^+$ are both four $\pi$-electron antiaromatic compounds, with X = $CH^+$ being the most antiaromatic of the series. Therefore, the expected order of aromaticity for these five-membered organic heterocycles is $CH^- > NH > O > CH_2 > BH > CH^+$. Results of 10 descriptors of

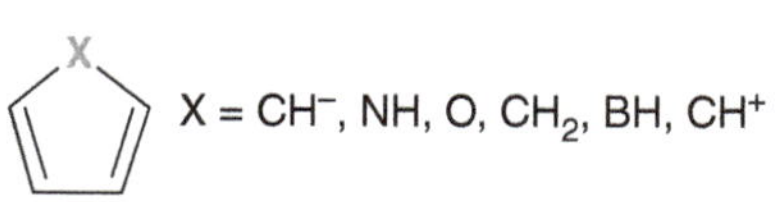


**Scheme 9.5** Five-membered heterocycles $C_4H_4X$ (X = $CH^-$, NH, O, $CH_2$, BH, $CH^+$).

**Table 9.4** $FLU^{1/2}$, MCI, $I_{NB}$, $I_{ring}$, $I_{NG}$, HOMA, NICS(0), NICS(1), and $NICS(1)_{zz}$ results for the $C_4H_4X$ (X = $CH^-$, NH, O, $CH_2$, BH, $CH^+$).

| | $FLU^{1/2}$ | MCI | $I_{NB}$ | $I_{ring}$ | $I_{NG}$ | HOMA | ASE[a)] | NICS(0) | NICS(1) | $NICS(1)_{zz}$ |
|---|---|---|---|---|---|---|---|---|---|---|
| $CH^-$ | 0.021 | 0.068 | 0.0467 | 0.0475 | 0.0447 | 0.810 | 22.1 | −12.61 | −9.63 | −33.68 |
| NH | 0.071 | 0.043 | 0.0426 | 0.0306 | 0.0410 | 0.854 | 20.6 | −13.64 | −10.12 | −30.99 |
| O | 0.091 | 0.028 | 0.0392 | 0.0224 | 0.0385 | 0.198 | 14.8 | −11.91 | −9.42 | −27.14 |
| $CH_2$ | 0.237 | 0.003 | 0.0251 | 0.0079 | 0.0312 | −0.903 | 0.0 | −3.12 | −4.95 | −12.64 |
| BH | N/A[b)] | −0.003 | −0.0379 | 0.0043 | 0.0416 | N/A[b)] | −22.5 | 19.99 | 11.29 | 37.30 |
| $CH^+$ | 0.218 | −0.038 | −0.0625 | 0.0001 | 0.0211 | −1.346 | −56.7 | 86.04 | 64.91 | 199.70 |

a) From Ref. [80] in kcal/mol.
b) There is no reference parameter for the B—C bond.
Source: Data from Ref. [78]

aromaticity, namely, FLU, MCI, $I_{NB}$, $I_{ring}$, $I_{NG}$, HOMA, ASE, NICS, NICS(1), and $NICS(1)_{zz}$ obtained at the B3LYP/6-311++G(d,p) level of theory are gathered in Table 9.4. As can be seen, electronic multicenter descriptors of aromaticity, that is, MCI, $I_{NB}$, $I_{ring}$, $I_{NG}$, perform well to reproduce the aromaticity trends along this series. The FLU index fails predicting $CH_2$ as the most antiaromatic of the series instead of $CH^+$, while HOMA values incorrectly predict that pyrrole ring is more aromatic than the cyclopentadienyl anion ($CH^-$). Among the three magnetic-descriptors of aromaticity used in this work, $NICS(1)_{zz}$ is the only one that reproduces the expected order, whereas NICS(0) and NICS(1) assign the stronger aromatic character of the series to the pyrrole ring. The order is consistent with ASE evaluations made by Cyrański and coworkers for this group of compounds [80].

Five-membered heterocycles have been used as model compounds for answering a fundamental question: Is the term aromaticity uniquely defined? A few important papers have postulated a statistically multidimensional character of aromaticity [119–122], but some other ideas have also been presented [57]. Analysis of mutual interrelations between energetic, magnetic, and geometric indexes of aromaticity for 75 five-membered π-electron systems as aza- and phospha- derivatives of furan, thiophene, pyrrole, and phosphole, and a set of 30 ring-monosubstituted compounds of them has led to the conclusion that aromaticity descriptors do not speak exactly with the same voice [80]. This important conclusion results from scatter plots of NICS, NICS(1), and HOMA versus ASE (Figure 9.9). Due to unsatisfactory correlations among these aromaticity indices, it was concluded that, in general, only a rough division of conjugated cyclic

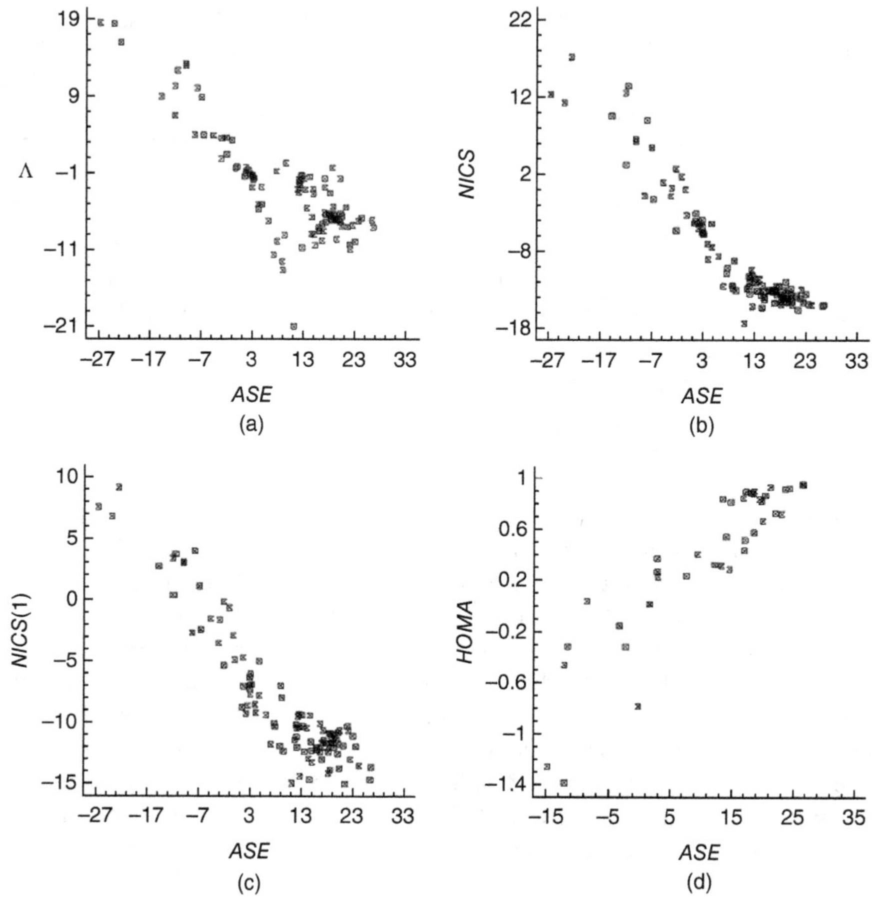


**Figure 9.9** Dependence between Λ, NICS, NICS(1), and HOMA versus ASE for 105 five-membered ring systems (aza and phospha derivatives of furane, thiophene, pyrrole, phosphole, and monoendosubstituted rings): (a) exaltation of magnetic susceptibility versus ASE; (b) NICS versus ASE; (c) NICS computed 1 Å above the ring centers versus ASE; (d) HOMA versus ASE. Source: Reprinted with permission from Ref. [80]. Copyright © 2002, American Chemical Society.

compounds into three major groups: aromatic, nonaromatic, and antiaromatic systems is possible.

Even variously defined ASE models applied to the same set of molecules [123], not always give good correlations. The previously presented conclusions do not exclude situations in which in series of structurally similar compounds various aromaticity characteristics can be well correlated.

## 9.5 Aromaticity of Nucleic Bases

Deoxyribonucleic acid (DNA) and ribonucleic acid (RNA) are the molecules that carry all genetic information in life on Earth. DNA is a macromolecule that has a structure of a double helix, whereas RNA is similar to DNA, but has only one strand of the helix and uses a slightly different set of nucleic bases. Constituents of DNA and RNA are the nucleic bases, which are a very important group of five heteroaromatic compounds. The heteroaromatics having a unique 6-MR, namely, cytosine, thymine, and uracil, are the three pyrimidine nucleic acid bases, whereas two other ones (the so-called purine nucleic acid bases), adenine and guanine, are bicyclic systems with fused 5- and 6-MRs. Due to functional groups as C=O or $NH_2$ as well as two kinds of nitrogen atoms in the ring, nucleic acid bases may exist in many tautomeric forms, differing in energy, electronic structure, and aromatic properties. Indeed, adenine may exist in 23 tautomeric forms, guanine in 36, uracil in 18, thymine in 13, and cytosine in 14. The most stable tautomers of adenine, guanine, uracil, thymine, and cytosine are shown in Figure 9.10. This Figure contains the relative stability among the different tautomers of each nucleic base.

The aromaticity of nucleic bases has been reviewed [124]. Nucleic bases are π-electron heterocyclic compounds that exhibit partial aromatic character. The first study of their aromaticity [125] already indicated that purine nucleobases are significantly more aromatic compared to pyrimidine ones and revealed the following sequence of decreasing aromaticity: adenine > guanine > cytosine > thymine ≈ uracil. This result was later confirmed by electronic and magnetic aromaticity indices [126, 127]. In general, the aromaticity of the nucleobases decreases significantly with the number of exocyclic groups attached to the ring with double bonds.

The main conclusion that can be drawn for the comparison of the relative stabilities of the different tautomers given in Figure 9.10 and the aromaticity analyzed with HOMA and NICS values in Table 9.5 is that the cyclic π-electron delocalization is not the only factor responsible for the relative stability of their tautomers. Noncovalent interactions (hydrogen bonding and π–π stacking) has a more important effect on the relative stability of the different tautomers than the π-electron delocalization of 5- and 6-MRs in individual nucleic bases. Uracil represents a paradigmatic example of this situation. Its most stable tautomer (**u1**) with two carbonyl groups is energetically more stable than all other tautomers by more than 10 kcal/mol. However, **u1** is the least aromatic of all five most stable tautomers of uracil. In this case, there is no direct relationship between the stability of uracil tautomers and their π-electron delocalization. The particular stability of **u1** comes from the three stabilizing N–H···O intramolecular hydrogen bonds.

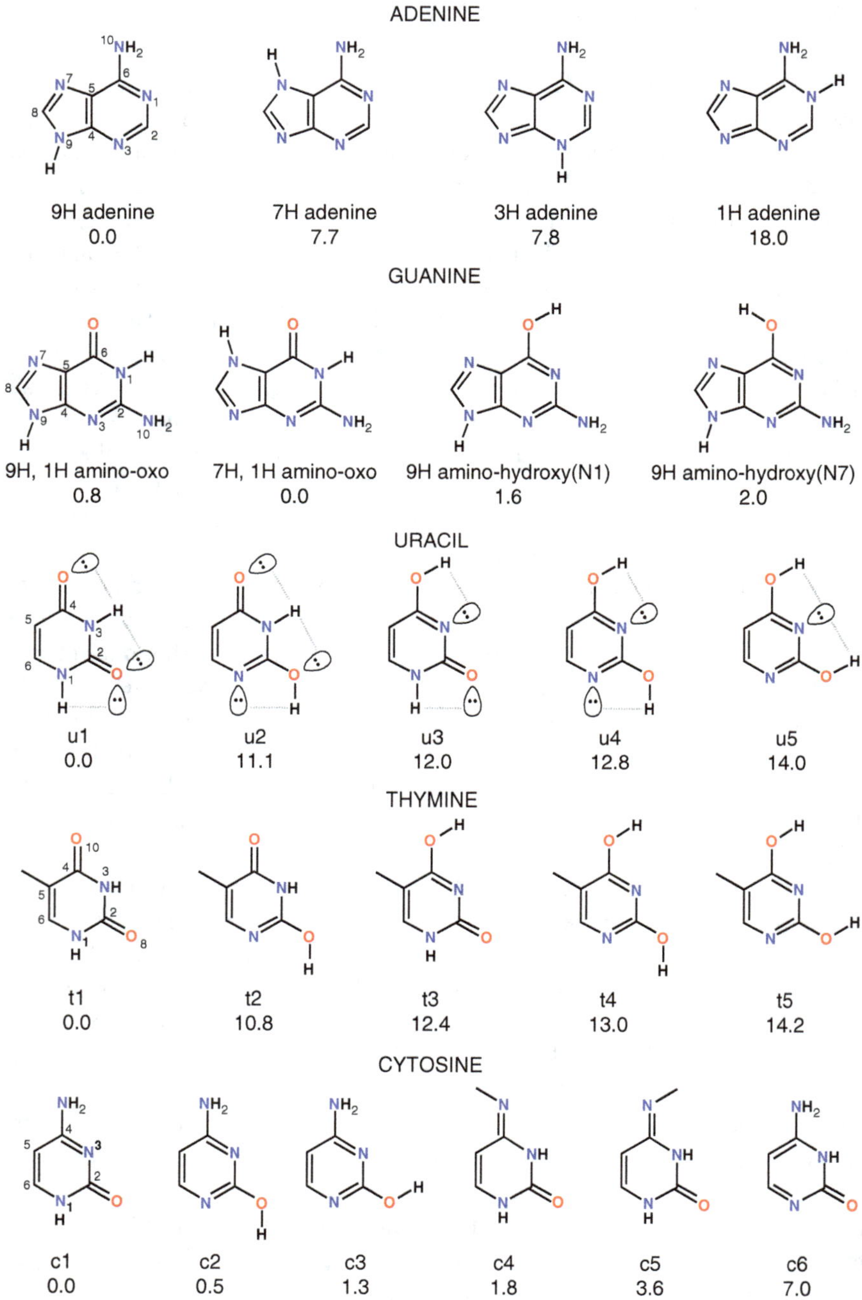


**Figure 9.10** Most stable tautomers of adenine, guanine, uracil, thymine, and cytosine. Relative energies given in kcal/mol. Source: Adapted from Ref. [124].

**Table 9.5** HOMA and NICS (units are ppm) indices for selected tautomers of adenine,[a] guanine,[b] uracil,[c] thymine,[d] and cytosine.[e]

| | $HOMA_5$ | $HOMA_6$ | $HOMA_{tot}$ [f] | $NICS(0)_5$ | $NICS(0)_6$ | $NICS(1)_{zz,5}$ | $NICS(1)_{zz,6}$ |
|---|---|---|---|---|---|---|---|
| **Adenine** | | | | | | | |
| 9H | 0.869 | 0.984 | 0.926 | −11.8 | −7.2 | −25.8 | −22.2 |
| 7H | 0.857 | 0.982 | 0.923 | −11.6 | −8.7 | −26.1 | −25.6 |
| 3H | 0.923 | 0.920 | 0.928 | −11.6 | −6.7 | −31.5 | −17.8 |
| 1H | 0.857 | 0.797 | 0.862 | −10.8 | −6.8 | −31.5 | −18.7 |
| **Guanine** | | | | | | | |
| 9H,1H amino-oxo | 0.878 | 0.704 | 0.761 | −12.4 | −3.1 | −25.7 | −6.3 |
| 7H,1H amino-oxo | 0.922 | 0.784 | 0.831 | −12.3 | −3.8 | −26.6 | −7.7 |
| 9H amino-hydroxy(N1) | 0.846 | 0.981 | 0.916 | −11.1 | −6.9 | −24.0 | −17.8 |
| 9H amino-hydroxy(N7) | 0.856 | 0.981 | 0.918 | −11.3 | −6.8 | −24.0 | −17.8 |
| **Uracil** | | | | | | | |
| u1 | — | 0.505 | — | — | — | — | — |
| u2 | — | 0.637 | — | — | — | — | — |
| u3 | — | 0.714 | — | — | — | — | — |
| u4 | — | 0.990 | — | — | — | — | — |
| u5 | — | 0.994 | — | — | — | — | — |
| **Thymine** | | | | | | | |
| t1 | — | 0.490 | — | — | −1.5 | — | −2.1 |
| t2 | — | 0.629 | — | — | −2.8 | — | −8.2 |
| t3 | — | 0.718 | — | — | −2.7 | — | −7.5 |
| t4 | — | 0.984 | — | — | −7.2 | — | −20.9 |
| t5 | — | 0.987 | — | — | −7.3 | — | −21.1 |
| **Cytosine** | | | | | | | |
| c1 | — | 0.699 | — | — | −1.3 | — | −5.7 |
| c2 | — | 0.976 | — | — | −5.3 | — | −18.1 |
| c3 | — | 0.981 | — | — | −5.4 | — | −18.3 |
| c4 | — | 0.534 | — | — | −0.6 | — | 0.5 |
| c5 | — | 0.512 | — | — | −0.6 | — | 0.4 |
| c6 | — | 0.800 | — | — | −1.8 | — | −6.7 |

a) Data taken from Ref. [128].
b) Data taken from Ref. [129].
c) Data taken from Ref. [130].
d) Data taken from Ref. [131].
e) Data taken from Ref. [132].
f) All bonds are taken into account.

**Table 9.6** NICS, PDI, and HOMA aromaticity descriptor of the five- and six-membered rings of guanine (G), and the six-membered ring of cytosine (C). NICS in ppm and PDI in electrons.[a)]

| | NICS | | | PDI | | HOMA | | |
|---|---|---|---|---|---|---|---|---|
| **System** | **G-5** | **G-6** | **C-6** | **G-6** | **C-6** | **G-5** | **G-6** | **C-6** |
| GC | −11.94 | −4.10 | −1.86 | 0.036 | 0.040 | 0.848 | 0.795 | 0.703 |
| $Ca^{2+}$-GC | −10.67 | −4.76 | −2.53 | 0.044 | 0.045 | 0.843 | 0.886 | 0.797 |
| $Cu^{+}$-GC | −10.64 | −4.59 | −2.25 | 0.040 | 0.043 | 0.869 | 0.898 | 0.761 |
| $Cu^{2+}$-GC | −7.37 | −2.00 | −3.07 | 0.022 | 0.040 | 0.915 | 0.760 | 0.822 |

a) Data taken from Ref. [135].

Finally, let us mention that aromaticity indices are useful for quantifying changes in aromaticity due to impact of some intramolecular (e.g., effect of functional group) or intermolecular (e.g., H-bonding, π-stacking, etc.) interactions [131, 133]. Interactions with metal cations and hydrogen bonding between nucleobases are very relevant because one of the possible mechanisms of DNA mutation [134] is associated with stabilization of certain tautomers of rare nucleobases by metal ions and mismatch in the base pairing.

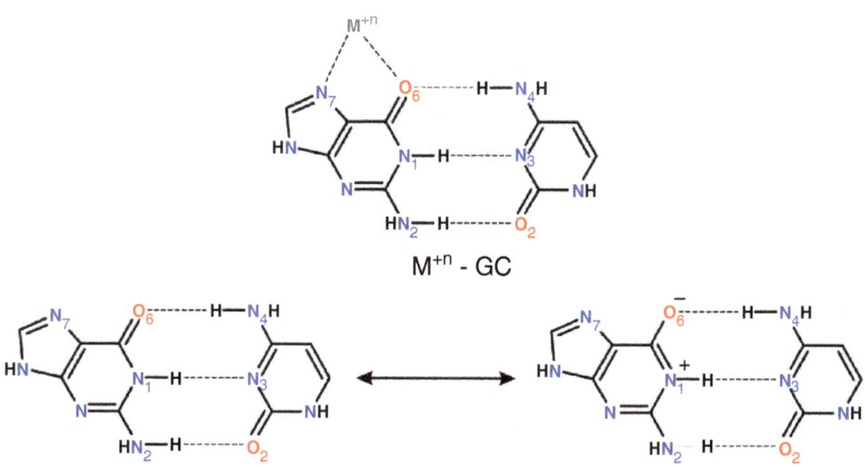


1

2

**Scheme 9.6** Schematic representation of the interaction of the guanine-cytosine (GC) complex with the metal cation (top) and the two main resonant structures of the guanine-cytosine base pair (bottom).

As an example of the effects of metal complexation, let us consider the changes in aromaticity due to interaction of GC pair with $Cu^{+}$, $Ca^{2+}$, and $Cu^{2+}$ cations [135]. The analysis was carried out with the HOMA, NICS, and PDI descriptors of aromaticity (Table 9.6). All indices point out an increase in the aromatic character of the 6-MRs when $Cu^{+}$ and $Ca^{2+}$ interact with the GC pair through $N_7$ and $O_6$ atoms of guanine (see Scheme 9.6 for numbering). This increase in aromaticity is in line with the fact that metal cations increase the weight of the resonance structure **2** in Scheme 9.6. As a result, the 6-MR of the guanine in the GC pair is more aromatic when GC interacts with a metal cation. In contrast, interaction of GC with $Cu^{2+}$ causes the removal of a π-electron from the guanine, thus reducing the aromaticity of its 6-MR [136].

## References


1. De Proft, F. and Geerlings, P. (ed.) (2014). *Structure, Bonding and Reactivity of Heterocyclic Compounds*. Berlin: Springer-Verlag.
2. Balaban, A.T., Oniciu, D.C., and Katritzky, A.R. (2004). *Chem. Rev.* 104: 2777–2812.
3. Katritzky, A.R. (2004). *Chem. Rev.* 104: 2125–2126.
4. Martín, N., Haley, M.M., and Tykwinski, R.R. (2012). *Chem. Commun.* 48: 10471–10471.
5. Martín, N. and Tobe, Y. (2017). *ChemPlusChem* 82: 943–944.
6. Fukazawa, A., Toyota, S., Tykwinski, R., and Wong, W. (2019). *ChemPlusChem* 84: 562–563.
7. Krygowski, T.M. and Cyrański, M.C. (2009). *Aromaticity in Heterocyclic Compounds*, vol. 19 (ed. R.R. Gupta), vii–xii. Berlin: Springer-Verlag.
8. King, B.T. (2012). *Pure Appl. Chem.* 84: iv.
9. Anderson, T. (1868). *Trans. R. Soc. Edinburgh* 25: 205–216.
10. Anderson, T. (1849). *Justus Liebigs Ann. Chem.* 70: 32–38.
11. Runge, F.F. (1834). *Ann. Phys. Chem.* 31: 65–78.
12. Stock, A. and Pohland, E. (1926). *Ber. Dtsch. Chem. Ges.* 59: 2215–2223.
13. Thorn, D.L. and Hoffmann, R. (1979). *Nouv. J. Chim.* 3: 39–45.
14. Cao, X.-Y., Zhao, Q., Lin, Z., and Xia, H. (2014). *Acc. Chem. Res.* 47: 341–354.
15. Elliott, G.P., Roper, W.R., and Waters, J.M. (1982). *J. Chem. Soc., Chem. Commun.* 811–813.
16. Frogley, B.J. and Wright, L.J. (2018). *Chem. Eur. J.* 24: 2025–2038.
17. Weller, K.J., Filippov, I., Briggs, P.M., and Wigley, D.E. (1998). *Organometallics* 17: 322–329.
18. Liu, B., Wang, H., Xie, H. et al. (2009). *Angew. Chem. Int. Ed.* 48: 5430–5434.

**19** Wang, T., Zhang, H., Han, F. et al. (2012). *Angew. Chem. Int. Ed.* 51: 9838–9841.

**20** Tsui, W.-K., Chung, L.-H., Tsang, W.-H. et al. (2015). *Organometallics* 34: 1005–1012.

**21** Chen, J., Lee, K.-H., Wen, T. et al. (2015). *Organometallics* 34: 890–896.

**22** Paneque, M., Posadas, C.M., Poveda, M.L. et al. (2003). *J. Am. Chem. Soc.* 125: 9898–9899.

**23** Zhu, C., Luo, M., Zhu, Q. et al. (2014). *Nature Commun.* 5: 3265.

**24** Frogley, B.J. and Wright, L.J. (2017). *Angew. Chem. Int. Ed.* 56: 143–147.

**25** Wang, T., Han, F., Huang, H. et al. (2015). *Sci. Rep.* 5: 9584.

**26** Zhu, Q., Lin, L., Qiu, R., and Zhu, J. (2020). *Chem. Eur. J.* 26: 5381–5387.

**27** Li, X., Kuznetsov, A.E., Zhang, H.-F. et al. (2001). *Science* 291: 859–861.

**28** Boldyrev, A.I. and Wang, L.-S. (2005). *Chem. Rev.* 105: 3716–3757.

**29** Li, X., Zhang, H.-F., Wang, L.-S. et al. (2001). *Angew. Chem. Int. Ed.* 40: 1867–1870.

**30** Huang, X., Zhai, H.-J., Kiran, B., and Wang, L.-S. (2005). *Angew. Chem. Int. Ed.* 44: 7251–7254.

**31** Zhai, H.-J., Averkiev, B.B., Zubarev, D.Y. et al. (2007). *Angew. Chem. Int. Ed.* 46: 4277–4280.

**32** Cheung, L.F., Kocheril, G.S., Czekner, J., and Wang, L.-S. (2020). *J. Am. Chem. Soc.* 142: 3356–3360.

**33** Minkin, V.I., Glukhovtsev, M.N., and Simkin, B.Y. (1994). *Aromaticity and Antiaromaticity: Electronic and Structural Aspects*. New York: Wiley.

**34** Gompper, R. and Wagner, H.-U. (1988). *Angew. Chem. Int. Ed. Engl.* 27: 1437–1455.

**35** Blanco, F., Alkorta, I., and Elguero, J. (2009). *Tetrahedron* 65: 6244–6250.

**36** Stępień, B.T., Krygowski, T.M., Cyrański, M.K. et al. (2004). *ARKIVOC* 185–201.

**37** Raczyńska, E.D., Kosińska, W., Ośmiałowski, B., and Gawinecki, R. (2005). *Chem. Rev.* 105: 3561–3612.

**38** Politzer, P. and Murray, J.S. (2017). *Struct. Chem.* 28: 1045–1063.

**39** Nein, Y.I. and Morzherin, Y.Y. (2012). *Russ. Chem. Bull.* 61: 1111–1116.

**40** Simas, A.M., Miller, J., and Filho, P.F.D.A. (1998). *Can. J. Chem.* 76: 869–872.

**41** Oziminski, W.P. and Ramsden, C.A. (2015). *Tetrahedron* 71: 7191–7198.

**42** Curutchet, C., Poater, J., Solà, M., and Elguero, J. (2011). *J. Phys. Chem. A* 115: 8571–8577.

**43** Poater, J., Solà, M., Alkorta, I., and Elguero, J. (2014). *Eur. J. Org. Chem.* 5370–5377.

**44** Sobczyk, L., Grabowski, S.J., and Krygowski, T.M. (2005). *Chem. Rev.* 105: 3513–3560.

**45** Szatyłowicz, H., Marek, P.H., Stasyuk, O.A. et al. (2020). *RSC Adv.* 10: 23350–23358.

**46** Wheeler, S.E. (2013). *Acc. Chem. Res.* 46: 1029–1038.

**47** Sung, Y.M., Oh, J., Cha, W.-Y. et al. (2017). *Chem. Rev.* 117: 2257–2312.

**48** Krygowski, T.M. and Szatyłowicz, H. (2015). *ChemTexts* 1: 12.

**49** Krygowski, T.M. and Stępień, B.T. (2005). *Chem. Rev.* 105: 3482–3512.

**50** Omelchenko, I.V., Shishkin, O.V., Gorb, L. et al. (2013). *Struct. Chem.* 24: 725–733.

**51** Cysewski, P. (2011). *J. Mol. Model.* 17: 2249–2257.

**52** Gellrich, U., Diskin-Posner, Y., Shimon, L.J.W., and Milstein, D. (2016). *J. Am. Chem. Soc.* 138: 13307–13313.

**53** Podsiadło, M., Jakóbek, K., and Katrusiak, A. (2010). *CrystEngComm* 12: 2561–2567.

**54** Cysewski, P., Jelinski, T., Krygowski, T.M., and Oziminski, W.P. (2012). *Curr. Org. Chem.* 16: 1920–1933.

**55** Gimarc, B.M. (1983). *J. Am. Chem. Soc.* 105: 1979–1984.

**56** Alkorta, I., Blanco, F., and Elguero, J. (2008). *J. Mol. Struct. (Theochem)* 851: 75–83.

**57** Subramanian, G., Schleyer, P.v.R., and Jiao, H. (1996). *Angew. Chem. Int. Ed. Engl.* 35: 2638–2641.

**58** Novak, I. (1997). *J. Mol. Struct. (Theochem)* 398–399: 315–323.

**59** Aihara, J.-I. (2005). *J. Phys. Org. Chem.* 18: 235–239.

**60** Elguero, E., Alkorta, I., and Elguero, J. (2018). *Heteroat. Chem.* 29: e21441.

**61** Mohajeri, A. and Shahamirian, M. (2011). *Comput. Theor. Chem.* 976: 19–29.

**62** Rajappa, S. and Gumaste, V.K. (2013). *Advances in Heterocyclic Chemistry*, vol. 108 (ed. A.R. Katritzky), 1–161. Academic Press.

**63** Cyrański, M.K., Krygowski, T.M., Krutošíková, A., and Sleziak, R. (2001). *Tetrahedron* 57: 8867–8873.

**64** Nyulászi, L. (2001). *Chem. Rev.* 101: 1229–1246.

**65** Bühl, M. and Hirsch, A. (2001). *Chem. Rev.* 101: 1153–1183.

**66** Bodwell, G.J., Bridson, J.N., Cyrański, M.K. et al. (2003). *J. Org. Chem.* 68: 2089–2098.

**67** Dobrowolski, M.A., Cyrański, M.K., Merner, B.L. et al. (2008). *J. Org. Chem.* 73: 8001–8009.

**68** Kawasumi, K., Zhang, Q., Segawa, Y. et al. (2013). *Nat. Chem.* 5: 739–744.

**69** Pascal, R.A. (2006). *Chem. Rev.* 106: 4809–4819.

**70** Dobrowolski, M.A., Cyrański, M.K., and Wróbel, Z. (2016). *Phys. Chem. Chem. Phys.* 18: 11813–11820.

**71** Stępień, M., Latos-Grażyński, L., Sprutta, N. et al. (2007). *Angew. Chem. Int. Ed.* 46: 7869–7873.

**72** Krygowski, T.M., Cyrański, M.K., Czarnocki, Z. et al. (2000). *Tetrahedron* 56: 1783–1796.

**73** Albert, A. (1968). *Heterocyclic Chemistry: An Introduction*, 2e. New York: Oxford University Press.

**74** Katritzky, A.R., Jug, K., and Oniciu, D.C. (2001). *Chem. Rev.* 101: 1421–1449.

**75** Bergman, E.D. (1971). *Aromaticity, Pseudo-aromaticity, Anti-aromaticity*. In: *Proceedings of an International Symposium held in Jerusalem 31 March – 3 April 1970* (ed. E.D. Bergmann and B. Pullman), 9–24. Jerusalem: Israel Academy of Science and Humanities.

**76** Poater, J., García-Cruz, I., Illas, F., and Solà, M. (2004). *Phys. Chem. Chem. Phys.* 6: 314–318.

**77** Bultinck, P. (2007). *Faraday Discuss.* 135: 347–365.

**78** Feixas, F., Matito, E., Poater, J., and Solà, M. (2008). *J. Comput. Chem.* 29: 1543–1554.

**79** Fias, S., Van Damme, S., and Bultinck, P. (2010). *J. Comput. Chem.* 31: 2286–2293.

**80** Cyrański, M.K., Krygowski, T.M., Katritzky, A.R., and Schleyer, P.v.R. (2002). *J. Org. Chem.* 67: 1333–1338.

**81** Poater, J., Duran, M., Solà, M., and Silvi, B. (2005). *Chem. Rev.* 105: 3911–3947.

**82** Wang, Y., Wu, J.I.C., Li, Q., and Schleyer, P.v.R. (2010). *Org. Lett.* 12: 4824–4827.

**83** Fabian, J. and Lewars, E. (2004). *Can. J. Chem.* 82: 50–69.

**84** Hou, B., Yi, P., Wang, Z. et al. (2014). *Comput. Theor. Chem.* 1046: 20–24.

**85** Cyrański, M.K. (2005). *Chem. Rev.* 105: 3773–3811.

**86** Cioslowski, J., Matito, E., and Solà, M. (2007). *J. Phys. Chem. A* 111: 6521–6525.

**87** Omelchenko, I.V., Shishkin, O.V., Gorb, L. et al. (2011). *Phys. Chem. Chem. Phys.* 13: 20536–20548.

**88** Bader, R.F.W. (1985). *Acc. Chem. Res.* 18: 9–15.

**89** Bader, R.F.W. (1990). *Atoms in Molecules: A Quantum Theory*. Oxford: Clarendon.

**90** Bader, R.F.W. (1991). *Chem. Rev.* 91: 893–928.

**91** Ebrahimi, A.A., Ghiasi, R., and Foroutan-Nejad, C. (2010). *J. Mol. Struct. (Theochem)* 941: 47–52.

**92** Chen, Z., Wannere, C.S., Corminboeuf, C. et al. (2005). *Chem. Rev.* 105: 3842–3888.

**93** Schleyer, P.v.R., Jiao, H.J., van Eikema Hommes, N.J.R. et al. (1997). *J. Am. Chem. Soc.* 119: 12669–12670.

**94** Fowler, P.W. and Steiner, E. (1997). *J. Phys. Chem. A* 101: 1409–1413.

**95** Ota, K. and Kinjo, R. (2020). *Chem. Asian J.* 15: 2558–2574.

96 Ota, K. and Kinjo, R. (2020). *Angew. Chem. Int. Ed.* 59: 6572–6575.

97 Islas, R., Chamorro, E., Robles, J. et al. (2007). *Struct. Chem.* 18: 833–839.

98 Baranac-Stojanović, M. and Stojanović, M. (2013). *RSC Adv.* 3: 24108–24117.

99 Kiran, B., Phukan, A.K., and Jemmis, E.D. (2001). *Inorg. Chem.* 40: 3615–3618.

100 Mohajeri, A. and Shahamirian, M. (2010). *J. Phys. Org. Chem.* 23: 440–450.

101 Mazurek, A. and Dobrowolski, J.C. (2012). *J. Org. Chem.* 77: 2608–2618.

102 Ozimiński, W.P. and Dobrowolski, J.C. (2009). *J. Phys. Org. Chem.* 22: 769–778.

103 Bleeke, J.R. (2001). *Chem. Rev.* 101: 1205–1227.

104 Bleeke, J.R. (2007). *Acc. Chem. Res.* 40: 1035–1047.

105 Fernández, I. and Frenking, G. (2007). *Chem. Eur. J.* 13: 5873–5884.

106 Fernandez, I., Frenking, G., and Merino, G. (2015). *Chem. Soc. Rev.* 44: 6452–6463.

107 Huang, Y.-Z., Yang, S.-Y., and Li, X.-Y. (2004). *J. Organomet. Chem.* 689: 1050–1056.

108 Iron, M.A., Lucassen, A.C.B., Cohen, H. et al. (2004). *J. Am. Chem. Soc.* 126: 11699–11710.

109 Havenith, R.W.A., De Proft, F., Jenneskens, L.W., and Fowler, P.W. (2012). *Phys. Chem. Chem. Phys.* 14: 9897–9905.

110 Periyasamy, G., Burton, N.A., Hillier, I.H., and Thomas, J.M.H. (2008). *J. Phys. Chem. A* 112: 5960–5972.

111 Psciuk, B.T., Lord, R.L., Winter, C.H., and Schlegel, H.B. (2012). *J. Chem. Theory Comput.* 8: 4950–4959.

112 El-Hamdi, M., El Bakouri El Farri, O., Salvador, P. et al. (2013). *Organometallics* 32: 4892–4903.

113 Zhu, J., Jia, G., and Lin, Z. (2007). *Organometallics* 26: 1986–1995.

114 Cyrański, M. and Krygowski, T.M. (1996). *Tetrahedron* 52: 13795–13802.

115 Mazurek, A. and Dobrowolski, J.C. (2015). *J. Phys. Org. Chem.* 28: 290–297.

116 Ghosh, D., Periyasamy, G., and Pati, S.K. (2011). *Phys. Chem. Chem. Phys.* 13: 20627–20636.

117 Zborowski, K.K. and Krygowski, T.M. (2014). *Tetrahedron Lett.* 55: 6359–6361.

118 Ozimiński, W.P., Dobrowolski, J.C., and Mazurek, A.P. (2003). *J. Mol. Struct.* 651-653: 697–704.

119 Katritzky, A.R., Barczynski, P., Musumarra, G. et al. (1989). *J. Am. Chem. Soc.* 111: 7–15.

120 Jug, K. and Köster, A.M. (1991). *J. Phys. Org. Chem.* 4: 163–169.

121 Krygowski, T.M., Ciesielski, A., Bird, C.W., and Kotschy, A. (1995). *J. Chem. Inf. Comput. Sci.* 35: 203–210.

122 Katritzky, A.R., Karelson, M., Sild, S. et al. (1998). *J. Org. Chem.* 63: 5228–5231.

123 Cyrański, M.K., Schleyer, P.v.R., Krygowski, T.M. et al. (2003). *Tetrahedron* 59: 1657–1665.

124 Szatyłowicz, H., Stasyuk, O.A., Solà, M., and Krygowski, T.M. (2021). *WIREs Comput. Mol. Sci.* 11: e1509.

125 Cyrański, M.K., Gilski, M., Jaskólski, M., and Krygowski, T.M. (2003). *J. Org. Chem.* 68: 8607–8613.

126 Cysewski, P. (2005). *J. Mol. Struct. (Theochem)* 714: 29–34.

127 Huertas, O., Poater, J., Fuentes-Cabrera, M. et al. (2006). *J. Phys. Chem. A* 110: 12249–12258.

128 Stasyuk, O.A., Szatyłowicz, H., and Krygowski, T.M. (2014). *Org. Biomol. Chem.* 12: 456–466.

129 Raczyńska, E.D., Makowski, M., Hallmann, M., and Kamińska, B. (2015). *RSC Adv.* 5: 36587–36604.

130 Raczyńska, E.D. (2017). *Comput. Theor. Chem.* 1121: 58–67.

131 Stasyuk, O.A., Szatyłowicz, H., and Krygowski, T.M. (2014). *Org. Biomol. Chem.* 12: 6476–6483.

132 Stasyuk, O.A., Szatyłowicz, H., and Krygowski, T.M. (2014). *Croat. Chem. Acta* 87: 335–342.

133 Stasyuk, O.A., Szatyłowicz, H., and Krygowski, T.M. (2016). *Struct. Chem.* 27: 111–118.

134 Pratviel, G. (2012). *Interplay between Metal Ions and Nucleic Acids in Metal Ions in Life Science* (ed. A. Sigel, H. Sigel and R.K.O. Sigel), 201–216. The Netherlands: Springer.

135 Poater, J., Sodupe, M., Bertran, J., and Solà, M. (2005). *Mol. Phys.* 103: 163–173.

136 Noguera, M., Bertran, J., and Sodupe, M. (2004). *J. Phys. Chem. A* 108: 333–341.

# 10

# Möbius Aromaticity

*"Chemistry: The craft and art of making stuff better than that given by nature"*
*W. H. Eugen Schwarz (the 2nd European Symposiumon Chemical Bonding 2018 held in Oviedo)*

## 10.1 Introduction

Möbius aromaticity is found in cyclic conjugated species having a molecular topology that resembles that of a Möbius strip. A Möbius strip is defined as a surface with only one side obtained by cutting a closed band into a single strip, twisting one of the two ends through 180°, and then reattaching the two ends. It was discovered independently by Johann Benedict Listing and August Ferdinand Möbius in 1858 [1]. As shown in Chapter 2, closed-shell electronic structures are reached for Möbius monocyclic conjugated systems having $4n$ π-electrons, whereas Möbius-type annulenes with $4n + 2$ π-electrons are antiaromatic. The linking number is used to describe topologically the number of times that a given strip winds (twist, $T_w$, and writhe, $W_r$, $L_k = T_w + W_r$, see Figure 2.4). Annulenes with an even $L_k$ value obey Hückel's $4n + 2$ π-electron rule of aromaticity for the closed-shell ground state, whereas those having an odd $L_k$ number follow the Möbius' $4n$ π-electron rule (see Figure 10.1). Möbius aromaticity was first described theoretically by Craig and Paddock [3] in 1958, and six years later was more elaborated by Heilbronner [4].

In pericyclic reactions, Zimmerman considered that a series of σ and π bonds involving $4n + 2$ electrons prefer to react in a pericyclic manner to form a Hückel aromatic ring and that a series of σ and π bonds with $4n$ electrons favor formation of a Möbius aromatic ring [5]. In 1993, Jiao and Schleyer [6] reported the first computational study of a Möbius transition state. They studied the [1, 7] sigmatropic hydrogen shift in (*Z*,*Z*)-1,3,5-heptatriene and they found that this pericyclic

*Aromaticity and Antiaromaticity: Concepts and Applications*, First Edition.
Miquel Solà, Alexander I. Boldyrev, Michał K. Cyrański, Tadeusz M. Krygowski, and Gabriel Merino.

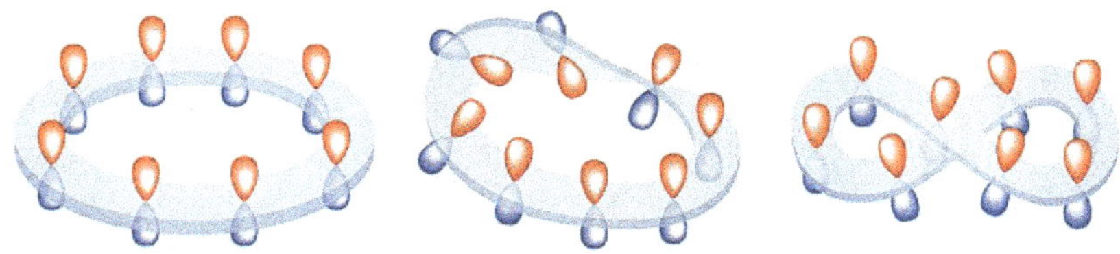


**Figure 10.1** Representation of different π-conjugation topologies with $L_k = 0, 1$, and 2 and their expected aromaticities as a function of the number of π-electrons. Source: Reprinted with permission from Ref. [2]. Copyright © 2020, American Chemical Society.

reaction proceeds via a Möbius transition state with the hydrogen moving to the antarafacial side. The Möbius character of this transition state was supported by the C—C bond lengths and the downfield $^1$H NMR chemical shifts, both similar to those of benzene, together with the stabilization energy due to resonance. Electrocyclic ring opening/ring closure pericyclic reactions occurring through Möbius transition states were also reported [7, 8].

The first computational study reporting a stable species with Möbius aromaticity was carried out in 1998 by Mauksch, Schleyer et al. [9]. They analyzed the cyclic 8π electrons [9]annulene cation, $C_9H_9^+$, which was detected experimentally some years before as a short live intermediate in the solvolysis of 9′-chloro-9-deuterobicyclo[6.1.0]nona-2,4,6-triene in aqueous acetone at 75 °C to give dihydroindenol in which deuterium was stochastically distributed over all nine carbon centers (see Figure 10.2) [10–12]. The geometry optimizations and single point calculations were performed at B3LYP/6-311+G(d,p) and CCSD(T)/DZP levels, respectively. Mauksch, Schleyer, and coworkers [9] found that the Möbius $C_9H_9^+$ isomer with a three-dimensional figure-of-eight and a *trans* bond was more stable than the all-*cis* Hückel one by about 1 kcal/mol. Moreover, this Möbius isomer presents small bond length alternation and negative nucleus-independent chemical shift (NICS) values, both typical of aromatic compounds. Ten years later, Herges et al. [13] performed high-level coupled-cluster calculations including corrections for solvent and vibrational contributions for $C_9H_9^+$ to confirm that the Hückel and the Möbius isomers are very close in energy. UV/Vis spectra measured in nanosecond laser flash photolysis experiments indicated the presence of only the Hückel isomer, which is likely to be in equilibrium with the Möbius one.

A way to enforce a twist in a cyclic conjugated system to form Möbius neutral aromatic species, is to introduce *trans* double bonds at suitable positions of annulenes. In this way, Rzepa et al. proposed the first Möbius [*n*]annulene for n = 16 [14]. However, not only in [16]annulene, but also in any attempt to produce

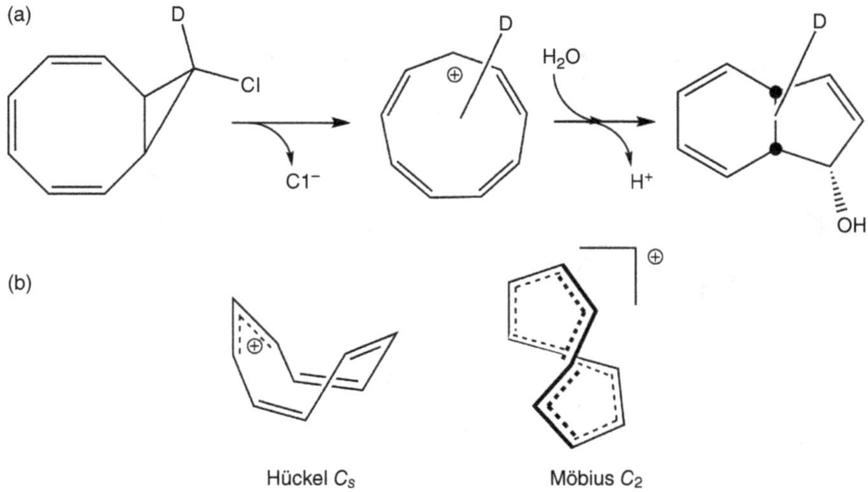


**Figure 10.2** (a) Generation of $C_9H_9^+$ by solvolysis of 9′-chloro-9-deuterobicyclo[6.1.0] nona-2,4,6-triene, and (b) Hückel and Möbius isomers of $C_9H_9^+$.

other Möbius [$n$]annulenes by introducing *trans* bonds, the most stable isomer has always presented Hückel topology (for $n = 16$, the Hückel isomer is most stable than the Möbius by 5.1 kcal/mol [13]). The reason must be found in the fact that stabilization by Möbius aromaticity cannot overcome the destabilization due to the torsional strain and the reduced overlap of neighboring $p$ orbitals in [$n$]annulenes.

Finally, in 2003, Ajami, Herges et al. [15] reported the synthesis of the first neutral and stable Möbius aromatic molecule by combining a Hückel out-of-plane aromatic structure with a "belt-like" in-plane aromatic structure. As shown in Figure 10.3, by cutting a normal π system and assembling it with a half-cylinder with in-plane aromaticity one can obtain a Möbius system less strained than the Hückel one. Experimentally, this situation was achieved by reacting dehydrodianthracene (TDDA) and cyclooctatetraene upon irradiation with a low-pressure mercury lamp. In this case, a ring enlargement metathesis reaction takes place and a Möbius aromatic system containing a bianthraquinodimethane unit (magenta) and a polyene bridge (blue) is formed (Figure 10.4). Five isomers were obtained from the reaction out of the 108 possible, one of which was the Möbius aromatic structure of Figure 10.3. Aromaticity of this structure was experimentally confirmed by NMR and X-ray crystallography and computationally with the magnetic properties of the molecule and the aromaticity stabilization energy. However, two years later, Castro et al. [16] questioned the classification of this compound as a Möbius aromatic species. By performing calculations of geometric ($\Delta r$, $\Delta r_m$, Julg A, HOMA – harmonic oscillator model of aromaticity) descriptors of aromaticity based on the published crystallographic data, and magnetic (NICS,

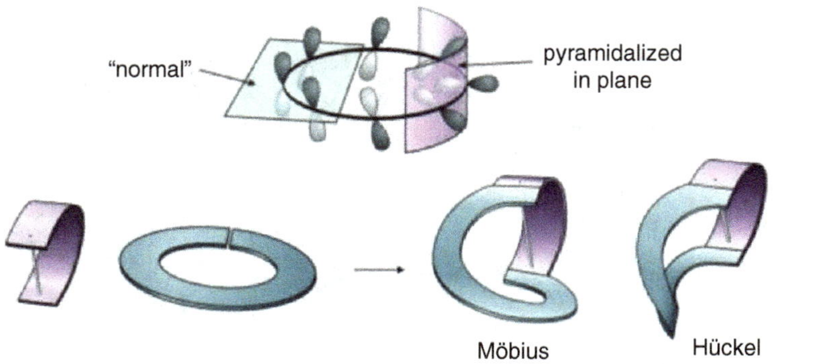


**Figure 10.3** Combination of out-of-plane and in-plane aromatic species as a synthetic strategy group to generate a Möbius aromatic compound. Source: Reprinted with permission from Ref. [13]. Copyright © 2006, Wiley.

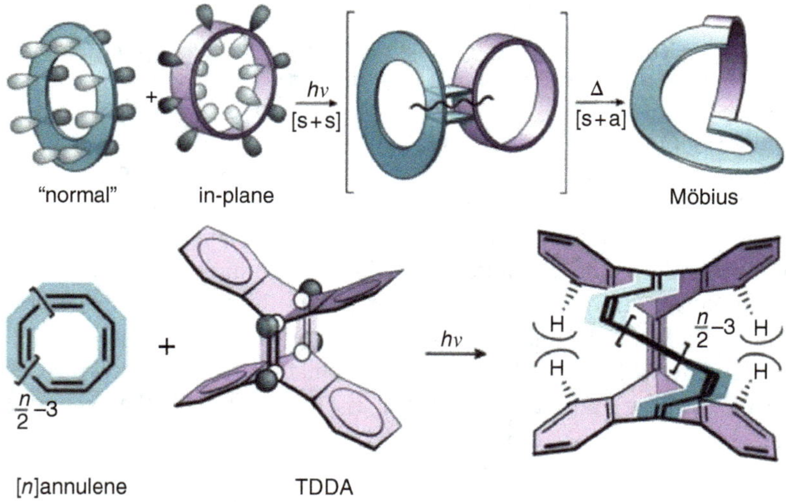


**Figure 10.4** Combination of out-of-plane and in-plane aromatic species as a synthetic strategy group to generate a Möbius aromatic compound. Source: Reprinted with permission from Ref. [13]. Copyright © 2006, Wiley.

magnetic susceptibility exaltation) indicators of aromaticity, as well as aromatic stabilization energies, the authors concluded that delocalization in this molecule is inhibited by large dihedral angles, which hinders effective $\pi$ overlap. Although it is obvious that there is a reduced overlap of neighboring $p$ orbitals in Möbius species, there is some controversy whether the reduction is enough to fully extinguish their aromaticity.

Since the work of Herges et al. [13, 15, 17], a number of Möbius aromatic molecules have been described. Möbius aromaticity has been mainly found in several planar metallacycles with $4n$ $\pi$-electrons and in macrocycles such as porphyrinoids or similar. These two types of Möbius aromatic species are concisely discussed in the next 10.2 and 10.3 subsections. More information can be found in the reviews by Rzepa [18], Herges [17], and Latos-Grażyński and coworkers [19].

## 10.2 Metallacyclic Möbius Aromatic Species

Two kinds of Möbius aromaticity classified as the Craig and Heilbronner types have been defined. The former was proposed by Craig in 1958, initially in cyclic molecules containing a S- or P-atom [3, 20], as "a novel type of aromaticity" by considering $p_\pi$–$d_\pi$–$p_\pi$ interactions that generates the typical phase-shift of Möbius aromatic compounds (see Figure 10.5, right). The latter (Figure 10.1, middle), proposed by Heilbronner [4], accomplishes the necessary phase-shift through twisting the nuclear framework without the need to involve *d* orbitals. Between these two possibilities, Craig-Möbius aromaticity is the most common in metallacyclic compounds.

Metalla-aromatic compounds are obtained when *sp*- or $sp^2$-hybridized carbon atoms in fully conjugated hydrocarbons are replaced by isolobal metal fragments [21]. If the metal is a transition metal, $p_\pi$–$d_\pi$–$p_\pi$ molecular orbitals (MOs) can have two different MO topologies (Figure 10.5), namely, Hückel (without such a phase inversion) and Craig-Möbius (with a single inversion of orbital phase) [3, 20]. In the particular case that a CH unit in benzene is substituted by a transition metal atom, a metallabenzene is obtained [22–24]. Thorn and Hoffmann predicted theoretically the existence of these species in 1979 [25]. Three years later, the first metallabenzene, an osmabenzene, was synthesized by Roper and coworkers [26] from the reaction of $Os(CO)(CS)(PPh_3)_3$ and two ethyne molecules. The structural analysis based on X-ray diffraction established the planarity of the ring.

Some authors [27, 28] suggested that these metallabenzenes are aromatic with 8 $\pi$-electrons. Other authors support this idea, arguing that some related compounds

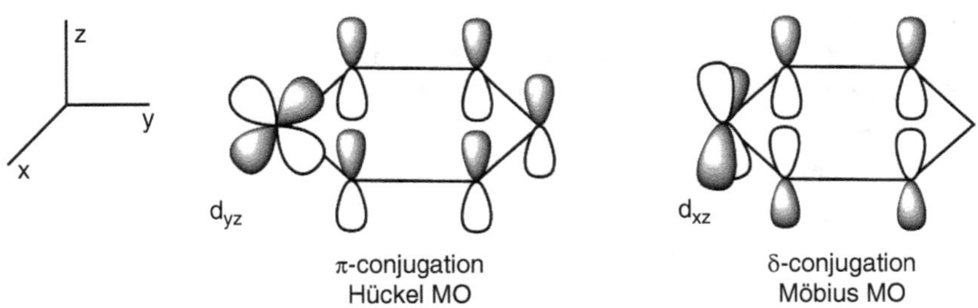


**Figure 10.5** Schematic representation of the two different topologies of $\pi$-molecular orbitals involving *d*-orbitals.

present also Möbius aromaticity with eight π-electrons [29, 30]. Other authors, however, classify these metallacycles as 6 or 10 π-electron systems and, therefore, Hückel aromatic [31–37]. The main problem is that the counting of π-electrons in metallacycles is cumbersome. For some occupied π-MOs, it is debatable whether the two electrons of a given π-MO have to be counted to decide the Möbius or Hückel aromatic character of a certain compound. This situation happens with occupied π-MOs that have a relatively large contribution of the metal and small coefficients of the *p* out-of-plane atomic orbitals of the C atoms of the ring or vice versa. This is why, for instance, in an aromatic dicupra[10]annulenes synthesized by Xi's group [38], Grande-Aztatzi et al. [39] concluded that this dicupra system is both 10 and 14 π-electron Hückel aromatic, whereas for Zhu and coworkers [40], this complex is Craig-type Möbius aromatic with eight occupied π MOs.

Not only, the π-electron counting, but also the fact that Möbius and Hückel orbitals co-exist in the same metalla-aromatic species poses the question of how to decide based on the MOs which type of aromaticity (Hückel or Möbius) is involved in a certain metalla-aromatic species. Mauksch and Tsogoeva proposed a solution to this problem. According to these authors, the metallacycle is aromatic (antiaromatic) when the number of π MOs is even and the π-HOMO is of Möbius (Hückel) topology and vice versa when the number of π-MOs is odd.

However, Szczepanik and Solà [41] concluded that the π-HOMO rule of Mauksch and Tsogoeva does not always hold. These authors found that it is very difficult in metallacycles to unambiguously associate their aromaticity with the $4n + 2$ (Hückel) and $4n$ (Möbius) π-electron rules. They used the electron density of delocalized bonds (EDDB) method [42–44] to quantify and visualize aromaticity in such difficult cases. With this method, it is possible to provide a detailed information on the real role of *d*-orbitals in metallacycles. Figure 10.6 shows the fifteen metalla-aromatic systems studied by Szczepanik and Solà [41] together with their EDDB values (as a reference the EDDB of benzene is 5.3 e). As can be seen, most of the metallacycles studied are aromatic or at least partially aromatic. When decomposed the aromaticity for the different *d* components, it is found that in some cases, for instance in complex 1, the Hückel aromaticity predominates, whereas in other cases like complex 11, the Möbius aromatic character prevails. However, most of the metallacycles studied cannot be classified exclusively as Hückel or Möbius because they have a hybrid Hückel–Möbius character. Therefore, the message was that a rigid classification of the aromatic metallacycles in Hückel or Möbius character is not possible since all these compounds have a certain percentage of Hückel and Möbius electron delocalization.

$ReB_4$ and $ReB_4^{-1}$ were the first discovered planar monocyclic metallaboron systems with Möbius aromaticity [45]. $ReB_4^{0/-1}$ were found to be planar pentagonal rings. Chemical bonding analyses showed that both $ReB_4^{0/-1}$ possess four delocalized π-electrons that suggest an antiaromatic system according to Hückel's

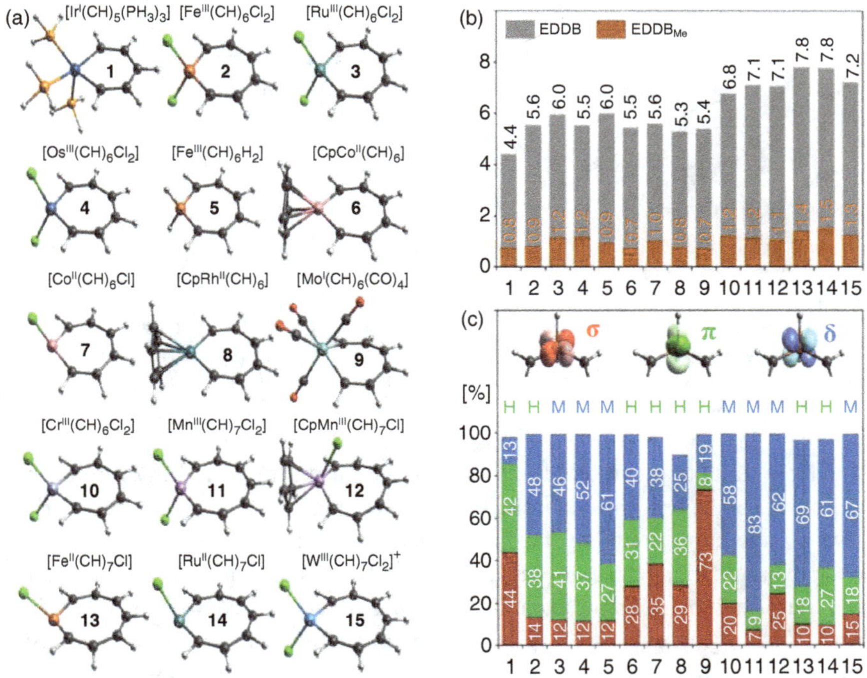


**Figure 10.6** (a) Optimized molecular structures of d-metallacycles 1-15. (b,c) Graphical representation of the EDDB results; labels EDDB and $EDDB_{Me}$ refer to the total EDDB and the EDDB due to the metal, respectively, and labels H (Hückel) and M (Möbius) refer to the topology of π-HOMO. Source: Reprinted with permission from Ref. [41]. Copyright © 2018, Wiley.

rule. However, its planar molecular structure with almost equal B—B bonds and negative NICS values indicate its aromatic character. Since two π-electrons are located in an orbital of Möbius topology, the authors concluded that $ReB_4^{0/-1}$ possesses Möbius aromaticity.

## 10.3 Macrocyclic Möbius Aromaticity

Over the past two decades, significant advances have been achieved in the synthesis of expanded porphyrins, which are higher homologues of porphyrins. The name of these molecules comes from the larger number of pyrrole rings they have compared to porphyrins, which have four pyrrole rings connected by *meso*-methine bridges (species “tetra” in Figure 10.7) [46, 47]. These compounds have a large and flexible annulene-like π-conjugated network with a number

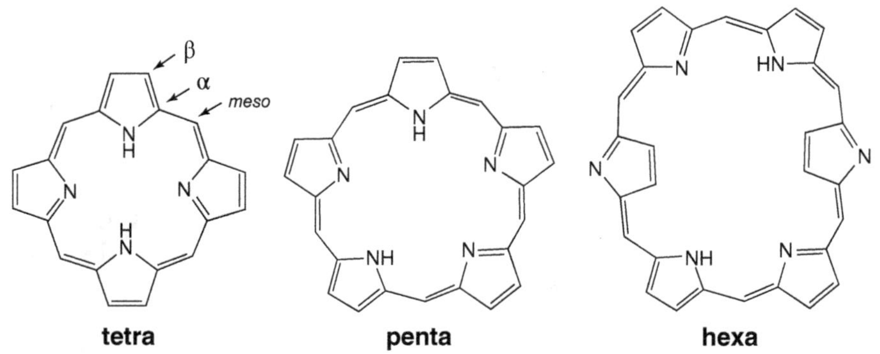


**Figure 10.7** Molecular structure of tetraporphyrin (porphyrin, N = 4), pentaphyrin (N = 5), and hexaphyrin (N = 6).

of pyrrole units that can be (de)protonated, can form intra- and intermolecular hydrogen-bonds, or can coordinate different metals. Expanded porphyrins have a large conformational flexibility that makes them attractive compounds to study aromaticity changes in different molecular topologies. Interestingly, the conformation of some expanded porphyrins can be controlled by adding peripheral substituents at the β and/or *meso* positions (Figure 10.7), by metal complex formation and anion binding, or by changing different physicochemical parameters such as redox and protonation states, solvent, temperature, and so on. Several reviews have discussed the recent advances in this field [48–55].

The first dynamic Hückel–Möbius system was described by Latos-Grażyński and coworkers in 2007 [56]. They synthesized the compound A,D-di-benzi[28] hexaphyrin(1.1.1.1.1.1) (hereafter di-*p*-benzihexaphyrin, 1 in Figure 10.8) in an acid-catalyzed condensation of dicarbinol (2 in Figure 10.8) with pyrrol and benzaldehyde followed by oxidation with 2,3-dichloro-5,6-dicyano-1,4-benzoquinone (DDQ). According to the authors, di-*p*-benzihexaphyrin has a 28-π-electron conjugation pathway as shown in Figure 10.8.

The X-ray structure of di-*p*-benzihexaphyrin has $C_2$ symmetry with a figure-eight conformation (Figure 10.8, right) corresponding to a Möbius band with a single half twist. The two phenylene rings are not equivalent, one is parallel and the other perpendicular to the $C_2$ symmetry axis. However, $^1$H NMR spectrum in solution shows that this conformation is not rigid in solution and the nonequivalence of the phenylene rings is lost because they can rotate freely. In solution, di-*p*-benzihexaphyrin exists in two conformations that are in equilibrium, the Möbius one and the Hückel one shown in Figure 10.8. The most stable is the Möbius conformation, the enthalpy difference between both conformers being 4.6 kcal/mol. Interestingly, the Möbius conformation exhibits chemical shifts characteristic of macrocyclic aromatic molecules, as expected from the fact that it

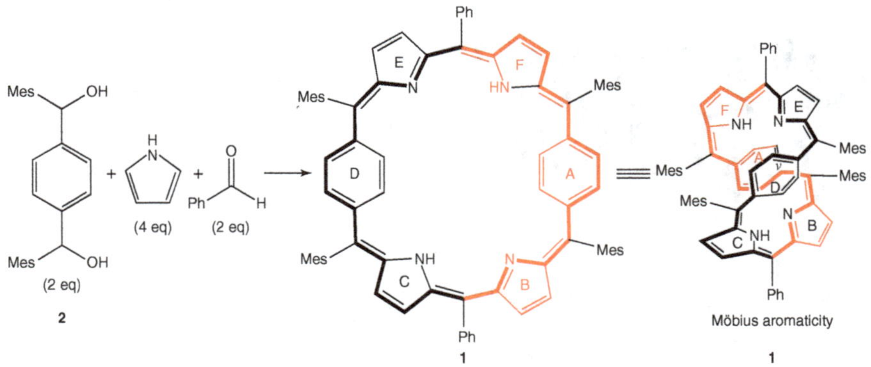


**Figure 10.8** Synthesis and molecular structure of di-*p*-benzihexaphyrin. Bold lines indicate the 28-π-electron conjugation pathway.

has a conjugated circuit with 28 π-electrons and Möbius topology. On the other hand, the Hückel conformer displays the typical paratropicity found in monocyclic $4n$-π-electron species. This di-*p*-benzihexaphyrin molecule represented the first example of dynamic switching between Hückel and Möbius topologies in a conjugated molecule.

In 2008, Osuka and coworkers [57] found an example of a $4n$ π expanded porphyrin that after metalation twist their molecular structure to form a Möbius aromatic compound. This behavior was observed in bis-Pd$^{II}$ [36]octaphyrin (Figure 10.9). The singly twisted Möbius structure was confirmed by X-ray diffraction analysis and the diatropic character of the ring current was clear from the $^1$H NMR spectrum (Figure 10.9). The inner β protons of inverted pyrroles were deshielded, while the outer β protons were significantly shielded. Moreover, the small bond length alternation and the large and negative NICS values inside the macrocycle provide enough evidence on the aromatic nature of this species.

$(4n+2)\pi$ Möbius antiaromatic species are much less common. The first structurally characterized Möbius antiaromatic expanded porphyrin was a [30]hexaphyrin bisphosphorus complex reported by Osuka and coworkers in 2010 [58]. The X-ray crystal structure of this rigid, neutral, and stable compound revealed a 30 π macrocyclic conjugation with a twisted Möbius topology. The inner pyrrolic β protons are deshielded, whereas the outer β protons are shielded, indicating the presence of a paratropic ring current as expected in antiaromatic molecules.

There are other examples of conformational switches involving Möbius aromatic expanded porphyrins. For instance, protonation of [32]heptaphyrins triggers a radical conformational change leading to a Möbius aromatic conformation [59, 60]. Protonation-induced conformational change is also effective

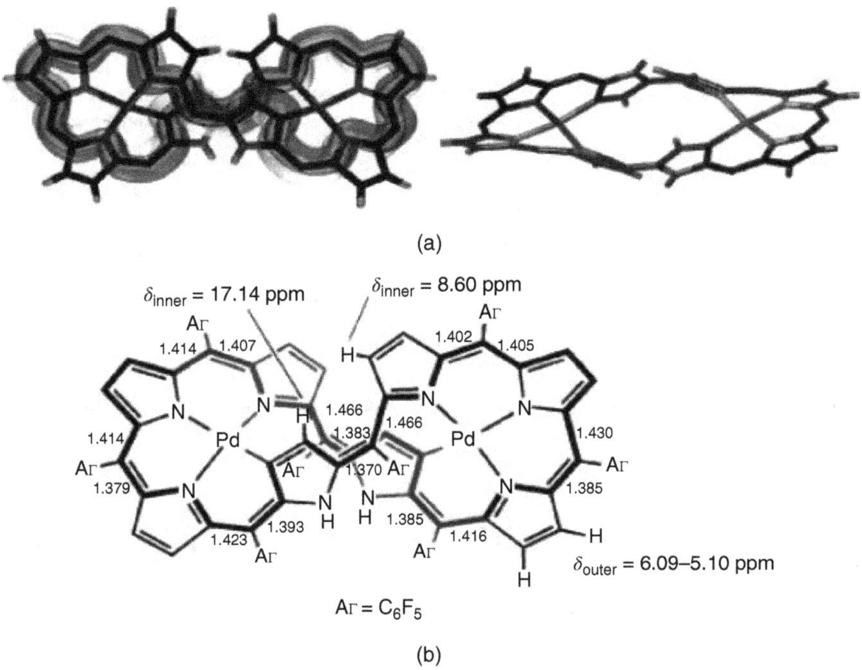


**Figure 10.9** (a) X-ray crystal structure of bis-Pd$^{II}$ [36]octaphyrin (*meso*-$C_6F_5$ substituents; left: top view; right: side view), (b) C(*meso*)−C(α) bond lengths and $^1H$ NMR chemical shifts of bis-Pd$^{II}$ [36]octaphyrin. Source: Reprinted with permission from Ref. [51]. Copyright © 2011, Wiley.

for [36]octaphyrin [61]. It has been also shown that *meso*-aryl-substituted [28]hexaphyrins exist at room temperature as an equilibrium of several interconverting twisted Möbius conformations with distinct aromaticities [62, 63].

Experimental evidences of aromaticity or antiaromaticity in expanded porphyrins comes from different sources. In X-ray data, aromatic expanded porphyrins present smaller bond length alternation in its π-conjugated structure. An important property of expanded porphyrins is that they fluorescence in the near-infrared (NIR) region. For aromatic expanded porphyrins a sharp and intense Soret-like band in the blue wavelength region of the visible spectrum (550–750 nm) and Q-like bands at about 900 nm is observed (Figure 10.10) [64, 65]. Moreover, aromatic expanded porphyrins display two-photon absorption (TPA) cross sections ($\sigma^{(2)}$ = 1000 up to 13 000 GM) larger than their antiaromatic counterparts. Finally, aromatic expanded porphyrins show a $^1H$ nuclear magnetic resonance (NMR) spectrum with β-CH signals in the deshielded region, while inner pyrrolic signals appear shielded due to the diatropic ring current generated by the external magnetic field in the annulene-like structure (Figure 10.11).

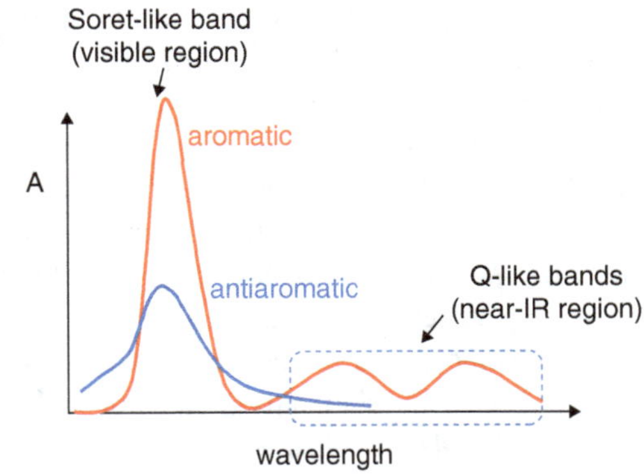


**Figure 10.10** UV/Vis spectrum of aromatic and antiaromatic compounds with the Soret-like and Q-like bands.

Antiaromatic expanded porphyrins exhibit opposite features to those previously described.

As to the computational diagnostics of aromaticity in expanded porphyrins, one can list structural bond length equalization due to cyclic delocalization (HOMA), energetic-enhanced stability (aromatic stabilization energy, ASE), diatropic magnetic ring currents, negative NICS values, electron delocalization indices such as EDDB, FLU, and ELF (see Chapter 8) and, more recently, a new electron

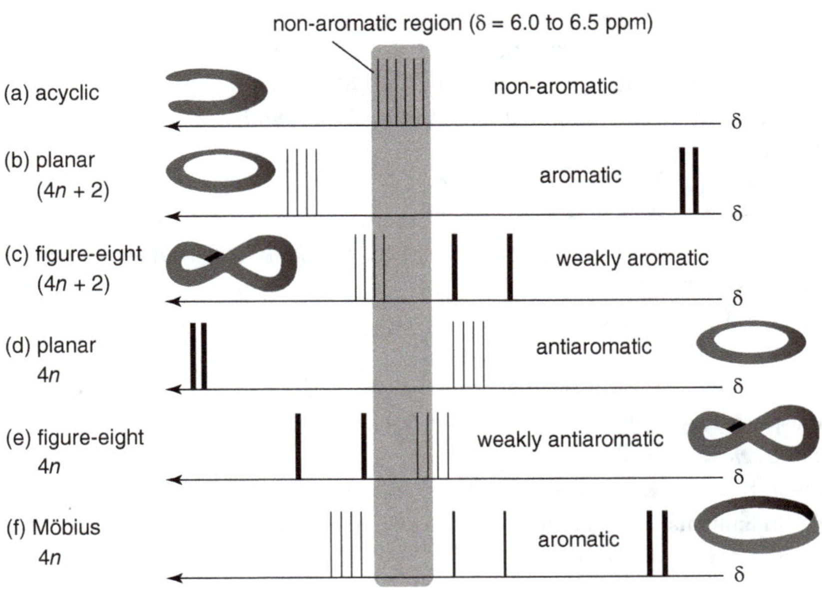


**Figure 10.11** Typical $^{1}$H NMR spectra of (a) acyclic $\pi$-conjugated oligopyrrole and (b–f) expanded porphyrins with different topologies and number of $\pi$ electrons. Source: Reprinted with permission from Ref. [51]. Copyright © 2011, Wiley.

delocalization descriptors especially designed to quantify aromaticity in large aromatic rings called AV1245 [66, 67], which is obtained by averaging the 4-center multicenter indices along a given closed circuit. In some cases, there is a controversy on which is the most conjugated pathway in a given expanded porphyrin and also in the relative importance of the macrocyclic aromaticity versus the local aromaticity of the pyrrole units. Interestingly, there are evidences showing that the Hückel's and Baird's rules of aromaticity in expanded porphyrins fade away as the ring structure increases and an optimal overlap between $p$ orbitals is no longer possible due to geometrical restrictions [68] (Figure 10.11).

Finally, let us mention that examples of triple twisted Möbius topologies and aromaticity [69, 70] or quadruple twisted Hückel aromaticity have been reported [71].

## References


1 Pickover, C.A. *The Möbius strip: Dr. August Möbius's marvelous band in mathematics, games, literature, art, technology, and cosmology*, vol. 2006. New York: Thunder's Mouth Press.

2 Sylvetsky, N., Banerjee, A., Alonso, M., and Martin, J.M.L. (2020). *J. Chem. Theory Comput.* 16: 3641–3653.

3 Craig, D.P. and Paddock, N.L. (1958). *Nature* 181: 1052–1053.

4 Heilbronner, E. (1964). *Tetrahedron Lett.* 5: 1923–1928.

5 Zimmermann, H. (1971). *Acc. Chem. Res.* 4: 272–280.

6 Jiao, H.J. and Schleyer, P.v.R. (1993). *Angew. Chem. Int. Ed. Engl.* 32: 1763–1765.

7 Rzepa, H.S. (2005). *Chem. Commun.* 5220–5222.

8 Mauksch, M. and Tsogoeva, S.B. (2009). *Angew. Chem. Int. Ed.* 48: 2959–2963.

9 Mauksch, M., Gogonea, V., Jiao, H., and Schleyer, P.v.R. (1998). *Angew. Chem. Int. Ed.* 37: 2395–2397.

10 Schleyer, P.v.R., Barborak, J.C., Su, T.M. et al. (1971). *J. Am. Chem. Soc.* 93: 279–281.

11 Anastassiou, A.G. and Yakali, E. (1971). *J. Am. Chem. Soc.* 93: 3803–3805.

12 Anastassiou, A.G., Eachus, S.W., Elliott, R.L., and Yakali, E. (1972). *J. Chem. Soc., Chem. Commun.* 531–532.

13 Ajami, D., Hess, K., Köhler, F. et al. (2006). *Chem. Eur. J.* 12: 5434–5445.

14 Martín-Santamaría, S., Lavan, B., and Rzepa, H.S. (2000). *J. Chem. Soc., Perkin Trans. 2* 1415–1417.

15 Ajami, D., Oeckler, O., Simon, A., and Herges, R. (2003). *Nature* 426: 819–821.

16 Castro, C., Chen, Z., Wannere, C.S. et al. (2005). *J. Am. Chem. Soc.* 127: 2425–2432.

17 Herges, R. (2006). *Chem. Rev.* 106: 4820–4842.

18 Rzepa, H.S. (2005). *Chem. Rev.* 105: 3697–3715.

19 Stępień, M., Szyszko, B., and Latos-Grażyński, L. (2010). *J. Am. Chem. Soc.* 132: 3140–3152.

20 Craig, D.P. (1959). *J. Chem. Soc. (Resumed)* 997–1001.

21 Hoffmann, R. (1982). *Angew. Chem. Int. Ed. Engl.* 21: 711–724.

22 Bleeke, J.R. (2001). *Chem. Rev.* 101: 1205–1227.

23 Wright, L.J. (ed.) (2017). *Metallabenzenes: An Expert View*. Chichester: Wiley.

24 Fernandez, I., Frenking, G., and Merino, G. (2015). *Chem. Soc. Rev.* 44: 6452–6463.

25 Thorn, D.L. and Hoffmann, R. (1979). *Nouv. J. Chim.* 3: 39–45.

26 Elliott, G.P., Roper, W.R., and Waters, J.M. *J. Chem. Soc., Chem. Commun.* 1982: 811–813.

27 Zhu, J., Jia, G., and Lin, Z. (2007). *Organometallics* 26: 1986–1995.

28 Zhu, C., Luo, M., Zhu, Q. et al. (2014). *Nat. Commun.* 5: 3265.

29 Zhu, C., Li, S., Luo, M. et al. (2013). *Nat. Chem.* 5: 698–703.

30 Mauksch, M. and Tsogoeva, S.B. (2010). *Chem. Eur. J.* 16: 7843–7851.

31 Periyasamy, G., Burton, N.A., Hillier, I.H., and Thomas, J.M.H. (2008). *J. Phys. Chem. A* 112: 5960–5972.

32 Iron, M.A., Lucassen, A.C.B., Cohen, H. et al. (2004). *J. Am. Chem. Soc.* 126: 11699–11710.

33 Fernández, I. and Frenking, G. (2007). *Chem. Eur. J.* 13: 5873–5884.

34 El-Hamdi, M., El Bakouri El Farri, O., Salvador, P. et al. (2013). *Organometallics* 32: 4892–4903.

35 Havenith, R.W.A., De Proft, F., Jenneskens, L.W., and Fowler, P.W. (2012). *Phys. Chem. Chem. Phys.* 14: 9897–9905.

36 De Proft, F. and Geerlings, P. (2004). *Phys. Chem. Chem. Phys.* 6: 242–248.

37 Psciuk, B.T., Lord, R.L., Winter, C.H., and Schlegel, H.B. (2012). *J. Chem. Theory Comput.* 8: 4950–4959.

38 Wei, J., Zhang, Y., Chi, Y. et al. (2016). *J. Am. Chem. Soc.* 138: 60–63.

39 Grande-Aztatzi, R., Mercero, J.M., Matito, E. et al. (2017). *Phys. Chem. Chem. Phys.* 19: 9669–9675.

40 An, K., Shen, T., and Zhu, J. (2017). *Organometallics* 36: 3199–3204.

41 Szczepanik, D.W. and Solà, M. (2019). *ChemistryOpen* 8: 219–227.

42 Szczepanik, D.W., Żak, E., Dyduch, K., and Mrozek, J. (2014). *Chem. Phys. Lett.* 593: 154–159.

43 Szczepanik, D.W. (2016). *Comput. Theor. Chem.* 1080: 33–37.

44 Szczepanik, D.W., Andrzejak, M., Dominikowska, J. et al. (2017). *Phys. Chem. Chem. Phys.* 19: 28970–28981.

45 Cheung, L.F., Kocheril, G.S., Czekner, J., and Wang, L.-S. (2020). *J. Am. Chem. Soc.* 142: 3356–3360.

46 Sessler, J.L. and Seidel, D. (2003). *Angew. Chem. Int. Ed.* 42: 5134–5175.

47 Neves, M.G.P.M.S., Martins, R.M., Tomé, A.C. et al. (1999). *Chem. Commun.* 385–386.

48 Shin, J.-Y., Kim, K.S., Yoon, M.-C. et al. (2010). *Chem. Soc. Rev.* 39: 2751–2767.

49 Osuka, A. and Saito, S. (2011). *Chem. Commun.* 47: 4330–4339.

50 Stępień, M., Sprutta, N., and Latos-Grażyński, L. (2011). *Angew. Chem. Int. Ed.* 50: 4288–4340.

51 Saito, S. and Osuka, A. (2011). *Angew. Chem. Int. Ed.* 50: 4342–4373.

52 Pawlicki, M. and Latos-Grażyński, L. (2015). *Chem. Asian J.* 10: 1438–1451.

53 Tanaka, T. and Osuka, A. (2017). *Chem. Rev.* 117: 2584–2640.

54 Szyszko, B., Białek, M.J., Pacholska-Dudziak, E., and Latos-Grażyński, L. (2017). *Chem. Rev.* 117: 2839–2909.

55 Tanaka, T. and Osuka, A. (2018). *Chem. Eur. J.* 24: 17188–17200.

56 Stępień, M., Latos-Grażyński, L., Sprutta, N. et al. (2007). *Angew. Chem. Int. Ed.* 46: 7869–7873.

57 Tanaka, Y., Saito, S., Mori, S. et al. (2008). *Angew. Chem. Int. Ed.* 47: 681–684.

58 Higashino, T., Lim, J.M., Miura, T. et al. (2010). *Angew. Chem. Int. Ed.* 49: 4950–4954.

59 Saito, S., Shin, J.-Y., Lim, J.M. et al. (2008). *Angew. Chem. Int. Ed.* 47: 9657–9660.

60 Alonso, M., Geerlings, P., and De Proft, F. (2013). *Chem. Eur. J.* 19: 1617–1628.

61 Karthik, G., Lim, J.M., Srinivasan, A. et al. (2013). *Chem. Eur. J.* 19: 17011–17020.

62 Sankar, J., Mori, S., Saito, S. et al. (2008). *J. Am. Chem. Soc.* 130: 13568–13579.

63 Marcos, E., Anglada, J.M., and Torrent-Sucarrat, M. (2012). *J. Phys. Chem. C* 116: 24358–24366.

64 Hong, Y., Oh, J., Sung, Y.M. et al. (2017). *Angew. Chem. Int. Ed.* 56: 2932–2936.

65 Kim, J.O., Hong, Y., Kim, T. et al. (2018). *J. Phys. Chem. Lett.* 9: 4527–4531.

66 Matito, E. (2016). *Phys. Chem. Chem. Phys.* 18: 11839–11846.

67 Casademont-Reig, I., Woller, T., Contreras-García, J. et al. (2018). *Phys. Chem. Chem. Phys.* 20: 2787–2796.

68 Casademont-Reig, I., Ramos-Cordoba, E., Torrent-Sucarrat, M., and Matito, E. (2020). *Molecules* 25: 711.

69 Schaller, G.R., Topić, F., Rissanen, K. et al. (2014). *Nat. Chem.* 6: 608–613.

70 Naulet, G., Sturm, L., Robert, A. et al. (2018). *Chem. Sci.* 9: 8930–8936.

71 Soya, T., Mori, H., and Osuka, A. (2018). *Angew. Chem. Int. Ed.* 57: 15882–15886.

# 11

# σ-, π-, δ-, and φ-Aromaticity

*"Work, look for peace and calm in work; you will find it nowhere else."*
Dmitri Mendeleev

## 11.1 Introduction

The concept of aromaticity was initially introduced in organic chemistry by Kekulé for describing low reactivity of benzene derivatives compared to other unsaturated hydrocarbons and high stability [1–3]. The next milestone in the development of the aromaticity concept came from Hückel, who formulated his famous $4n+2$ rule for π-electrons based on the quantum mechanical consideration of benzene [4–6]. More than 30 years later, Breslow introduced the concept of antiaromaticity [7, 8], which was the antonym of aromaticity referring to cyclic hydrocarbons with $4n$ π-electrons, which have high reactivity and low stability. Until recently, aromaticity and antiaromaticity were confined to π-electron organic molecules.

Dewar proposed that the conjugation patterns in benzene and cyclopropane might be very similar, and that certain attributes of cyclopropane might be explained as manifestations of σ-aromaticity [9–11]. Later, Schleyer and coworkers proved that cyclopropane is not a σ-aromatic molecule [12], but the possibility of σ-aromaticity started floating in chemistry. Indeed, the $4n+2$ rule was based on symmetry of π molecular orbitals (MOs) (Figure 11.1), and we should expect that if we find cyclic molecular systems with $4n+2$ delocalized σ-electrons, they will be σ-aromatic and those with the $4n$ delocalized σ-electrons will be σ-antiaromatic. We can imagine systems that have $4n+2$ delocalized both σ-electrons and π-electrons. Such systems should be named as double σ- and π-aromatic. The first doubly aromatic system, the 3,5-dehydrophenyl cation, was identified by Chandrasekhar et al. [13] and introduced double (σ- and π-) aromaticity in chemistry. Finally, we could think about chemical systems, which

*Aromaticity and Antiaromaticity: Concepts and Applications*, First Edition.
Miquel Solà, Alexander I. Boldyrev, Michał K. Cyrański, Tadeusz M. Krygowski, and Gabriel Merino.

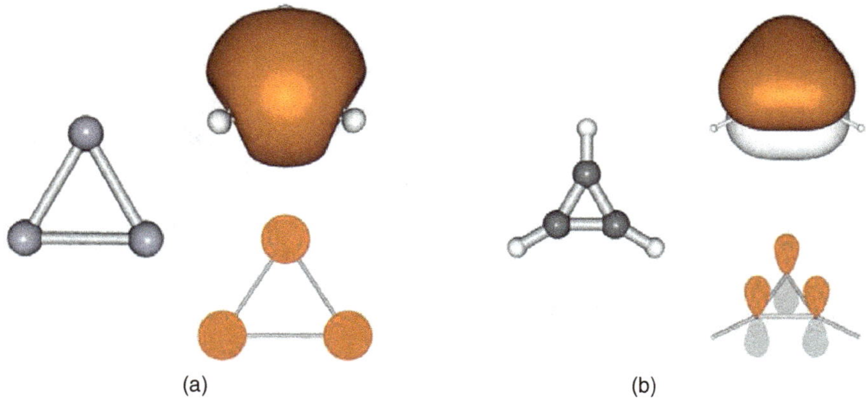

Figure 11.1 (a) Structure of the $Li_3^+$ cluster, HOMO ($1a_1'$) of $Li_3^+$ and schematic representation of HOMO as a linear combination of *2s*-AOs of Li atoms. (b) Structure of the $C_3H_3^+$ cation, HOMO ($1a_2''$) of $C_3H_3^+$ and its schematic representation as a linear combination of $2p_z$-AOs C atoms.

have either delocalized $4n+2$ π-electrons and $4n$ σ-electrons, or delocalized $4n$ π-electrons and $4n+2$ σ-electrons, and such systems are called molecules with conflicting aromaticity [14].

In recent years, concepts of aromaticity, antiaromaticity, double aromaticity, and conflicting aromaticity were extended into inorganic chemistry. Boldyrev and Wang suggested that for cyclic systems made out of *d*- and *f*-metals, there are possibilities for δ-aromaticity and φ-aromaticity because now *d*- and *f*-atomic orbitals (AOs) are available for delocalized bonding [14]. Multiple aromaticity and antiaromaticity in inorganic systems leads to new complications for electron counting rules (see detailed discussion about that in Ref. [15]). The electron counting rules for *s*-AO based σ-aromaticity are the same as the Hückel's $4n+2/4n$ rules for π-aromaticity/antiaromaticity for all cyclic structures. The counting rules for *p*-AO based σ-aromaticity are $4n+4$ (aromaticity) and $4n+6$ (antiaromaticity) for cyclic structures with even numbers of atoms, and $4n+2$ (aromaticity) and $4n$ (antiaromaticity) for cyclic structures with odd numbers of atoms because there are two types of σ-orbitals, namely, radial $p_{\sigma\text{-}r}$- and tangential $p_{\sigma\text{-}t}$-molecular orbitals (MOs) that should be considered together. In the simplest case of the occupation of just one completely bonding $p_{\sigma\text{-}r}$- or $p_{\sigma\text{-}t}$-MO, the system is also aromatic in spite of 4 σ-electrons participating in delocalized bonding. For *p*-AO based π-aromaticity the counting rules are $4n+2/4n$ for aromaticity/antiaromaticity for all cyclic structures. For *d*-AO based σ- and π-aromaticity, the counting rules are $4n+4$ (aromaticity) and $4n+6$ (antiaromaticity) for cyclic structures with even numbers of atoms, and $4n+2$ (aromaticity) and $4n$ (antiaromaticity) for cyclic structures with odd numbers of atoms because there are two types of σ-orbitals, namely, $d_{\sigma\text{-}r}$- and $d_{\sigma\text{-}t}$-MOs, and two types of π-orbitals, namely, $d_{\pi\text{-}r}$- and $d_{\pi\text{-}t}$-MOs. For *d*-AO based δ-aromaticity the electron counting rule is $4n+2/4n$ for aromaticity/antiaromaticity, respectively. There are a number of reviews on aromaticity/antiaromaticity in inorganic chemistry [14–29]. We do not discuss *p*-AO based π-aromaticity and antiaromaticity because they are well known in chemistry; instead, we focus on other new types of aromaticity and antiaromaticity. The number of systems with new types of aromaticity and antiaromaticity and conflicting aromaticity is very large. In order to keep discussion on these types of aromaticity compact, we consider only few repetitive molecules for every kind of aromaticity. More extended discussion of new types of aromaticity can be found in reviews [14–29].

## 11.2 σ-Aromatic and σ-Antiaromatic Species

### 11.2.1 σ-Aromatic Species

If only *s*-AOs are involved in chemical bonding, the chemical species may be σ-aromatic or σ-antiaromatic with the electron counting rule $4n+2/4n$ for singlet coupling σ-aromaticity/antiaromaticity [15]. Such chemical species should be built out of atoms of the I and II rows. For example, $M_3^+$ (M = Li…Cs) and $H_3^+$ are aromatic ($n = 0$ for aromaticity) and $M_3^-$ (M = Li…Cs) are antiaromatic ($n = 1$ for antiaromaticity) based on the $4n+2/4n$ rules. The simplest and the smallest aromatic compound is the $H_3^+$ cluster, which is the most abandoned aromatic species in the universe. It has a triangular global minimum structure and it is σ-aromatic since it has the three-center two-electron (3c-2e) bond occupied by two electrons. A calculated current density map shows that $H_3^+$ shows a market diatropic ring-current and the zz-component of the nucleus-independent chemical shift ($NICS_{zz}$) value calculated at z = 0.0 is very negative (−36.9 ppm) [30]. Thus, both aromaticity criteria confirm its σ-aromaticity. The smallest aromatic metal cluster is $Li_3^+$, which is known to have a triangular global minimum energy structure [31]. Aromaticity in $Li_3^+$ was initially assessed by Alexandrova and Boldyrev [32]. Similarly to $H_3^+$, it has only one completely delocalized bonding σ-MO (Figure 11.1a).

The highest occupied MO (HOMO) in $Li_3^+$, which is solely responsible for bonding in this cation, is similar to the HOMO in $C_3H_3^+$, where it is solely responsible for π-bonding. Since we accept π-aromaticity in $C_3H_3^+$, we should accept σ-aromaticity in $Li_3^+$ since both are completely bonding MOs and both systems have delocalized MOs confined to the $4n+2$ rule for aromaticity. Alexandrova and Boldyrev calculated that the resonance energy if $Li_3^+$ is rather high, 36 kcal/mol [32], and that $NICS_{zz}$ calculated at the center of this cation is −8.8 ppm. Both of these criteria are signaling that this cation is aromatic. However, Fowler and coworkers concluded that $Li_3^+$ shows no global current and therefore it is nonaromatic according to this criterion. A chemical bonding pattern for other $M_3^+$ (M = Na – Cs) cations was found to be similar to $Li_3^+$, and thus all of them may be considered as aromatic.

### 11.2.2 σ-Antiaromatic Species

The tetramers $M_4$ (M = Li–Cs) all have a rhombus structure with 4 σ-electrons and thus can be viewed as σ-antiaromatic (Figure 11.2).

At the perfect square structure, $M_4$ tetramers have $1a_{1g}^2 1e_u^2$ electronic configuration, which is in the singlet state is unstable and undergoes the Jahn-Teller distortion towards the rhombus structure, whereas π-antiaromaticity in cyclobutadiene leads to rectangular structure. The $M_4$ clusters allow us to introduce the phenomenon of local aromaticity. Although $M_4$ rhombic clusters are globally σ-antiaromatic (4 σ-electrons and rhombic distortion), they form two areas of delocalized bonding (2 σ-electrons each) inside of each triangle leading to regions of local or island aromaticity. Indeed, the Adaptive Natural Density Partitioning (AdNDP) analysis performed for the $Li_4$ clusters revealed these three-center regions of local σ-aromaticity as two 3c-2e bonds with the occupation numbers (ONs) equal to 2.0 |e| (Figure 11.2). More discussion on σ-aromaticity and σ-antiaromaticity can be found in Refs. [14–29, 32].

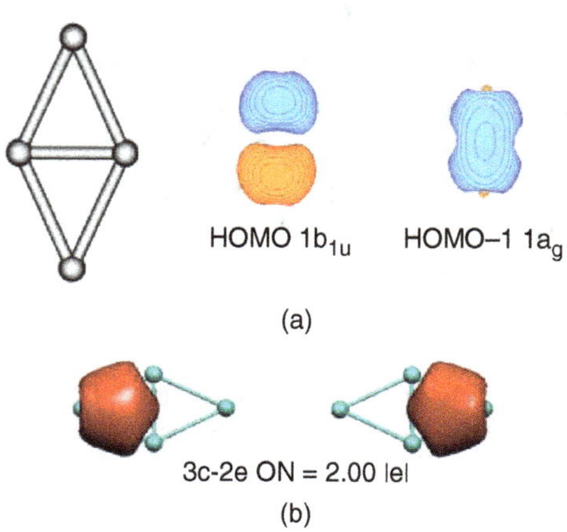


**Figure 11.2** (a) Structure and CMOs of $Li_4$ $D_{2h}$, ($^1A_g$) cluster; (b) results of the AdNDP localization.

## 11.3 σ-, π-Doubly Aromatic, and σ-, π-Doubly Antiaromatic Species and Species with σ-, π-Conflicting Aromaticity

### 11.3.1 σ-, π-Doubly Aromatic Species

When *p*-AOs are involved in delocalized bonding, we may encounter σ-, π-double aromaticity, σ-, π-double antiaromaticity, and σ-, π-conflicting aromaticity. The best way to discuss these cases is to consider chemical bonding in boron clusters [16, 26]. Let us first consider a σ-, π-double aromatic $B_4$ cluster [32]. It is convenient to use the AdNDP method proposed by Zubarev and Boldyrev [33] for chemical bonding analysis, where delocalized bonding is involved. The AdNDP method is an extension of the Natural Bond Orbital (NBO) method proposed by Foster and Weinhold [34]. The NBO method is very efficient in obtaining lone pairs and two-center two-electron (2c-2e) bonds from canonical molecular orbitals (CMOs). Thus, it can recover a Lewis structure if it is a good description of the molecule. However, if bonding in a molecule is multicenter, the AdNDP method is an appropriate method for chemical bonding analysis since it is capable of recovering 3c-2e, 4c-2e,..., nc-2e bonds. The AdNDP analysis of chemical bonding in $B_4$ cluster is presented in Figure 11.3 together with the CMOs.

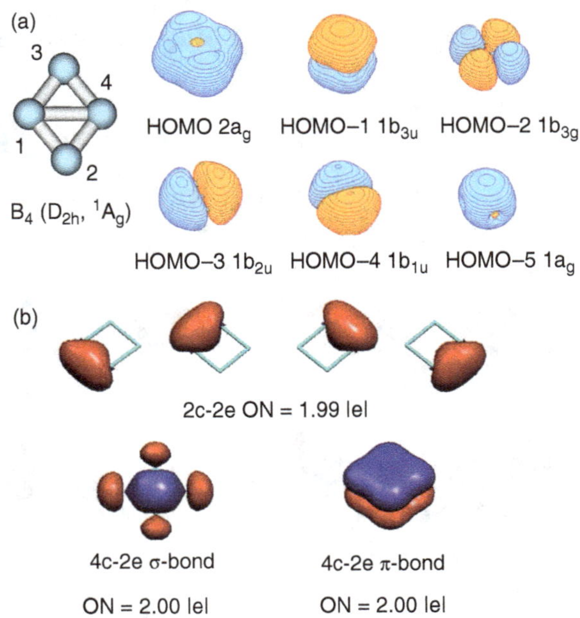


**Figure 11.3** (a) Structure and CMOs of $B_4$ $D_{2h}$ ($^1A_g$) cluster; (b) results of the AdNDP analysis.

The AdNDP analysis reveals that the $B_4$ cluster has four 2c-2e peripheral B–B σ-bonds, one completely delocalized 4c-2e σ-bond and one completely delocalized 4c-2e π-bond. The completely delocalized bonds are similar to CMOs, as it should be in this case. Thus, 4c-2e σ-bond and 4c-2e π-bond constitute double aromaticity in this cluster. The double aromatic cluster supposes to have a high symmetry ($D_{4h}$), but the small deviation from it is due to the second order Jahn-Teller effect [35], which does not affect σ-, π-double aromaticity.

### 11.3.2 σ-, π-Doubly Antiaromatic Species

The $B_6^{2-}$ cluster is σ-, π-doubly antiaromatic. The structure of this cluster is shown in Figure 11.4a [33, 36].

From the CMOs (Figure 11.4a), it is clear that this cluster has two π-MOs (HOMO-4 is completely bonding and HOMO is partially antibonding), which makes this cluster π-antiaromatic. The σ-bonding is more difficult to rationalize and that is where AdNDP analysis is really helpful. Figure 11.4b shows results of the AdNDP localization. There are six 2c-2e peripheral B–B peripheral σ-bonds,

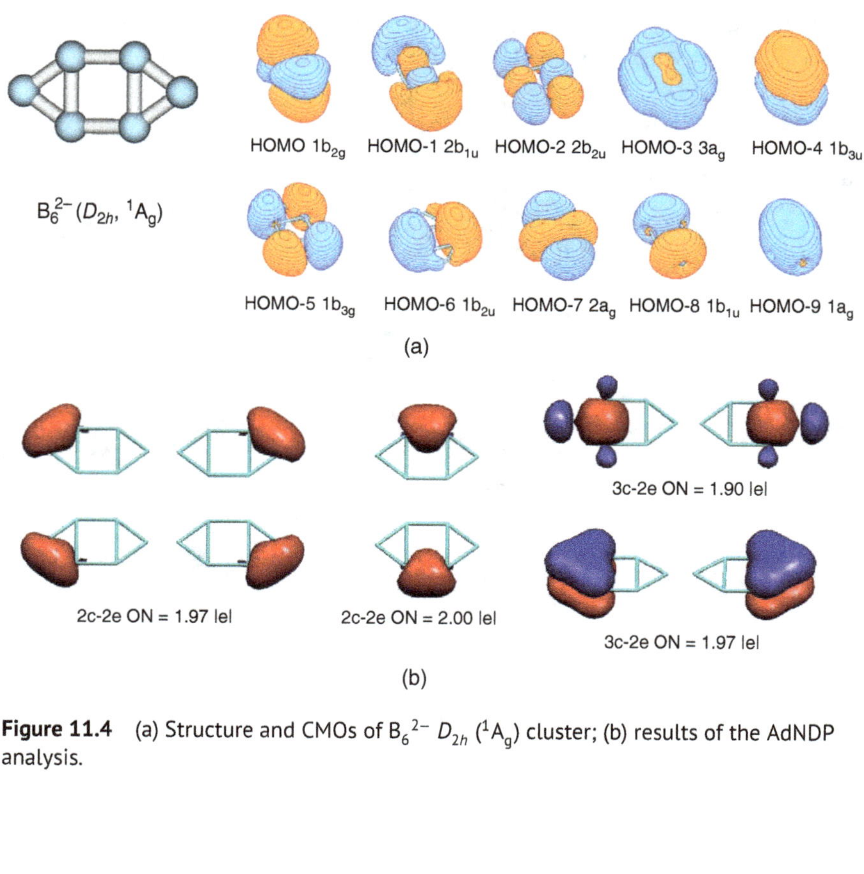


**Figure 11.4** (a) Structure and CMOs of $B_6^{2-}$ $D_{2h}$ ($^1A_g$) cluster; (b) results of the AdNDP analysis.

two 3c-2e σ-bonds, and two 3c-2e π-bonds. The AdNDP analysis revealed 3c-2e σ- and π-bonds in this case instead of 6c-2e bonds because completely bonding (HOMO-3 and HOMO-4) together with partially antibonding (HOMO-1 and HOMO) led to island aromaticity results in 3c-2e σ- and π-bonds. That is similar to cyclobutadiene, when two MOs (completely bonding and partially π-antibonding MOs) lead to the formation of two 2c-2e π-bonds. Thus, two 3c-2e σ- and two 3c-2e π-bonds constitutes σ- and π-double antiaromaticity. Distorted hexagon structure is also consistent with the antiaromatic nature of chemical bonding in this cluster. Our conclusion on antiaromaticity of the $B_6^{2-}$ $D_{2h}$ ($^1A_g$) structure is supported by the results of Havenith et al. [37], who studied the induced ring current maps. They demonstrated that in π-only and (σ + π) maps, a strong paramagnetic current is discernible in the inner square of the molecule, consistent with the expected antiaromatic nature of this system.

### 11.3.3 Species with σ-Antiaromaticity and π-Aromaticity

The low symmetry ($C_{2v}$) of the global minimum structure of the $B_5^-$ cluster [38] (Figure 11.5a) is a clear indication that this cluster is probably antiaromatic.

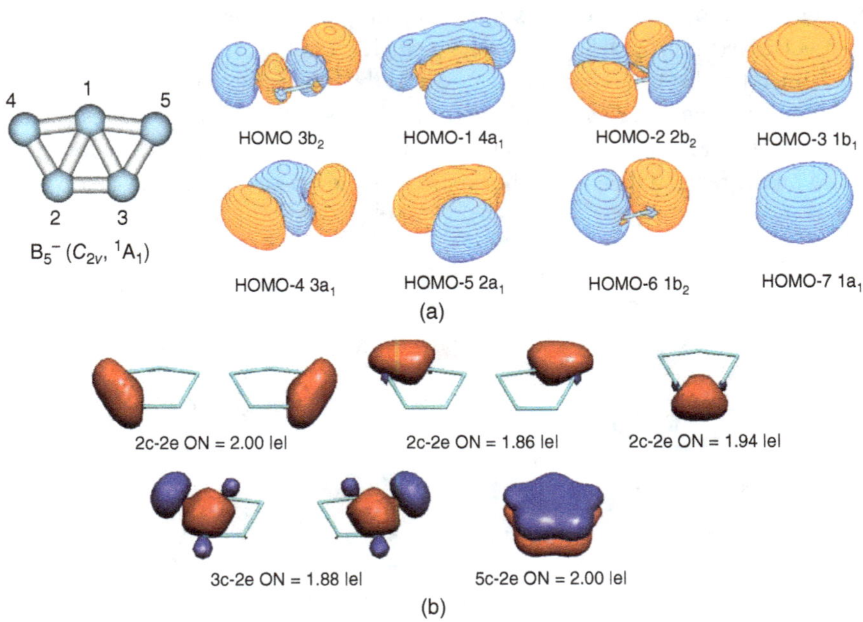


**Figure 11.5** (a) Structure and CMOs of $B_5^-$ $C_{2v}$ ($^1A_1$) cluster; (b) results of the AdNDP analysis.

However, inspection of CMOs (Figure 11.5a) clearly shows that this cluster is π-aromatic (HOMO-3). Since it is hard to interpret σ-CMOs, we used the AdNDP method [33]. Figure 11.5b shows results of the AdNDP localization. Again, one can see that $B_5^-$ has five peripheral 2c-2e B–B σ-bonds, one 5c-2e π-bond, and two 3c-2e σ-bonds. The 5c-2e π-bond and two 3c-2e σ-bonds constitute π-aromaticity and σ-antiaromaticity. Thus, the $B_5^-$ cluster has a conflicting aromaticity, which explains its low symmetry structure since σ-antiaromaticity wins versus π-aromaticity in this case.

### 11.3.4 Species with σ-Aromaticity and π-Antiaromaticity

The $Li_3Al_4^-$ cluster is an example of a system with σ-aromaticity and π-antiaromaticity [39]. It was shown that the $LiAl_4^-$ cluster (pyramidal structure with the planar square $Al_4^{2-}$ base capped with $Li^+$ cation) is a doubly (σ- and π-) aromatic system [40]. Indeed, CMOs shown for just the base $Al_4^{2-}$ dianion in Figure 11.6 show one completely bonding π-CMO indicating π-aromaticity in this cluster and two completely bonding σ-CMOs indicating σ-aromaticity (separately for σ-radial and σ-tangential aromaticity, different counting rule $4n+4$ is applied in this case).

In the $Li_3Al_4^-$ cluster, we have two more electrons compared to the $LiAl_4^-$ cluster, and those two electrons are occupying one of the doubly degenerate π-1$e_u$-LUMO (in $LiAl_4^-$). Since we have a singlet coupling, the system undergoes the Jahn-Teller distortion and the $Al_4^{4-}$ base in this cluster has somewhat distorted rectangular shape (the distortion from the perfect rectangular shape comes due to asymmetrical capping by $Li^+$ cations) similar to the antiaromatic cyclobutadiene. Based on the rectangular shape, the presence of the 4 *p*-electrons and appreciably

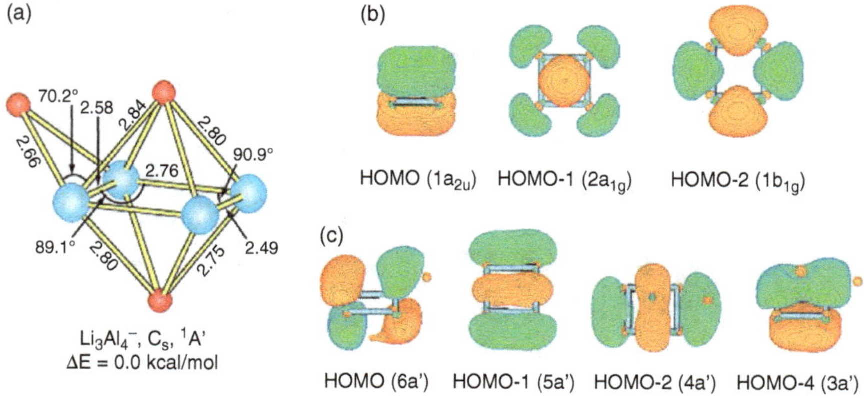


**Figure 11.6** (A) Optimized global minimum structure of $Li_3Al_4^-$; bond lengths in Å. (B) CMOs for $Al_4^{2-}$; (C) CMOs for $Li_3Al_4^-$.

low vertical electron detachment energy (1.39 eV in $Li_3Al_4^-$ versus 2.15 eV in $LiAl_4^-$) we concluded that this anion being σ-aromatic and π-antiaromatic is net antiaromatic. The assignment of this cluster to net antiaromaticity created a controversy in the literature [14]. Everybody agrees that the $Li_3Al_4^-$ cluster is σ-aromatic and π-antiaromatic, but some people disagree on the net antiaromaticity (see the discussion about that in Ref. [14]). Other examples of conflicting aromaticity can be found in Ref. [14–29].

## 11.4 δ-Aromaticity

δ-aromaticity can be found only in transition metal clusters or clusters made out of *f*-elements because it requires a specific overlap of *d*- or *f*-orbitals shown in Figures 11.7 and 11.8 for triatomic atomic cluster built out of transition metal atoms or for triatomic clusters built out of *f*-elements, respectively [15].

As we can see from the model system in Figure 11.7, we can compose a completely bonding $1a_1'$-CMO and two partially bonding/antibonding $1e'$-CMOs. Subsequently, for a system with two electrons on the $1a_1'$-CMO, we have a δ-aromaticity, and for a system with four singlet coupled electrons ($1a_1'^2 1e'^2$), we have a δ-antiaromaticity.

*f*-AO based δ-aromaticity/antiaromaticity in a cyclic triatomic system is based on $f_{\delta-r}$-radial *f*-AOs and $f_{\delta-t}$-tangential *f*-AOs (Figure 11.8). In the case of *f*-AO based δ-aromaticity/antiaromaticity, there are six δ-CMOs: $1a_1'$, $1e'$, and $2e'$, and $1a_2'$ which are completely bonding, partially bonding/antibonding, and completely antibonding, respectively (see Ref. [15] for possible electronic configurations and the counting rules for aromaticity/antiaromaticity).

The first experimentally observed example of δ-aromaticity was found by Zhai et al. [41] in joint photoelectron spectroscopy and theoretical study. They showed that the $Ta_3O_3^-$ cluster possesses a global minimum structure with a perfect $D_{3h}$ ($^1A_1'$) planar triangular structure and that it's doubly π- and δ-aromatic species (Figure 11.9).

Out of the 34 valence electrons in $Ta_3O_3^-$, 24 belong to either pure oxygen lone pairs or those polarized towards the Ta atoms (responsible for the covalent contributions to the Ta-O bonding). The remaining 10 valence electrons

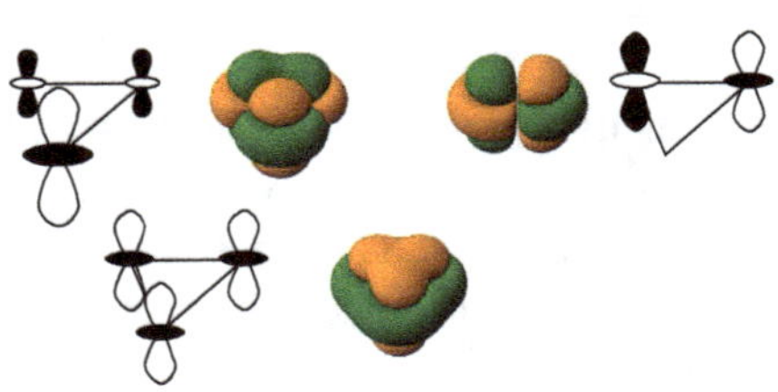


**Figure 11.7** *d*-AO based δ-CMOs for a model triatomic system plotted by Orbital Viewer 1.04.

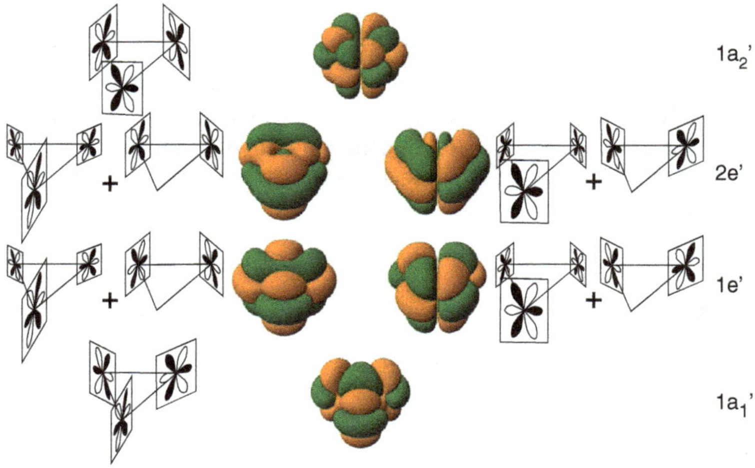


**Figure 11.8** *f*-AO based δ-CMOs for a model triatomic system plotted by Orbital Viewer 1.04.

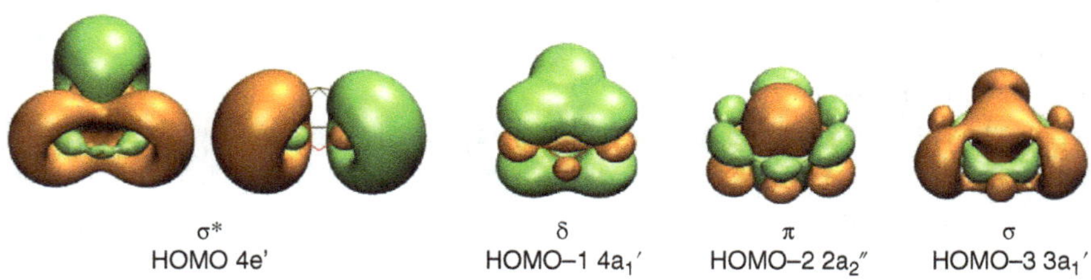


**Figure 11.9** The five valence MOs responsible in the $Ta_3O_3^-$ ($D_{3h}$, $^1A_1'$) cluster.

were responsible for direct metal–metal bonding between the Ta atoms. The corresponding occupied MOs are shown in Figure 11.9. The doubly degenerate bonding/antibonding-HOMO (4e′) and the completely bonding HOMO-3 (3a$_1$′) are of σ-type. The bonding character of HOMO-3 was cancelled by the antibonding nature of the HOMO and thus the σ orbitals did not contribute significantly to direct metal-metal bonding. The HOMO-1 and HOMO-2 are completely bonding δ and π orbitals, and, thus, this cluster is doubly (δ and π) aromatic according to the $4n + 2$ rule for aromaticity, with $n = 0$ applied separately to the δ and π systems.

δ-aromaticity provides us with the possibility to discover molecules with a triple σ-, π-, and δ-aromaticity. Averkiev and Boldyrev [42] theoretically predicted the first example of such aromaticity in $Hf_3$ in the $D_{3h}$, $^1A_1'$ $1a_1'^{(2)}2a_1'^{(2)}$ $1e'^{(4)}1a_2''^{(2)}3a_1'^{(2)}$ state. The results of the AdNDP analysis for this system are presented in Figure 11.10.

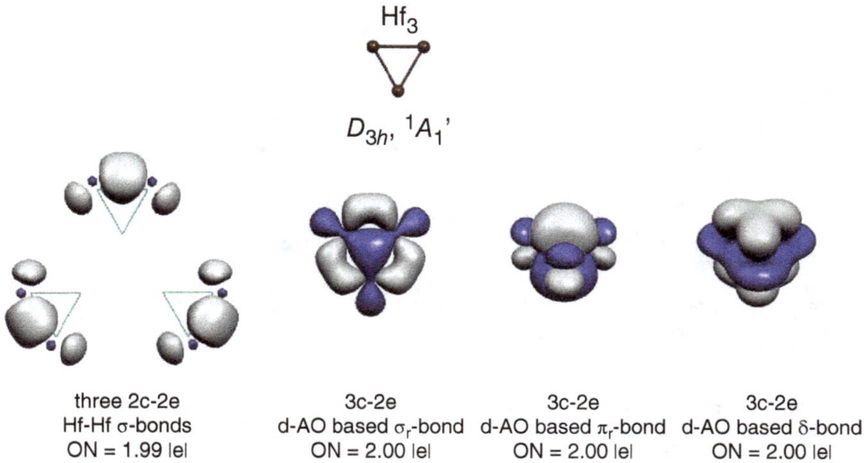


**Figure 11.10** Geometric structure, three 2c-2e Hf–Hf σ-bonds, 3c-2e d-AO based σ-bond, 3c-2e d-AO based π-bond, and 3c-2e d-AO based δ-bond of triply σ-, π- and δ-aromatic $Hf_3$ cluster.

The following bonding pattern is revealed for the $Hf_3$ species: three 2c-2e Hf–Hf σ-bonds are formed out of hybrid *6s,5d*-AOs of Hf atoms, and three completely delocalized bonds. They are: one completely bonding 3c-2e *d*-radial based σ-bond with ON = 2.00 |e|, one completely bonding 3c-2e *d*-radial based π-bond ON = 2.00 |e|, and one completely bonding 3c-2e *d*-AO based δ-bond ON = 2.00 |e|. Assignment of the triple (σ-, π-, and δ-) aromaticity can be made on the basis of the $4n + 2$ rule (odd number of atoms, see Ref. [15]). Though it is not clear at this point if the singlet state is a ground electronic state for the $Hf_3$ cluster or not, it is certain that the lowest singlet state for this cluster is an example of the system with triple aromaticity.

We may think that δ-aromaticity will not play any significant role in chemistry because the overlap between *d*-AOs will be significantly smaller than the overlap between *p*-AOs. However, Sergeeva and Boldyrev have shown that the stability of the $Pd_4^{2+}$ cluster is primarily due to δ-aromaticity [43] in a remarkable $[Pd_4(\mu_4\text{-}C_9H_9)\ (\mu_4\text{-}C_8H_8)]^+$ triple-decker sandwich compound. Murahashi et al. [44] have synthesized this compound, where the $Pd_4^{2+}$ square cluster is sandwiched between aromatic ($C_9H_9^-$) and antiaromatic ($C_8H_8$) molecules. The AdNDP analysis (Figure 11.11) revealed four lone pairs on each Pd atom based on *4d* AOs (not shown in Figure 11.11) and three 4c-2e *d*-orbitals (Figure 11.11), which were solely responsible for the Pd–Pd bonding, rendering it δ-aromaticity.

Therefore, the $[Pd_4(\mu_4\text{-}C_9H_9)(\mu_4\text{-}C_8H_8)]^+$ compound is the first synthesized compound in the solid state with δ-aromaticity. Even though δ-aromaticity is

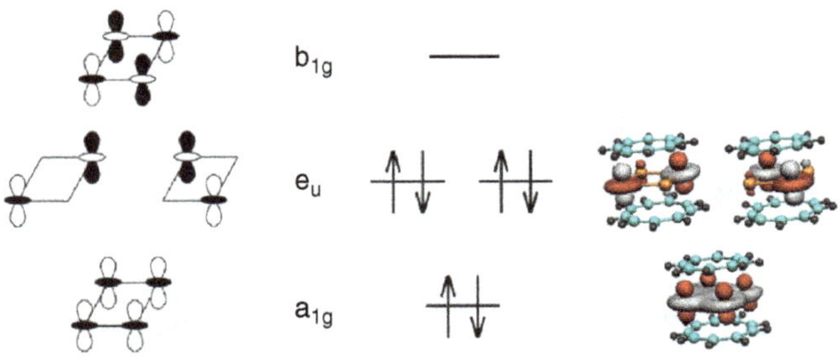


**Figure 11.11** The δ-bonding in the $Pd_4^{2+}$ cluster in the $[Pd_4(\mu_4\text{-}C_9H_9)(\mu_4\text{-}C_8H_8)]^+$ complex revealed by the AdNDP analysis.

admittedly weaker than σ- and π-aromaticity, it can play an important role in chemistry.

## 11.5 ϕ-Aromaticity

*f*-elements provide us with an opportunity to discover ϕ-aromaticity and ϕ-antiaromaticity. Starting with the triatomic model, we can see that the *f*-AO based ϕ-aromaticity/antiaromaticity in a triatomic system is based on completely ϕ-bonding $1a_2''$-MO and two partially bonding/antibonding $1e''$-MOs (Figure 11.12).

Applying Hückel's rules of aromaticity/antiaromaticity in cyclic triatomic systems, we can say that the system with the electronic configurations $1a_2''^{(2)}1e''^{(0)}$ (singlet coupling) and $1a_2''^{(2)}1e''^{(2)}$ (triplet coupling) are ϕ-aromatic, and with the electronic configurations $1a_2''^{(2)}1e''^{(2)}$ (singlet coupling) and $1a_2''^{(1)}1e''^{(1)}$ (triplet coupling) are ϕ-antiaromatic. As for *f*-AO based ϕ-aromaticity/antiaromaticity in tetratomic systems, there are four ϕ-MOs (Figure 11.13): one completely bonding

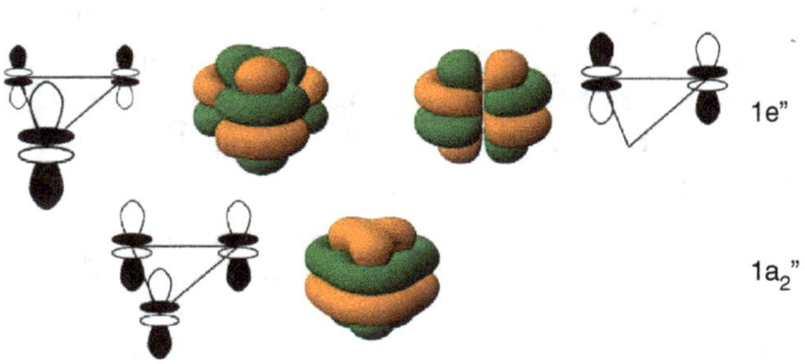


**Figure 11.12** *f*-AO based ϕ-MOs for model triatomic system plotted by Orbital Viewer 1.04.

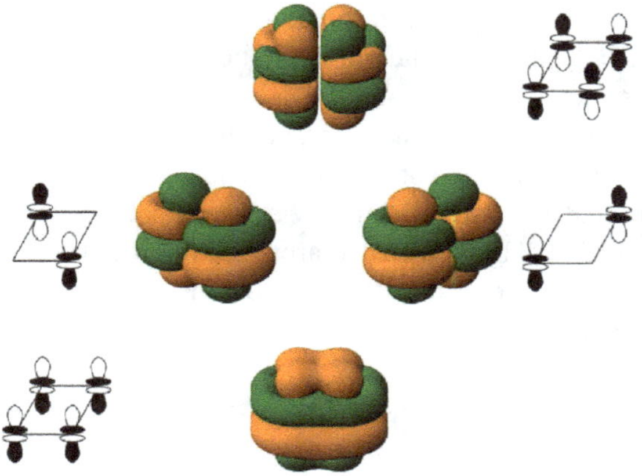


**Figure 11.13** *f*-AO based ϕ-MOs for model tetratomic system plotted by Orbital Viewer 1.04.

$1a_{2u}$, two degenerate partially bonding/antibonding $1e_g$, and one completely antibonding $1b_{2u}$-MO.

The preceding chemical bonding analysis on model triatomic and tetratomic systems can be used for assessing aromaticity/antiaromaticity in real molecules and clusters. However, hybridization may complicate this analysis.

Tsipis et al. [45] presented the first evidence of *f*-AO participation in delocalized bonding. They studied computationally a series of prototypical planar isocyclic cyclo-$U_nX_n$ ($n = 3, 4, X = O$, NH) and heterocyclic cyclo-$U_n(\mu_2\text{-}X)_n$ ($n = 3, 4$; X = C, CH, NH) clusters as well as E@[c-$U_4(\mu_2\text{-}C)_4$], (E = $H^+$, C, Si, Ge), and U@[c-$U_5(\mu_2\text{-}C)_5$] molecules, including a planar tetracoordinate element E and pentacoordinate U at the ring center. They demonstrated that *5f* orbitals play a key role in the bonding of these *f*-block metal systems with a significant contribution to the cyclic electron delocalization. The aromaticity in these clusters was verified by $NICS_{zz}(0)$ and $NICS_{zz}(1)$ calculations. The high negative values of $NICS_{zz}(0)$ (−135 to −353 ppm) and $NICS_{zz}(1)$ (−40 to −171 ppm) indicate the high aromatic character of cyclo-$U_nX_n$ ($n$ = 3,4, X = O, NH) and heterocyclic cyclo-$U_n(\mu_2\text{-}X)_n$ ($n$ = 3,4; X = C, CH, NH) species. This pioneering work is just the first step towards more extended discussions of φ-antiaromaticity in chemistry.

## 11.6 Conclusions

The large variety of different types of aromaticity/antiaromaticity, conflicting aromaticity in inorganic chemistry, which are briefly discussed in this chapter and which are due to participation of the s-, p-, d-, and f-AOs in delocalized bonding, provides us with unique tools in explaining of chemical bonding in new complex chemical species. It is clear that aromaticity and antiaromaticity could be very useful concepts in explaining structure, stability, and other molecular properties of isolated and embedded clusters of transition metals and transition metal oxide clusters. In spite of some success in application of aromaticity and antiaromaticity beyond organic chemistry, this field is still in its infancy. We expect that the number of chemical compounds, which will be classified as aromatic or antiaromatic, will grow tremendously. Catalysis, biochemistry, 0D-, 1D-, 2D-, and 3D-materials will be new frontiers for aromaticity and antiaromaticity in coming future.

## References


1 Kekule, A. (1865). *Bull. Soc. Chim. Fr. (Paris)* 3: 98–110.
2 Kekulé, A. (1965). *Bull. Acad. R. Belg.* 119: 551–563.
3 Kekulé, A. (1865). *Ann. Chem.* 137: 129–136.
4 Hückel, E. (1931). *Z. Physik* 70: 104–186.
5 Hückel, E. (1931). *Z. Physik* 72: 310–337.
6 Hückel, E. (1932). *Z. Physik* 76: 628–648.
7 Breslow, R. (1965). *Chem. Eng. News* 43: 90–100.
8 Breslow, R. (1973). *Acc. Chem. Res.* 6: 393–398.
9 Dewar, M.J.S. (1979). *Bull. Soc. Chim. Belg.* 88: 957–967.
10 Dewar, M.J.S. and McKee, M.L. (1980). *Pure Appl. Chem.* 52: 1431–1441.
11 Dewar, M.J.S. (1984). *J. Am. Chem. Soc.* 106: 669–682.
12 Wu, W., Ma, B., Wu, J.I.-C. et al. (2009). *Chem. Eur. J.* 15: 9730–9736.
13 Chandrasekhar, J., Jemmis, E.D., and Schleyer, P.v.R. (1979). *Tetrahedron Lett.* 20, 3707–20, 3710.
14 Boldyrev, A.I. and Wang, L.S. (2005). *Chem. Rev.* 105: 3716–3757.
15 Sergeeva, A.P., Averkiev, B.B., and Boldyrev, A.I. (2010). *Metal-Metal Bonding. Structure and Bonding* (ed. G. Parkin), 275–306. Berlin, Heidelberg: Springer.
16 Alexandrova, A.N., Boldyrev, A.I., Zhai, H.-J., and Wang, L.S. (2006). *Coord. Chem. Rev.* 250: 2811–2866.
17 Tsipis, C.A. (2005). *Coord. Chem. Rev.* 249: 2740–2762.
18 Tsipis, C.A. (2010). *Metal-Metal Bonding*, Structure and Bonding, vol. 136 (ed. G. Parkin), 217–274. Berlin, Heidelberg: Springer.
19 Zubarev, D.Y., Sergeeva, A.P., and Boldyrev, A.I. (2009). *Chemical Reactivity Theory. A Density Functional View* (ed. P.K. Chattaraj), 439–452. New York: CRC Press, Taylor & Francis Group.
20 Zubarev, D.Y. and Boldyrev, A.I. (2009). *Computational Inorganic and Bioinorganic Chemistry* (ed. E.I. Solomon, R.A. Scott and R.B. King), 551–562. Chistester, UK: Wiley.

**21** Sergeeva, A.P. and Boldyrev, A.I. (2010). *Aromaticity and Metal Clusters* (ed. P.K. Chattaraj), 55–68. Boca Raton: CRC Press.

**22** Galeev, T.R. and Boldyrev, A.I. (2011). *Annu. Rep. Prog. Chem. Sect. C: Phys. Chem.* 107: 124–147.

**23** Galeev, T.R. and Boldyrev, A.I. (2013). Reference module in chemistry, molecular sciences and chemical engineering. In: *Comprehensive Inorganic Chemistry II (Second Edition)* (ed. J. Reedijk and K. Poeppelmeier), 245–275. Elsevier, Amsterdam.

**24** Romanescu, C., Galeev, T.R., Li, W.-L. et al. (2013). *Acc. Chem. Res.* 46: 350–358.

**25** Popov, I.A. and Boldyrev, A.I. (2014). *The Chemical Bond*, Chemical Bonding Across the Periodic Table (ed. G. Frenking and S. Shaik), 421–444. Weinheim: Wiley-VCH.

**26** Sergeeva, A.P., Popov, I.A., Piazza, Z.A. et al. (2014). *Acc. Chem. Res.* 47: 1349–1358.

**27** Mercero, M., Boldyrev, A.I., Merino, G., and Ugalde, J.M. (2015). *Chem. Soc. Rev.* 44: 6519–6534.

**28** Feixas, F., Matito, E., Poater, J., and Solà, M. (2013). *WIREs Comput. Mol. Sci.* 3: 105–122.

**29** Boldyrev, A.I. and Wang, L.S. (2016). *Phys. Chem. Chem. Phys.* 18: 11589–11605.

**30** Havenith, R.W.A., De Proft, F., Fowler, P.W., and Geerlings, P. (2005). *Chem. Phys. Lett.* 407: 391–396.

**31** Bishop, D.M., Chaillet, M., Larrieu, K., and Pouchan, C. (1984). *Mol. Phys.* 51: 179–183.

**32** Alexandrova, A.N. and Boldyrev, A.I. (2003). *J. Phys. Chem. A* 107: 554–560.

**33** Zubarev, D.Y. and Boldyrev, A.I. (2008). *Phys. Chem. Chem. Phys.* 10: 5207–5217.

**34** Foster, J.P. and Weinhold, F. (1980). *J. Am. Chem. Soc.* 102: 7211–7218.

**35** Martin, J.M.L., Francois, J.P., and Gijbels, R. (1992). *Chem. Phys. Lett.* 189: 529–536.

**36** Alexandrova, A.N., Boldyrev, A.I., Zhai, H.-J. et al. (2003). *J. Phys. Chem. A* 107: 1359–1369.

**37** Havenith, R.W.A., Fowler, P.W., and Steiner, E. (2002). *Chem. Eur. J.* 8: 1068–1073.

**38** Zhai, H.-J., Wang, L.S., Alexandrova, A.N., and Boldyrev, A.I. (2002). *J. Chem. Phys.* 117: 7917–7924.

**39** Kuznetsov, A.E., Birch, K.A., Boldyrev, A.I. et al. (2003). *Science* 300: 622–625.

**40** Li, X., Kuznetsov, A.E., Zhang, H.-F. et al. (2001). *Science* 291: 859–861.

**41** Zhai, H.J., Averkiev, B.B., Zubarev, D.Y. et al. (2007). *Angew. Chem. Int. Ed.* 46: 4277–4280.

**42** Averkiev, B.B. and Boldyrev, A.I. (2007). *J. Phys. Chem. A* 111: 12864–12866.
**43** Sergeeva, A.P. and Boldyrev, A.I. (2010). *Phys. Chem. Chem. Phys.* 12: 12050–12054.
**44** Murahashi, T., Inoue, R., Usui, K., and Ogoshi, S. (2009). *J. Am. Chem. Soc.* 131: 9888–9889.
**45** Tsipis, A.C., Kefalidis, C.E., and Tsipis, C.A. (2008). *J. Am. Chem. Soc.* 130: 9144–9155.

# 12

# The Distortivity of π-Electrons

*"True science teaches, above all, to doubt, and to be ignorant."*
*Miguel de Unamuno*

## 12.1 Introduction

Planarity and bond length equalization (BLE) are two structural features of benzene that are linked to its aromatic character. Fusion of two benzene rings to form naphthalene keeps the planarity but destroys the perfect BLE. More significant C—C bond length alternation (BLA) in the benzene ring is achieved when this ring is annelated with clamping groups such as the cyclopropa, cyclobuta, and cyclobutadieno clamps [1, 2]. However, calculated ring currents [1] and different electronic, magnetic, and geometric indices of aromaticity [3–5] indicate that the aromatic character of the benzene ring changes only slightly. These results, together with resonance energy calculations of several BLA deformed benzene rings [6–8], show that benzene can suffer significant C—C BLA without a substantial loss of its cyclic electron delocalization and its aromaticity [9]. Thus, the π-electron delocalization and the aromatic character of benzene is robust. Indeed, it resists not only BLA, but also large out-of-plane distortions [10–15], a fact that has been well proved both experimentally and theoretically.

Experimentally, the conformational ring flexibility of benzene is supported by the fact that deuterated benzene in its orthorhombic crystal structure suffers slight but nonnegligible distortions caused by crystal packing forces that change the symmetry of the molecule from $D_{6h}$ to $C_{3v}$ [16]. Moreover, the crystal structures of most acenes and their derivatives [17] as well as sterically overcrowded polycyclic aromatic hydrocarbons (PAHs) [18–20] show clear deviations from planarity. Additional support to the flexibility of the benzene ring comes

*Aromaticity and Antiaromaticity: Concepts and Applications*, First Edition.
Miquel Solà, Alexander I. Boldyrev, Michał K. Cyrański, Tadeusz M. Krygowski, and Gabriel Merino.

from [$n$]para- and metacyclophanes [10, 11] and pyrenophanes [12–14]. These species can undergo severe distortions of the aromatic ring(s) without significant loss of aromaticity in comparison with the undistorted benzene and pyrene molecules, respectively, as indicated by $^{1}$H nuclear magnetic resonance (NMR) signals, magnetic anisotropies, UV/Vis spectra, X-ray molecular structures, and theoretical calculations [14].

From a theoretical point of view, the flexibility of the benzene ring as well as the robustness of its π-electronic delocalization and aromatic character have been discussed in several works. The change in the aromatic character of the benzene ring due to BLA and BLE was analyzed using the harmonic oscillator model of aromaticity (HOMA) index and the nucleus-independent chemical shifts (NICSs) as indicators of aromaticity [21]. NICS values indicated slight changes in the aromatic character of benzene, while HOMA results predicted somewhat larger variations. Bader et al. [22] studied the effect of BLA and BLE distortions on the delocalization of π-electrons in benzene. They found that there is no significant change in the delocalization pattern due to BLE distortions, although the π-electronic delocalization between *para*-related carbon atoms is clearly reduced due to BLA deformations. Finally, different studies analyzed chair and boat distortions of the benzene ring [23–25] concluding that the benzene ring can suffer deviations from planarity up to 25° without a substantial loss of electron conjugation.

In summary, the aromatic character of the benzene rings remains quite unaltered despite substantial distortions. In this chapter, we discuss the particular features of the BLA distortion in benzene.

## 12.2 The Kekulean Distortion

Among the different distortions that a benzene ring can suffer, the Kekulean or BLA distortion of in-plane $b_{2u}$ symmetry that changes the $D_{6h}$ symmetry of benzene into a Kekulé-like $D_{3h}$ symmetry structure (Scheme 12.1) is particularly interesting for three reasons: (i) upon excitation to the first $^{1}B_{2u}$ excited state of benzene, there is an important (and unexpected) upshift of this $b_{2u}$ frequency mode; (ii) this $b_{2u}$ distortion localizes the π-electrons and, therefore, it represents one of the simplest way to reduce the aromaticity of the ring; and (iii) this $b_{2u}$

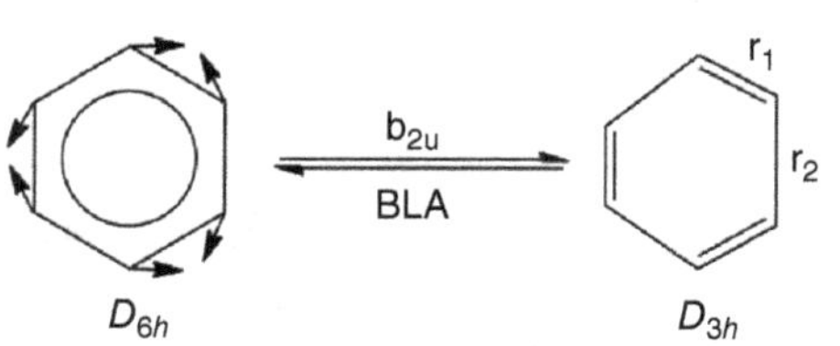


**Scheme 12.1** The Kekulé vibrational mode in benzene.

BLA vibrational mode breaks the maximum hardness principle (MHPs) and the minimum polarizability principle (MPPs). These three effects are discussed in more detail in the next subsections.

## 12.3 Frequencies of the Kekulé Vibrational Mode in Benzene

Berry [26] noted in 1961 that the frequency of the in-plane $b_{2u}$ vibration in benzene (Scheme 12.1) is surprisingly low as compared to that of other vibrational modes such as, for instance, the $a_{1g}$ breathing mode that shortens and elongates all C—C bonds, simultaneously keeping the BLE and $D_{6h}$ symmetry of the benzene ring. Experimental gas-phase two-photon spectroscopy measurements [27] provide a value of 1309 $cm^{-1}$ for this vibrational mode. However, assuming that both the σ- and π-electronic systems of benzene oppose resistance to this in-plane $b_{2u}$ distortion we would predict a larger frequency of about 1600 $cm^{-1}$ [28–31]. In addition, there is a remarkable up-shift of this low frequency from 1309 to 1570 $cm^{-1}$ upon excitation to the $1^1B_{2u}$ excited state [27, 32, 33]. This result, reinforced theoretically through coupled cluster calculations [34], is surprising in the context of a $\pi^* \leftarrow \pi$ transition from the $1^1A_{1g}$ ground state to the $1^1B_{2u}$ excited state. Considering that the π-system should be weakened in the $1^1B_{2u}$ excited state ($\pi^* \leftarrow \pi$ transition), we may anticipate a reduction in the frequency of this $b_{2u}$ vibrational mode upon excitation [35].

The solution to this paradox came from the work of Hiberty, Shaik, and coworkers [7, 31, 36–39], among others [26, 40, 41], who proved using different procedures that the π-electrons of benzene possess a distortive tendency away from the $D_{6h}$ symmetry structure (see next sections). Moving to the $1^1B_{2u}$ excited state reduces the π-distortive tendency of the $1^1A_{1g}$ ground state and, consequently, increases the frequency of the in-plane $b_{2u}$ vibration in benzene. Calculations of the second derivatives of orbital energies with respect to this $b_{2u}$ normal coordinate performed by Gobbi et al. [42] provided further support to this statement. According to these authors, the second derivative of the total energy with respect to normal coordinates can be estimated as the sum of the second derivatives of the orbital energies, thus allowing for a σ/π-separation of the force constants [42]. For the BLA $b_{2u}$ vibration mode of benzene, it was found that the π-force constant in the $1^1A_{1g}$ ground state is negative while the σ-force constant is positive [42].

An alternative physical basis for the large increase in the frequency of this $b_{2u}$ mode when going from the $1^1A_{1g}$ to the $1^1B_{2u}$ states of benzene is based on the pseudo-Jahn-Teller (PJT) effect [43–46], that is, the vibronic coupling between the ground and lowest-lying $^1B_{2u}$ excited state through the in-plane CC stretching modes [32, 47–49]. The PJT effect is usually formulated using second-order perturbation theory [43], where the energy is expanded about the point of minimum energy $Q_0$ as follows:

$$E = E_0 + Q\langle\Psi_0|\frac{\partial U}{\partial Q}|\Psi_0\rangle + \frac{Q^2}{2}\langle\Psi_0|\frac{\partial^2 U}{\partial Q^2}|\Psi_0\rangle + \sum_k \frac{\left[Q\langle\Psi_0|\frac{\partial U}{\partial Q}|\Psi_k\rangle\right]^2}{E_0 - E_k} + \ldots \tag{12.1}$$

In Eq. (12.1), $Q$ is a nuclear displacement coordinate, and $U$ the nuclear-electronic and nuclear-nuclear parts of the potential energy. If the original wave function $\Psi_0$ is nondegenerate, the linear term is zero at the minimum geometry, and the sum of the two second-order terms gives the force constant of the vibration $Q$. At the minimum, the first second-order term is positive. The last term corresponds to the PJT coupling. It only appears for nontotally symmetric vibrations Q with the same symmetry as the product $\Psi_0\Psi_k$ (in our example, the $\Psi_0\Psi_1$ and the vibrational mode have the same $b_{2u}$ symmetry). Its value is negative for the ground state of the molecule, because $E_0 < E_1$, and therefore it lowers the force constant of the ground state. There is an inverse effect on the excited state $\Psi_1$, leading in principle to an increase of the force constant. The effect of the coupling on the energy gap between the two states is shown in Figure 12.1, where the relevant

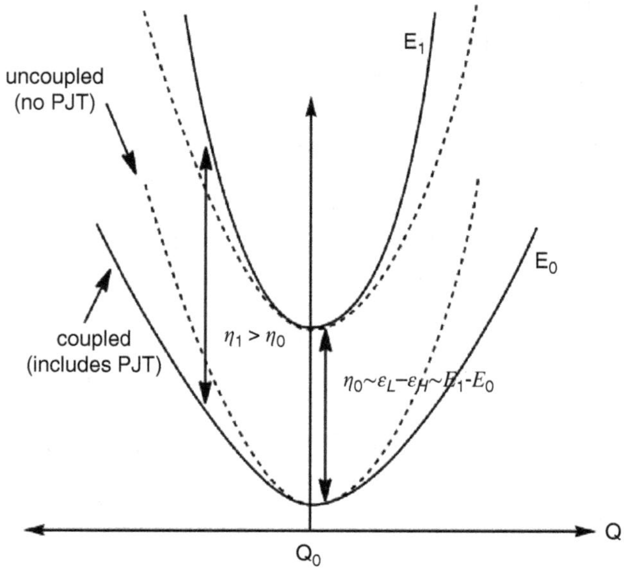


**Figure 12.1** Representation of the pseudo Jahn-Teller (PJT) effect between $\Psi_0$ and $\Psi_1$ along a nontotally symmetric vibrational mode Q. Dashed parabolas: "uncoupled" potential energy surfaces. Full-line parabolas: "coupled" potential energy surfaces. Source: Reproduced with permission from Ref. [49]. Copyright © 2006, American Chemical Society.

excited state is $\Psi_1$. Further couplings between $\Psi_0$ and higher-lying excited states of the symmetry of $\Psi_1$, as well as between $\Psi_1$ and other excited states of the symmetry of $\Psi_0$ may have a role, but the most important is, generally, the $\Psi_0\Psi_1$ coupling. As a result of the coupling, there is a lowering of the force constant of the ground state and an increase of the force constant of the excited state $\Psi_1$.

## 12.4 Changes in Aromaticity in the Kekulean Distortion

For the BLA distortion, Table 12.1 lists the changes in aromaticity after deformation with consecutive steps of 0.05 Å ($\Delta R = r_2 - r_1$, Scheme 12.1) without further reoptimization of the distorted benzene ring geometry [3, 4]. The change in aromaticity along this distortion was studied with the *para*-delocalization index (PDI), the aromatic fluctuation index (FLU), the multicenter MCI, $I_{ring}$, $I_{NB}$, and $I_{NG}$ indices, the harmonic oscillator model of aromaticity (HOMA), and the nucleus-independent chemical shift (NICS) in the form of NICS(0), NICS(1), and NICS(1)$_{zz}$ indicators of aromaticity (for definitions of the indicators, see Chapters 5–8). For the series of indices used, we have that the more negative the NICS, the lower the FLU$^{1/2}$ index, and the higher the HOMA, PDI, MCI, $I_{ring}$, $I_{NG}$, and $I_{NB}$ values, the more aromatic the rings are. As expected from the localization of the π-electrons during the BLA distortion, all descriptors of aromaticity indicate a reduction in the aromatic character of the benzene ring due to BLA distortion. With the exception of HOMA, descriptors of aromaticity show a minor decrease of aromaticity for distortions up to $\Delta R = 0.1$ Å. From comparison among the different indices, it is likely that the HOMA index overestimates the

**Table 12.1** PDI, FLU$^{1/2}$, MCI, $I_{NB}$, $I_{ring}$, $I_{NG}$, HOMA, NICS(0), NICS(1), and NICS(1)$_{zz}$ results for the BLA distortion of the benzene ring.[a)] NICS in ppm and PDI, MCI, $I_{NB}$, $I_{ring}$, and $I_{NG}$ in electrons.

| $\Delta R$[b)] | PDI | FLU$^{1/2}$ | MCI | $I_{NB}$ | $I_{ring}$ | $I_{NG}$ | HOMA | NICS (0) | NICS (1) | NICS(1)$_{zz}$ |
|---|---|---|---|---|---|---|---|---|---|---|
| 0 | 0.103 | 0.000 | 0.072 | 0.041 | 0.048 | 0.041 | 0.989 | −8.05 | −10.22 | −29.25 |
| 0.05 | 0.102 | 0.045 | 0.070 | 0.041 | 0.047 | 0.041 | 0.827 | −7.91 | −10.07 | −28.79 |
| 0.1 | 0.099 | 0.088 | 0.064 | 0.040 | 0.044 | 0.041 | 0.344 | −7.54 | −9.65 | −27.47 |
| 0.15 | 0.094 | 0.130 | 0.056 | 0.039 | 0.039 | 0.040 | −0.461 | −7.00 | −9.02 | −25.48 |
| 0.2 | 0.088 | 0.168 | 0.046 | 0.038 | 0.033 | 0.039 | −1.588 | −6.37 | −8.26 | −23.08 |
| 0.25 | 0.082 | 0.204 | 0.037 | 0.037 | 0.028 | 0.038 | −3.038 | −5.78 | −7.45 | −20.52 |

a) Data from Ref. [4].
b) Difference between the long and short bond in $b_{2u}$ distorted benzene ($\Delta R = r_2 - r_1$ in Å, Scheme 12.1).

loss of aromaticity for the BLA distortion. Interestingly, even after distortions of the aromatic ring as large as $\Delta R = 0.25$ Å, some aromatic character is preserved. This result conforms the already mentioned robustness of the π-electronic delocalization and aromaticity of the benzene ring.

## 12.5 The Maximum Hardness and Minimum Polarizability Principles

Two of the most important principles of reactivity derived in the framework of the conceptual Density Functional Theory [50] are the MHP and the MPP. The MHP [51] states that, at a given temperature and external potential, molecular systems evolve toward maximum hardness, where the hardness is given by:

$$\eta = \left(\frac{\partial^2 E}{\partial N^2}\right)_{v(\vec{r}),T} \tag{12.2}$$

$N$ is the number of electrons and $\nu(\vec{r})$ the potential of the nuclei plus any other present external potential. Calculations of hardness usually involve the following approximations:

$$\eta \cong IP - EA \cong \varepsilon_L - \varepsilon_H \tag{12.3}$$

That is, the hardness is approximated as the difference between ionization potential (IP) and electron affinity (EA) of the molecule, and these quantities are, in turn, estimated as the negative of the energies of the highest occupied molecular orbital (HOMO, $\varepsilon_H$) and lowest unoccupied molecular orbital (LUMO, $\varepsilon_L$), respectively, following Koopmans' theorem [52]. A formal proof of the MHP was given by Parr and Chattaraj [53] under the constraints that the chemical potential ($\mu$) and the so-called external potential ($\nu[\vec{r}]$) must remain constant upon distortion of the molecular structure. Here $\nu(\vec{r})$ is the potential acting on an electron at $\vec{r}$ due to the nuclear attraction plus such other external forces as may be present. These are two severe constraints that are usually not fulfilled. On the other hand, the MPP was formulated on the basis of the MHP and an inverse relationship between hardness and polarizability ($\alpha$) [54]. This principle states that the natural direction of evolution of any system is towards a state of minimum polarizability.

Consider now a molecule in their equilibrium geometry and make a displacement from equilibrium along a nontotally symmetric normal mode. For this particular molecular motion and using symmetry arguments, Pearson and Palke [55] showed that the values of the average external potential, hardness, polarizability, and chemical potential for positive deviation will be the same as those for the negative deviation from equilibrium. Then, if Q represents a nontotally

**Table 12.2** Chemical polarizability ($\alpha$) and hardness ($\eta$) for the molecular distortion of benzene along the $b_{2u}$ normal vibrational modes depicted in Scheme 12.1 calculated at the Hartree-Fock (HF) and B3LYP levels of theory with the aug-cc-pVTZ basis set.[a),b)]

| Method | Property | Equilibrium | $b_{2u} \pm 0.04$[c)] |
|---|---|---|---|
| HF | $\alpha$ | 66.74075 | 66.74038 |
| | $\eta$ | 0.368479 | 0.368495 |
| B3LYP | $\alpha$ | 69.33744 | 69.32336 |
| | $\eta$ | 0.242999 | 0.243114 |

a) From Ref. [61].
b) $\eta$ calculated using Eq. (12.3) using HOMO and LUMO energies. $\alpha$ and $\eta$ are given in au.
c) The distortion along the $b_{2u}$ mode is given in bohr.

symmetric normal mode coordinate, it follows that $(\delta\mu/\delta Q) = 0$ and $(\delta v_{en}/\delta Q) = 0$ at the equilibrium geometry, $v_{en}$ being the electron-nuclear attraction potential. Hence, the chemical and external potential, $\nu(\vec{r})$, are roughly constant for small distortions along nontotally symmetric normal modes, thus nearly following the two mentioned conditions of Parr and Chattaraj [53]. As a consequence, the MHP and MPP are expected to be obeyed for nontotally symmetric vibrations, as confirmed by most numerical calculations of hardness and polarizability along the nontotally symmetric normal modes [56–59].

Interestingly, the in-plane BLA $b_{2u}$ vibrational mode in benzene represents one of the few examples [60–62] of a nontotally symmetric normal mode that disobeys the MPP and the MHP. Results of hardness and polarizability for this vibrational mode are given in Table 12.2 at different levels of theory. Results in Table 12.2 show that the hardness increases and the polarizability decreases when going from the equilibrium structure to a $b_{2u}$ distorted benzene. Figure 12.1 helps to explain the reason of this behavior. As can be seen in Figure 12.1, the hardness of the equilibrium structure ($\eta_0$), estimated as the energy difference between the ground and the excited state, is smaller than that of a $b_{2u}$ distorted structure ($\eta_1$, $\eta_0 < \eta_1$). Therefore, the PJT helps to understand the origin of the breakdown of the MHP and MPP for this particular vibrational mode [63].

## 12.6 The Distortive Nature of $\pi$-Electrons

As discussed in Section 12.5, the Kekulean or BLA distortion of in-plane $b_{2u}$ symmetry that changes the $D_{6h}$ of benzene into a $D_{3h}$ symmetry structure

(Scheme 12.1) has several interesting features that can be explained if one takes into account the distortive nature of π-electrons. This was proposed first by Longuet-Higgins and Salem [64] when treating the BLA in long polyenes and cyclic annulenes. The same concept was also used by Berry [26] in an attempt to explain the observed increased frequency of the $b_{2u}$ Kekulé vibrational mode when going from the ground $^1A_{1g}$ to the first $^1B_{2u}$ excited state [27, 32, 33]. This idea was reinforced later on by the work of Haas and Zilberg [65, 66] and the theory about the distortive character of π-electrons in aromatic systems gained support over the years, especially thanks to the work of Hiberty, Shaik, and coworkers [7, 36, 37, 66, 67], and others [42, 49]. More recently, Pierrefixe and Bickelhaupt [68] provided further support to the idea that the regular geometry of benzene is due to the σ- and not to the π-electrons, the latter having a slight tendency to localize double-bonds [68–70]. In the next paragraphs, we discuss these two approaches.

The approach by Shaik, Hiberty, et al. [7, 31, 38, 66] is mainly based on the Valence Bond (VB) method. Scheme 12.2 shows the distortion $D_{6h}$ benzene (with r = 1.40 Å) into a $D_{3h}$ benzene (with $r_1$ = 1.4627 and $r_2$ = 1.34 Å) with the same total nuclear repulsion between carbon atoms for the two different structures. In the ground state, the distortion energy of 7.2 kcal/mol (at the π-CI/6-31G level) contains σ- and π-contributions. In the lowest-lying septuplet state and in the quasiclassical state (QS), π-delocalization is turned off during the distortion. In the high spin (HS) septuplet state, with all π-electrons unpaired, π-bonding and π-delocalization are absent. In this HS state, the benzene $D_{6h}$ to $D_{3h}$ distortion needs 14.5 kcal/mol. The QC state corresponds to a situation in which the π-electrons are unpaired and uncoupled using the VB method. In this state, in which the π-bonding and π-delocalization is turned off, the distortion requires 12.5 kcal/mol. Therefore, Shaik, Hiberty, et al. [7, 71] concluded that the energy required to distort the $D_{6h}$ benzene into a $D_{3h}$ benzene decreases significantly

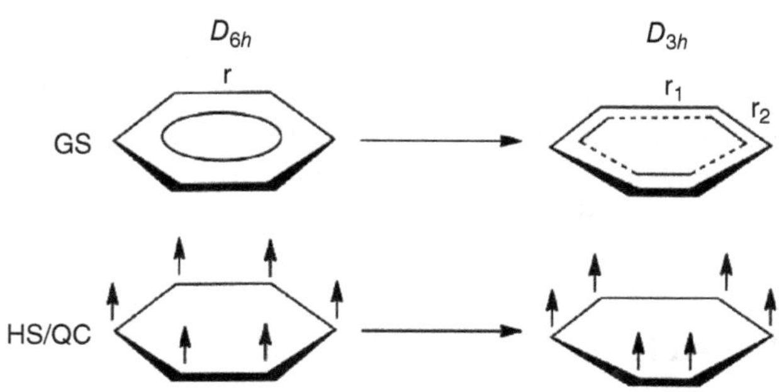


**Scheme 12.2** Benzene distorted from $D_{6h}$ to $D_{3h}$ in its ground state (GS) and in its lowest-lying septuplet state (HS) or in a quasiclassical state (QS).

when both π-bonding and π-delocalization are present. Therefore, this result gave support to the distortive tendency of π-electrons in benzene.

With the VB approach, it is possible to compute the energy of each Kekulé resonance structure of benzene ($K_1$ and $K_2$ in Scheme 12.3) as well as of the combination of the two resonance structures to give the bonding $\Psi_0$ ($K_1 + K_2$, $A_{1g}$) and antibonding $\Psi_1$ ($K_1 - K_2$, $B_{2u}$) states. The energies of the two Kekulé resonance structures along the $b_{2u}$ distortion mode together with the resulting ground and excited states are shown in the VB correlation diagram of Figure 12.2.

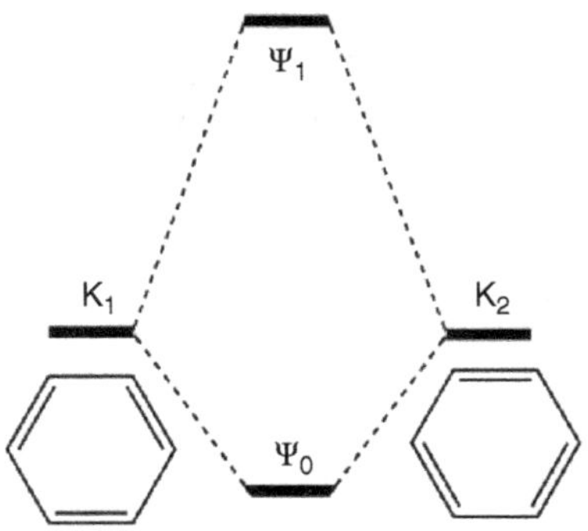


**Scheme 12.3** Generation of the $A_{1g}$ ground state and $B_{2u}$ excited state in benzene by combination of the Kekulé resonance structures $K_1$ and $K_2$.

The avoided crossing of the two Kekulé structures leads to their resonance mixing and formation of the $\Psi_0$ and $\Psi_1$ states. As it is apparent from Figure 12.2a, the frequency of this $b_{2u}$ mode is larger in the $\Psi_1$ than in the $\Psi_0$ state as discussed in the previous subsection. One can also calculate the energies of the $K_1$, $K_2$, $\Psi_0$, and $\Psi_1$ with the QC approach in order to disconnect the π-bonding and π-delocalization. The difference between the complete and QC approaches gives the π-contribution, so Figure 12.2b shows the π-energy along the same $b_{2u}$ distortion. The VB diagram clearly shows that the π-electrons in the ground state ($^1A_{1g}$) are distortive, whereas in the $1^1B_{2u}$ excited state they favor BLE and $D_{6h}$ symmetry.

Using the MO theory, different approaches were used to prove that the π-system in benzene is distortive [68, 72]. Here, we refer to the procedure followed by Pierrefixe and Bickelhaupt [68]. In this procedure, the authors used the energy decomposition analysis (EDA) of Morokuma-Ziegler [73–75]. The EDA is a fragment-based analysis [76–79]. In the case of benzene, we can take the two identical fragments formed by two $(CH)_3^{9\cdot}$ moieties with six unpaired σ-electrons and three unpaired π-electrons as depicted in Scheme 12.4.

The interaction of the two fragments, one with nine α-electrons and the other with nine β-electrons, generates the six σ 2-electron 2-center (2c-2e) bonds and the delocalized π-system. The interaction energy $\Delta E_{\text{int}}$ corresponds to the energy

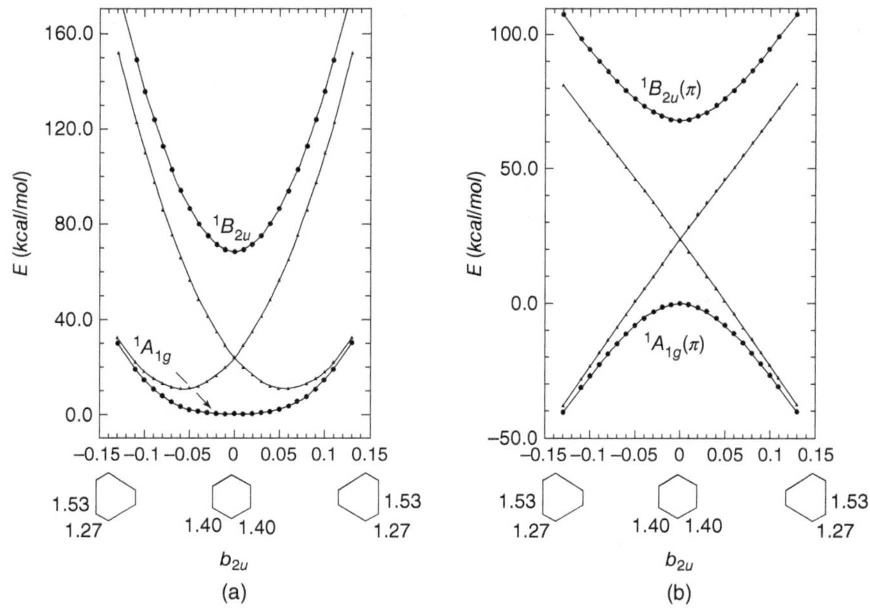


**Figure 12.2** VB correlation diagrams with the avoided crossing for benzene along the $b_{2u}$ mode for (a) the full VB calculation and (b) only the π-electron system. Regular lines correspond to the energies of the Kekulé resonance structures, and bold lines refer to the resulting ground and excited states. Energies in kcal/mol and bond lengths in Å. Source: Reproduced with permission from Ref. [7]. Copyright © 2001, American Chemical Society.

change when the two separated fragments at the geometry they have in optimized benzene are combined to form benzene. This $\Delta E_{\text{int}}$ can be decomposed into:

$$\Delta E_{\text{int}} = \Delta V_{\text{elstat}} + \Delta E_{\text{Pauli}} + \Delta E_{\text{oi}} \tag{12.4}$$

In Eq. (12.4), $\Delta V_{\text{elstat}}$ corresponds to the classical electrostatic interaction between the unperturbed charge distributions of the fragments and is usually attractive. The Pauli-repulsion $\Delta E_{\text{Pauli}}$ comprises the destabilizing interactions between occupied orbitals and is responsible for the steric repulsions. The orbital interaction term, $\Delta E_{\text{oi}}$, accounts for electron-pair bonding (interactions between half-filled orbitals in the two fragments), charge transfer (donor–acceptor interactions between occupied orbitals on one moiety with unoccupied orbitals of the other) and polarization (empty/occupied orbital mixing on one fragment due to the presence of another fragment). Moreover, using the extended transition state (ETS) scheme [80], the $\Delta E_{\text{oi}}$ term can be divided into the contributions of orbitals with different symmetry. For planar systems, like the ones under analysis in the present work, the σ/π separation is possible:

$$\Delta E_{\text{oi}} = \Delta E_{\sigma} + \Delta E_{\pi} \tag{12.5}$$

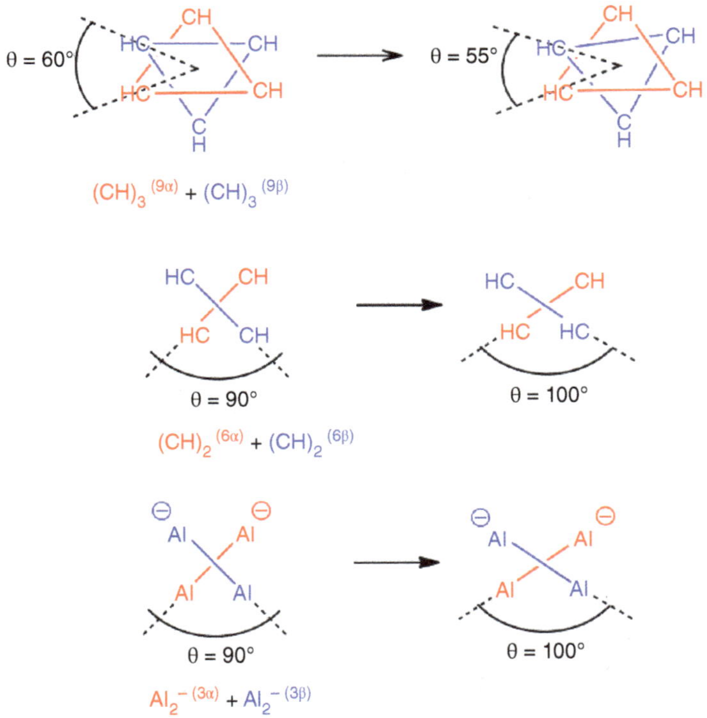


**Scheme 12.4** The two fragments (red and blue) considered in the energy decomposition analysis along the bond length alternation distortion of benzene (top), cyclobutadiene (center), and $Al_4^{2-}$ (bottom). The degree of bond length alternation is controlled by the angle θ.

Because by construction there are no doubly occupied π-orbitals in the fragments, all $\Delta E_{\text{Pauli}}$ is of σ-origin and one can write:

$$\Delta E_{\text{int}} = \Delta V_{\text{elstat}} + \text{total}\,\sigma + \text{total}\,\pi \tag{12.6}$$

where "total σ" = $\Delta E_\sigma + \Delta E_{\text{Pauli}}$ and "total π" = $\Delta E_\pi$.

The result of the analysis is shown in Figure 12.3. As can be seen in Figure 12.3, the energy of $D_{6h}$ benzene raises on distortion ($\Delta E_{\text{int}}$ is more positive). However, the π-system and the electrostatic interaction favor the $D_{3h}$ distorted structure. Only the σ-system opposes the distortion and prefers the BLE structure. Therefore, as found with the VB approach, the MO method also finds that the π-electrons in benzene are distortive and promote the localization of the double bonds.

For butadiene, the trends of the $\Delta V_{\text{elstat}}$, "total σ", and "total π" terms are similar. Now the fragments taken for the analysis are $(CH)_2^{6\cdot}$ moieties in the diagonal of

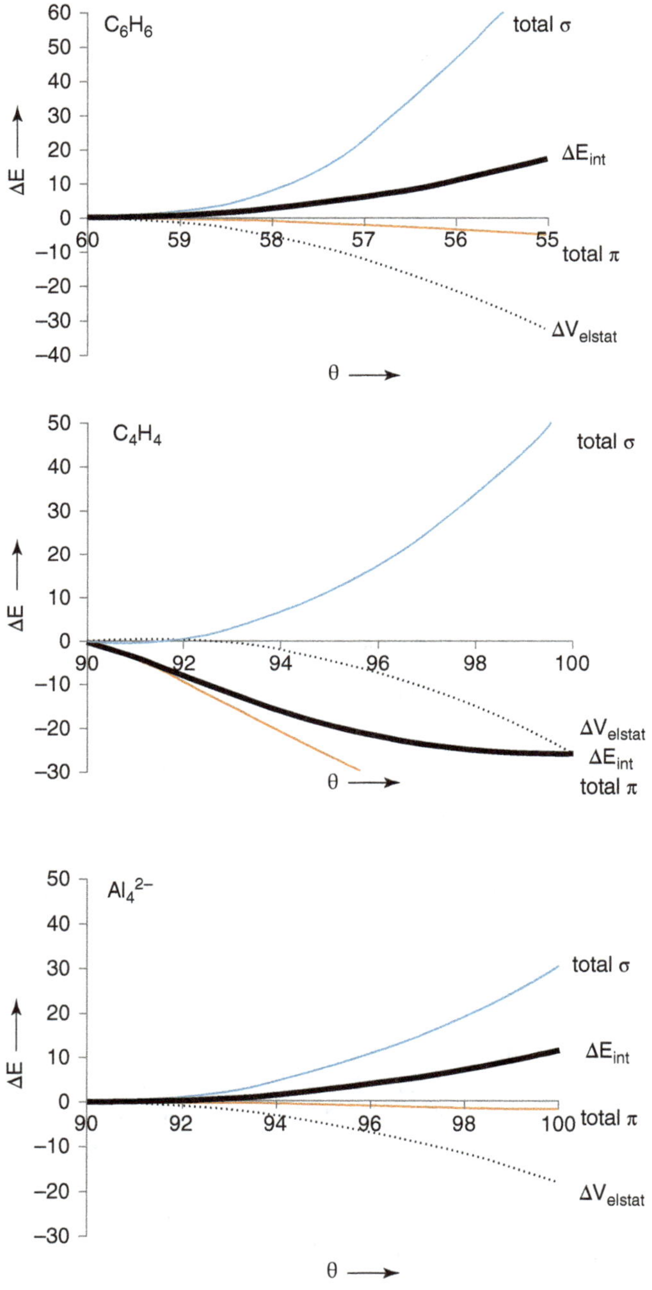


**Figure 12.3** Energy decomposition analysis (kcal/mol) of benzene (top), cyclobutadiene (center), and $Al_4^{2-}$ (bottom) constructed from two equivalent rigid fragments as a function of the distortion mode (degrees) from delocalized to localized situation. Energies in kcal/mol and angles in degrees. Source: Adapted from Refs. [68, 81].

$C_4H_4$ (see Scheme 12.4). However, for this antiaromatic molecule, the $\pi$-electrons are much more distortive than in benzene and, as a result, $\Delta E_{int}$ favors BLA ("energy goes down"), that is, localization of the double bonds. The reason for the higher distortive character of the $\pi$-electrons in butadiene as compared to benzene can be understood by looking at Scheme 12.5. In the BLA distortion of benzene from $D_{6h}$ to $D_{3h}$ there is some minor stabilization of the $\pi$-electrons. During the distortion, the $\pi$-MOs gain and loose bonding overlap and the net effect is a minor gain in bonding. However, in cyclobutadiene, when moving from $D_{4h}$ to $D_{2h}$, one of the nonbonding e″ degenerate orbitals gains stabilization in every C—C bond whereas the opposite situation is found for the other nonbonding e″ degenerate orbital. As a whole, the cyclobutadiene is stabilized markedly along the BLA distortion. So, basically, the $\pi$-system of the aromatic and antiaromatic molecules behave similarly, both are distortive, but the $\pi$-system of the antiaromatic molecules has a larger distortive character than that of the aromatic molecules.

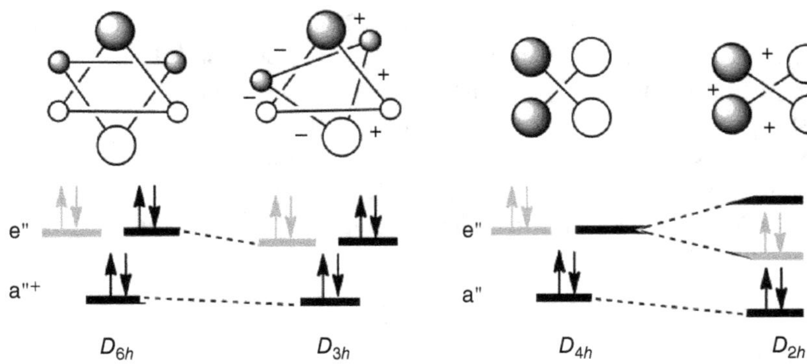


**Scheme 12.5** Effect of BLA distortion on the $\pi$-molecular orbitals of benzene (left) and cyclobutadiene (right). The molecular orbital plotted at top refers to the grey level. Adapted from Scheme 2 of Ref. [68].

The $Al_4^{2-}$ cluster of $D_{4h}$ symmetry discussed in Chapter 11 is a doubly $\sigma$- and $\pi$-aromatic species. In the case of the $\sigma$-system, it has both $\sigma$-radial and $\sigma$-tangential aromaticities. The result of the EDA of this system [81] taking equivalent $Al_2^-(\alpha\alpha\alpha)$ and $Al_2^-(\beta\beta\beta)$ fragments (see Scheme 12.4) is shown in Figure 12.3. The interaction energy of the two fragments to give $Al_4^{2-}$, $\Delta E_{int}$, becomes less stabilizing along the distortion mode from $D_{4h}$ to $D_{2h}$, that is, $\Delta E_{int}$ is against localization. This behavior originates from the $\sigma$-system. The $\pi$-electrons prefer localization, but only very slightly in this case. The $\Delta V_{elstat}$ term also favors localization. However, all together, the $\sigma$-electrons dominate and the result is the $D_{4h}$ equilibrium geometry of $Al_4^{2-}$ with delocalized $\sigma$-radial, $\sigma$-tangential, and $\pi$-bonds.

## 12.7 Conclusions

As we have seen in the previous paragraphs, the distortive character of the π-system is widely accepted and it has been proved using arguments coming from VB and Molecular Orbital (MO) theories. This distortive nature of π-electrons leads to, at least, two paradoxes: (i) aromaticity in benzene comes from the π-electron system, but this system goes against one of the most fundamental properties of aromatic benzene, which is BLE; and (ii) if the π-system is destabilized in the $D_{6h}$ structure as compared to the Kekulé-like $D_{3h}$ symmetry structure, we would expect high reactivity in benzene, but this molecule is quite unreactive in its ground state. As to the first paradox, it is important to note that, as pointed out by Shaik, Hiberty, and coworkers [7], the π-distortivity and resonance stabilization are two coexisting features of classical aromatic species. This means that the aromatic molecule gains stability through resonance stabilization despite of π-distortivity. As to the second, we have shown that the HOMO of benzene is stabilized by a Kekulé distortion (Scheme 12.5). This stabilization is responsible for the breakdown of the MHP for this nonsymmetry distortion [61] and should make $D_{6h}$ benzene more reactive than a hypothetical $D_{3h}$ one. However, the stabilization of the HOMO in the distortion is minor. Moreover, the reactivity depends not only on the energy of the frontier MOs but also on the overlap between the frontier MOs of the reactants. It is likely that more delocalized π-orbitals in $D_{6h}$ benzene leads to poorer overlaps and, consequently, to lower reactivity.

## References


1 Soncini, A., Havenith, R.W.A., Fowler, P.W. et al. (2002). *J. Org. Chem.* 67: 4753–4758.

2 Frank, N.L., Baldridge, K.K., and Siegel, J.S. (1995). *J. Am. Chem. Soc.* 117: 2102–2103.

3 Feixas, F., Matito, E., Poater, J., and Solà, M. (2007). *J. Phys. Chem. A* 111: 4513–4521.

4 Feixas, F., Matito, E., Poater, J., and Solà, M. (2008). *J. Comput. Chem.* 29: 1543–1554.

5 Solà, M., Feixas, F., Jiménez-Halla, J.O.C. et al. (2010). *Symmetry* 2: 1156–1179.

6 Mo, Y., Wu, W., and Zhang, Q. (1994). *J. Phys. Chem.* 98: 10048–10053.

7 Shaik, S.S., Shurki, A., Danovich, D., and Hiberty, P.C. (2001). *Chem. Rev.* 101: 1501–1539.

8 Dijkstra, F., van Lenthe, J.H., Havenith, R.W.A., and Jenneskens, L.W. (2003). *Int. J. Quantum Chem.* 91: 566–574.

9 Fleischer, U., Kutzelnigg, W., Lazzeretti, P., and Mühlenkamp, V. (1994). *J. Am. Chem. Soc.* 116: 5298–5306.

10 Kraakman, P.A., Valk, J.M., Niederlander, H.A.G. et al. (1990). *J. Am. Chem. Soc.* 112: 6638–6646.

11 Gready, J.E., Hambley, T.W., Kakiuchi, K. et al. (1990). *J. Am. Chem. Soc.* 112: 7537–7540.

12 Bodwell, G.J., Bridson, J.N., Houghton, T.J. et al. (1996). *Angew. Chem. Int. Ed. Engl.* 35: 1320–1321.

13 Bodwell, G.J., Bridson, J.N., Houghton, T.J. et al. (1999). *Chem. Eur. J.* 5: 1823–1827.

14 Bodwell, G.J., Bridson, J.N., Cyrański, M.K. et al. (2003). *J. Org. Chem.* 68: 2089–2098.

15 Dobrowolski, M.A., Cyrański, M.K., Merner, B.L. et al. (2008). *J. Org. Chem.* 73: 8001–8009.

16 Jeffrey, G.A., Ruble, J.R., McMullan, R.K., and Pople, J.A. (1987). *Proc. R. Soc. London, Ser. A* 414: 47–57.

17 Pascal, R.A. Jr., (2006). *Chem. Rev.* 106: 4809–4819.

18 Moran, D., Stahl, F., Bettinger, H.F. et al. (2003). *J. Am. Chem. Soc.* 125: 6746–6752.

19 Boschi, R., Clar, E., and Schmidt, W. (1974). *J. Chem. Phys.* 60: 4406–4418.

20 Poater, J., Visser, R., Solà, M., and Bickelhaupt, F.M. (2007). *J. Org. Chem.* 72: 1134–1142.

21 Cyrański, M.K. and Krygowski, T.M. (1999). *Tetrahedron* 55: 6205–6210.

22 Bader, R.F.W., Streitwieser, A., Neuhaus, A. et al. (1996). *J. Am. Chem. Soc.* 118: 4959–4965.

23 Dijkstra, F. and van Lenthe, J.H. (1999). *Int. J. Quantum Chem.* 74: 213–221.

24 Jenneskens, L.W., Vaneenige, E.N., and Louwen, J.N. (1992). *New J. Chem.* 16: 775–779.

25 Datta, A. and Pati, S.K. (2006). *Chem. Phys. Lett.* 433: 67–70.

26 Berry, R.S. (1961). *J. Chem. Phys.* 35: 2253–2254.

27 Wunsch, L., Neusser, H.J., and Schlag, E.W. (1975). *Chem. Phys. Lett.* 31: 433–440.

28 Califano, S. and Crawford, B. (1960). *Spectrochim. Acta* 16: 889–899.

29 Scherer, J.R. and Overend, J. (1961). *Spectrochim. Acta* 17: 719–730.

30 Scherer, J.R. (1962). *J. Chem. Phys.* 36: 3308–3321.

31 Hiberty, P.C. and Shaik, S.S. (2004). *Phys. Chem. Chem. Phys.* 6: 224–231.

32 Mikami, N. and Ito, M. (1976). *J. Chem. Phys.* 64: 3077–3078.

33 Wunsch, L., Metz, F., Neusser, H.J., and Schlag, E.W. (1977). *J. Chem. Phys.* 66: 386–400.

34 Christiansen, O., Stanton, J.F., and Gauss, J. (1998). *J. Chem. Phys.* 108: 3987–4001.

35 Robey, M.J. and Schlag, E.W. (1977). *J. Chem. Phys.* 67: 2775–2781.
36 Shaik, S.S. and Hiberty, P.C. (1985). *J. Am. Chem. Soc.* 107: 3089–3095.
37 Shaik, S.S., Hiberty, P.C., Le Four, J.-M., and Ohanessian, G. (1987). *J. Am. Chem. Soc.* 109: 363–374.
38 Hiberty, P.C., Danovich, D., Shurki, A., and Shaik, S. (1995). *J. Am. Chem. Soc.* 117: 7760–7768.
39 Jug, K., Hiberty, P.C., and Shaik, S.S. (2001). *Chem. Rev.* 101: 1477–1500.
40 Stanger, A. and Tkachenko, E. (2001). *J. Comput. Chem.* 22: 1377–1386.
41 Kovacevic, B., Baric, D., Maksic, Z.B., and Müller, T. (2004). *ChemPhysChem* 5: 1352–1364.
42 Gobbi, A., Yamaguchi, Y., Frenking, G., and Shaefer, H.F. III, (1995). *Chem. Phys. Lett.* 244: 27–31.
43 Pearson, R.G. (1969). *J. Am. Chem. Soc.* 91: 4947–4955.
44 Bersuker, I.B. (2001). *Chem. Rev.* 101: 1067–1114.
45 Bersuker, I.B., Balabanov, N.B., Pekker, D., and Boggs, J.E. (2002). *J. Chem. Phys.* 117: 10478–10486.
46 Bearpark, M.J., Blancafort, L., and Robb, M.A. (2002). *Mol. Phys.* 100: 1735–1739.
47 Metz, F., Robey, M.J., Schlag, E.W., and Dörr, F. (1977). *Chem. Phys. Lett.* 51: 8–12.
48 Orlandi, G. and Zerbetto, F. (1986). *Chem. Phys. Lett.* 131: 409–413.
49 Blancafort, L. and Solà, M. (2006). *J. Phys. Chem. A* 110: 11219–11222.
50 Geerlings, P., De Proft, F., and Langenaeker, W. (2003). *Chem. Rev.* 103: 1793–1873.
51 Pearson, R.G. (1999). *J. Chem. Educ.* 76: 267–275.
52 Koopmans, T. (1933). *Physica (Utrecht)* 1: 104–137.
53 Parr, R.G. and Chattaraj, P.K. (1991). *J. Am. Chem. Soc.* 113: 1854–1855.
54 Ghanty, T.K. and Ghosh, S.K. (1993). *J. Phys. Chem.* 97: 4951–4953.
55 Pearson, R.G. and Palke, W.E. (1992). *J. Phys. Chem.* 96: 3283–3285.
56 Pearson, R.G. (1997). *Chemical Hardness: Applications from Molecules to Solids.* Oxford: Wiley-VCH.
57 Pal, S., Naval, N., and Roy, R. (1993). *J. Phys. Chem.* 97: 4404–4406.
58 Makov, G. (1995). *J. Phys. Chem.* 99: 9337–9339.
59 Chattaraj, P.K., Fuentealba, P., Jaque, P., and Toro-Labbé, A. (1999). *J. Phys. Chem. A* 103: 9307–9312.
60 Torrent-Sucarrat, M., Luis, J.M., Duran, M., and Solà, M. (2001). *J. Am. Chem. Soc.* 123: 7951–7952.
61 Torrent-Sucarrat, M., Luis, J.M., Duran, M., and Solà, M. (2002). *J. Chem. Phys.* 117: 10651–10570.
62 Torrent-Sucarrat, M., Duran, M., Luis, J.M., and Solà, M. (2005). *J. Phys. Chem. A* 109: 615–621.

**63** Blancafort, L., Torrent-Sucarrat, M., Luis, J.M. et al. (2003). *J. Phys. Chem. A* 38: 7337–7339.

**64** Longuet-Higgins, H.C. and Salem, L. (1959). *Proc. Roy. Soc. London, Ser. A* 251: 172–185.

**65** Haas, Y. and Zilberg, S. (1995). *J. Am. Chem. Soc.* 117: 5387–5388.

**66** Shaik, S., Zilberg, S., and Haas, Y. (1996). *Acc. Chem. Res.* 29: 211–218.

**67** Shaik, S., Shurki, A., Danovich, D., and Hiberty, P.C. (1996). *J. Am. Chem. Soc.* 118: 666–671.

**68** Pierrefixe, S.C.A.H. and Bickelhaupt, F.M. (2007). *Chem. Eur. J.* 13: 6321–6328.

**69** Pierrefixe, S.C.A.H. and Bickelhaupt, F.M. (2008). *J. Phys. Chem. A* 112: 12816–12822.

**70** Pierrefixe, S.C.A.H. and Bickelhaupt, F.M. (2008). *Aust. J. Chem.* 61: 209–215.

**71** Shaik, S.S., Shurki, A., Danovich, D., and Hiberty, P.C. (1997). *J. Mol. Struct. THEOCHEM* 389-399: 155–167.

**72** Fernández, I. and Frenking, G. (2007). *Faraday Discuss.* 135: 403–421.

**73** Kitaura, K. and Morokuma, K. (1976). *Int. J. Quantum Chem.* 10: 325–340.

**74** Morokuma, K. (1977). *Acc. Chem. Res.* 10: 294–300.

**75** Ziegler, T. and Rauk, A. (1977). *Theor. Chim. Acta* 46: 1–10.

**76** Bickelhaupt, F.M. and Baerends, E.J. (2000). *Reviews in Computational Chemistry*, vol. 15 (ed. K.B. Lipkowitz and D.B. Boyd), 1–86. New York: Wiley-VCH.

**77** te Velde, G., Bickelhaupt, F.M., Baerends, E.J. et al. (2001). *J. Comput. Chem.* 22: 931–967.

**78** Hopffgarten, M. and Frenking, G. (2012). *WIREs Comput. Mol. Sci.* 2: 43–62.

**79** Fernández, I. and Bickelhaupt, F.M. (2014). *Chem. Soc. Rev.* 43: 4953–4967.

**80** Ziegler, T. and Rauk, A. (1979). *Inorg. Chem.* 18: 1755–1759.

**81** Poater, J., Feixas, F., Bickelhaupt, F.M., and Solà, M. (2011). *Phys. Chem. Chem. Phys.* 13: 20673–20681.

# 13

# Three-Dimensional Aromaticity

*"For a chemist, concepts like aromaticity, acid-base behavior, functional groups and substituent effects wilt at the edges when they are defined too closely. They cannot be mathematicised, they cannot be defined unambiguously, yet they are of fantastic utility to our science."*

*Roald Hoffman* (The Same and Not the Same, *Columbia University Press,* ***1995****, New York, p. 20*)

## 13.1 Introduction

The concept of aromaticity was originally developed to account for the unusual thermodynamic and kinetic stability of two-dimensional planar benzene derivatives. The determination in 1959 of the structure of the first *closo* borane $B_{10}H_{10}{}^{2-}$ by Lipscomb et al. [1], and the synthesis in 1960 of the dodecaborate(12) anion by Hawthorne and Pitochelli [2], and the first derivatives of *closo*-dodecaborate and *closo*-decaborate by Muetterties and coworkers [3] in 1962 move the concept of aromaticity from two to three dimensions. It is now well established that certain molecules such as *closo* boranes with the general formula $B_nH_n{}^{2-}$ $(5 \leq n \leq 14)$ [4] and carborane clusters in which some BH units of *closo* boranes have been replaced by isoelectronic $CH^+$ groups [5, 6] as well as some charged fullerenes [7–9], among other species are examples of molecules with aromaticity in three dimensions. This kind of aromaticity requires electron delocalization in closed three-dimensional circuits. As in the case of classical planar organic aromatic molecules, three-dimensional aromatic molecules, in general, have closed-shell electronic structure or have the highest-energy occupied shell only half-filled with same-spin electrons, display enhanced chemical stability compared to similar nonaromatic molecules, react through substitution rather than addition, exhibit diatropic ring currents in the presence of an external magnetic field, and have a tendency to bond length equalization, which makes these compounds, in general,

*Aromaticity and Antiaromaticity: Concepts and Applications*, First Edition.
Miquel Solà, Alexander I. Boldyrev, Michał K. Cyrański, Tadeusz M. Krygowski, and Gabriel Merino.

highly symmetric. From the shape of the molecules, three-dimensional aromaticity has been divided into spherical, cubic, octahedral, tetrahedral, or cylindrical aromaticity. Examples of species with all these types of three-dimensional aromaticity are described in the next sections. There are no molecules with a perfect spherical shape. Aromatic molecules with icosahedral symmetry are considered usually spherical aromatic species. In some cases, cubic, octahedral, or tetrahedral aromatic species are also considered as a special case of spherical aromaticity. In other cases, as in this chapter, they are treated separately.

## 13.2 Spherical Aromaticity

Spherical aromaticity is formally used to justify the unusually stable nature of some spherical compounds such as charged fullerenes or some polyhedral *closo* boranes. In the case of fullerenes, the electrons that are delocalized are those located in the out-of-plane *2p* orbital of the C atoms, and, consequently, delocalization takes place on the surface of the sphere. In polyhedral *closo* boranes, the situation is different. All valence electrons, except those that forming external B—H bonds, are delocalized inside the approximately spherical cage. Therefore, we distinguish between two types of spherical aromaticity, namely, the spherical aromaticity that occurs on the surface of the sphere and spherical aromaticity that involves valence electrons inside the sphere.

### 13.2.1 Aromaticity on the Surface of the Sphere

Fullerenes are a class of carbon allotropes formed by carbon atoms in the shape of a hollow sphere or ellipsoid. The first discovered fullerene was $C_{60}$. It was discovered by Kroto et al. [10] when they tried to simulate the conditions of cool red giant stars in the laboratory and a 720 mass peak (i.e., $C_{60}$) appeared to be extremely intense in the mass spectrum. Because of its soccer-ball-like structure, $C_{60}$ was named Buckminsterfullerene as homage to the geodesic dome architect, Buckminster Fuller. In the original paper that reported the discovery of $C_{60}$ [10], the authors already suggest the possible aromatic character of this molecule. One year later, Klein et al. [11] proposed a superaromatic character of buckminsterfullerene based on the large number (K = 12 500) of Kekulé resonance structures that can be depicted for this species that has 60 $\pi$-electrons. However, not all of these large number of resonance structures have the same weight in the description of $C_{60}$. The one that represents the 20 six-membered rings (6-MRs) as cyclohexatriene species and the 12 5-MRs as a [5]radialene system (Figure 13.1) has the most important contribution as evidenced by the two different C—C bonds of $C_{60}$. The so-called [6] bonds have a bond length of 1.355 Å [12] and connect two hexagons, whereas the

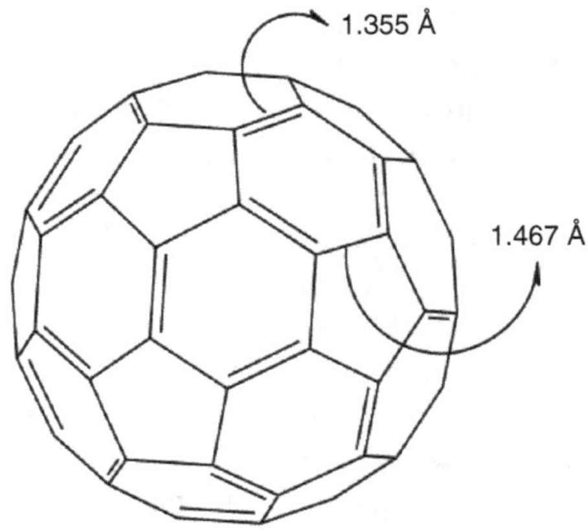


**Figure 13.1** The Kekulé resonance structure out of the existing 12 500 Kekulé resonance structures that better represents the molecular structure of $C_{60}$ together with the C−C bond lengths of the short [6] and long [5, 6] bonds.

bonds in the ring junction between a hexagon and a pentagon, named [5, 6] bonds, are longer (1.467 Å [12]).

From the very beginning, the aromatic character of $C_{60}$ was controversial. It is now widely accepted that fullerenes have ambiguous aromatic character [9, 13, 14], with some properties that are typical of aromatic compounds and others that are not. For instance, their rather considerable stability is in favor of having aromatic character [14], whereas its rich chemical reactivity [15–19] is against its possible aromaticity. Most of the usual aromatic descriptors point out a low aromatic character for $C_{60}$. Its aromatic stabilization energy (ASE) per $\pi$ electron of about 0.6 kJ/mol [20] is much less than that of classical organic aromatic compounds. For instance, the ASE per $\pi$ electron of benzene is 20.1 kJ/mol [21]. The harmonic oscillator model of aromaticity (HOMA) of $C_{60}$ of only 0.263 also points out a relatively low aromaticity [20]. The nucleus-independent chemical shift (NICS) computed in the center of the fullerenic cage is −11.2 ppm, indicating a weak aromatic character [22, 23]. If we consider only the radial set of molecular orbitals (MOs) with $\pi$ character to compute the NICS in the center [24], we obtain a value of +21.2 ppm. Taking into account this result, Chen et al. [23] classified $C_{60}$ as a spherically $\pi$-antiaromatic molecule. In endohedral $Ng@C_{2n}$ fullerenes (Ng = noble gas), the chemical shift of the Ng (δ(Ng), usually Ng = $^{3}He$ or $^{129}Xe$) nucleus measured relative to free Ng acts as a probe for the ring currents and magnetic properties of the fullerene cage [25, 26]. According to δ($^{3}He$) values, $C_{60}$ (−6.3 ppm) is one of the less aromatic fullerenes [8, 27, 28]. Ring current maps of $C_{60}$ show strong paramagnetic currents associated with 5-MRs and relatively weak diamagnetic currents in 6-MRs [29–33]. These diamagnetic and paramagnetic currents in $C_{60}$ partially cancel leading to its vanishingly small magnetic susceptibility, which is much less than that of benzene [34, 35]. The negative NICS values of the weakly aromatic 6-MRs and positive NICS of antiaromatic or nonaromatic 5-MRs concur with the observed ring currents [8, 9, 22, 36, 37]. In line with NICS results, HOMA values calculated from experimental bond lengths for 5- and 6-MRs of $C_{60}$ are 0.10 and 0.55, respectively [38]. The multicenter MCI index also agrees with the 6-MRs of $C_{60}$ being more aromatic than the 5-MRs [39]. Finally, the *para*-delocalization (PDI) index and the aromatic fluctuation (FLU) index for $C_{60}$ again indicate relatively weak local aromatic character of the 6-MRs [36, 40].

Fullerenes have large electron affinities [41]. When fullerenes are reduced, their local and global aromatic properties change. The negative charge accumulates mainly in the 5-MRs, which become more cyclopentadienyl anion-like, increasing their aromaticity. Thus, when going from $C_{60}$ to $C_{60}{}^{2-}$ the ring currents of all 6-MRs become paramagnetic, whereas those of the 5-MRs change from strong paramagnetic into diamagnetic. Calculated NICS(0) values of the 5- and 6-MRs confirm this change [42]. Diamagnetic ring currents in the 5-MRs are also found in $C_{60}{}^{6-}$ and $C_{70}{}^{6-}$ [43]. Despite the charge transfer may decrease the aromaticity of 6-MRs in some cases, the global aromatic character of the negatively charged fullerene cages tends to increase. Thus, whereas $C_{60}$ is considered spherically $\pi$-antiaromatic or at least nonaromatic, its hexaanion form, $C_{60}{}^{6-}$, is found to be $\pi$-aromatic [23].This result is confirmed by the strongly shifted to higher field $\delta(^3He)$ in $^3He@C_{60}{}^{6-}$ [44, 45] and by the values of the NICS/NICS$_\pi$ in the center of the cage that are decreased from −2.8/21.2 to −50.0/−24.4 ppm when going from $C_{60}$ to $C_{60}{}^{6-}$ [23], respectively.

It is widely accepted that endohedral metallofullerenes (EMFs), $M@C_{2n}$, can be formally described as $M^{q+}@C_{2n}{}^{q-}$ [46]. The number of electrons formally transferred from the metallic cluster, M, to the cage, $C_{2n}$, ranges from 2 (e.g., when having a Sm atom incarcerated) to 6 (for example, in the case of encapsulated $Sc_3N$). The reduction of the fullerenic cages in EMFs results in an increase of their aromaticity that has important consequences already described in Chapter 4. From a structural point of view, the most stable isomers differ when going from neutral to negatively charged cages. As is discussed in Section 4.2, in neutral hollow fullerenes, aromaticity does not play an important role and the strain energy is the key factor that determines the stability of the isomeric carbon cages. This is translated into the isolated pentagon rule (IPR) obeyed by all hollow fullerenes known to date. This IPR rule states that all pentagons must be surrounded by hexagons to alleviate the strain produced by two fused pentagons [47]. On the other hand, in negatively charged fullerenes or EMFs, maximizing the total aromaticity is the main stabilizing element. Consequently, the IPR does not apply and is substituted by the maximum aromaticity rule [48]. This rule affirms that the most stable anionic fullerene isomeric cage is the one whose total aromaticity is maximized irrespective of whether the cage follows or not the IPR rule. From a reactivity point of view, the regioselectivity of cycloaddition reactions (Diels-Alder, Bingel-Hirsch, ...) changes when going from hollow fullerenes to EMFs as is discussed in Section 4.3.2.4.

Rabilloud [49] studied $Li_{12}C_{60}$ showing that charge transfer from Li to $C_{60}$ leads to significant changes in the structure of the $C_{60}$ cage. In the $Li_{12}C_{60}$ most stable structure, Li atoms are located above the 12 pentagonal sites and the carbon–carbon distances are found to be approximately equal for both the single and the double bonds, 1.458 and 1.449 Å, respectively (Figure 13.2). Although aromaticity was not discussed in that work, the low bond length alternation observed seems to prove that the formally formed $C_{60}{}^{12-}$ cage is particularly aromatic. This $C_{60}{}^{12-}$ cage with 72 π-electrons has a closed-shell configuration and follows the Hirsch's $2(n+1)^2$ rule for predicting the aromaticity of spherical systems [50] that is discussed in Section 2.9. Hirsch considered that the π-electron system of a roughly spherical fullerene can be approximated by a spherical electron gas of uniform density surrounding the surface of a sphere. The wave functions of this electron gas are characterized by the angular momentum quantum number $l$ ($l = 0, 1, 2, 3 \ldots$) similar to the situation found for the atomic orbitals, each energy level being $2l+1$ times degenerated. Consequently, all π-shells are totally filled and spherical systems are aromatic when we have $2(n+1)^2$ electrons (Figure 2.12). According to Hirsch's rule, icosahedral $C_{20}{}^{+2}$, $C_{60}{}^{10+}$, $C_{60}{}^{12-}$, or $C_{80}{}^{8+}$ are aromatic fullerenes. This prediction has been confirmed by means of NICS and multicenter index (MCI) calculations [8, 39, 50, 51].

With the same philosophy that Baird used to extend the Hückel's rule to open-shell systems, it is possible to extend the Hirsch's rule to spherical open-shell systems [39]. By doing so, it is found that spherical compounds with a half-filled last energy level, that is, those with $2n^2+2n+1$ electrons and spin $n+\frac{1}{2}$, should be aromatic. Bond length alternation values, NICS, and MCI indices show that systems such as the $C_{60}{}^{19+}$ ($S = 9/2$) and $C_{60}{}^{1-}$ ($S = 11/2$) are indeed aromatic confirming the validity of this rule. It is worth noting that the $S = 9/2$ electronic state for $C_{60}{}^{19+}$ and the $S = 11/2$ for $C_{60}{}^{1-}$ are excited states for these species. This rule has been discussed in more detail in Section 2.10.

Homoaromaticity, like aromaticity, can also be found in three-dimensional systems. The first three-dimensional homoaromatic-described species was the

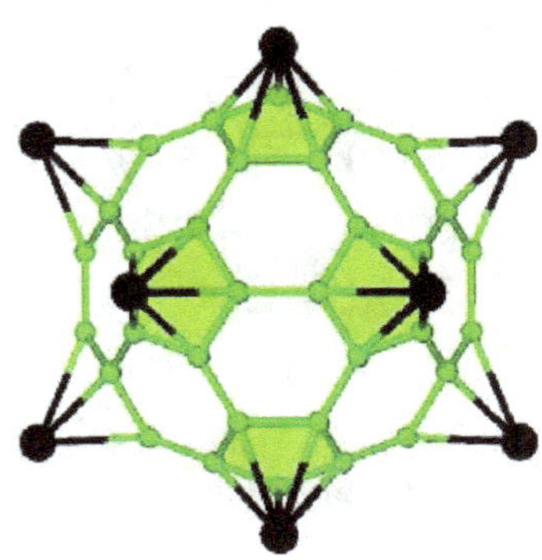

**Figure 13.2** The most stable molecular structure of $Li_{12}C_{60}$. Source: Reprinted with permission from Ref. [49]. Copyright 2010 American Chemical Society.

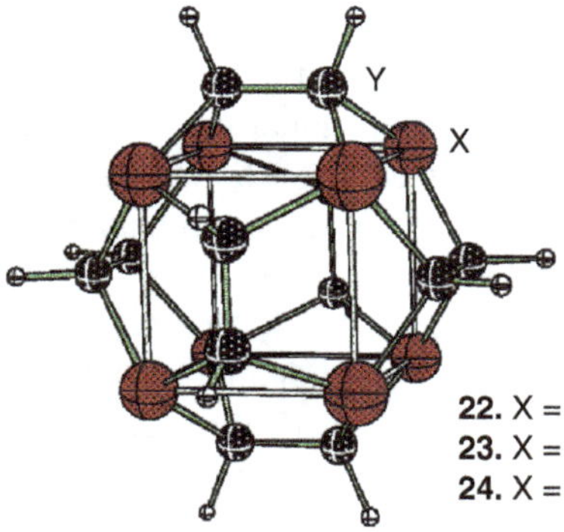


**Figure 13.3** The molecular structure of $C_{12}X_8H_{12}$ spherical homoaromatic species. Source: Reprinted with permission from Ref. [54]. Copyright 2003 Wiley.

1,3-dehydro-5,7-adamantanediyl cation [52]. In the case of spherical homoaromaticity, the same $2(n + 1)^2$ Hirsch's rule applies [53]. As an example, we can mention the $C_{12}X_8H_{12}$ (X = C, Si, Ge) species depicted in Figure 13.3. These neutral eight-center eight-π-electron stable species, which follow Hirsch's rule with $N = 1$, are highly aromatic, as denoted by their GIAO-B3LYP/6-31G(d) NICS values calculated at the cage center of about −33 ppm [54]. Based on the $2(n + 1)^2$ rule, Schleyer and coworkers designed a series of spherical sila- and germa-homoaromatic systems rich in group 14 elements [54]. Another example of π-spherical homoaromatic species [55] are fulleroids. These systems are formed when the addition to a certain C—C bond of a fullerene leads to the breaking of the attacked C—C bond and the formation of an open bond on the surface of the cage (see Figure 13.4) [56]. In $C_{60}$ fulleroids, the 60 π-electron systems remain more or less intact because all carbon atoms keep their $sp^2$ hybridization. Support

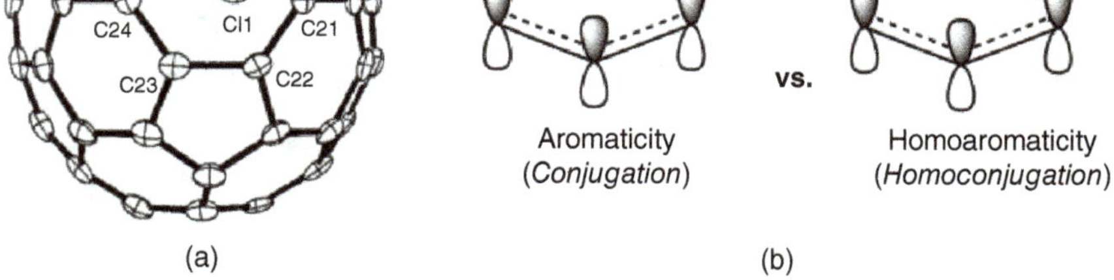


**Figure 13.4** (a) The X-ray of the $C_{71}Cl_2$ fulleroid structure [56] and (b) schematic representation of aromaticity versus homoaromaticity. Source: Reprinted with permission from Ref. [7]. Copyright 2014 Royal Society of Chemistry.

**Table 13.1** B3LYP/6-31G(d) $NICS(1)_{zz}$ (in ppm) and MCI (in electrons) values for three different $Ge_{12}$ anions and cations.[a)]

| Species | $NICS(1)_{zz}$ [b)] | MCI [c)] |
|---|---|---|
| $Ge_{12}^{2-}$ (S = 0) | −5.5 | 0.049 |
| $Ge_{12}^{4+}$ (S = 0) | −69.0 | 0.088 |
| $Ge_{12}^{1-}$ (S = 5/2) | −405.9 | 0.113 |

a) In all cases, the structure considered is that of the $I_h$ $Ge_{12}^{2-}$ B3LYP/6-31G(d) optimized geometry with d(Ge–Ge) = 2.685 Å. Data from ref. [39].
b) Calculated for ghost atoms located at 1 Å above the ring centers determined by the non-weighted mean of the heavy atom coordinates.
c) Calculated in a three-membered ring.

to the homoaromaticity of fulleroids comes from the similarity of the endohedral chemical shifts and UV spectra to that of the parent fullerenes [57–59], the significant π-overlap values in Hückel MO theory [56], and NICS(0) and PDI values [36].

As an example of inorganic compound with spherical aromaticity, let us to discuss here the germaniaspherene $Ge_{12}^{2-}$ species, a molecule of $I_h$ symmetry that belongs to the group of dianions called spherenes [60, 61]. As can be seen in Table 13.1, $Ge_{12}^{2-}$ species have a modest negative $NICS(1)_{zz}$ value and a small MCI, indicating weak aromatic character. On the contrary, both the closed-shell $Ge_{12}^{4+}$ that follows the Hirsch's rule with $n = 1$ and the open-shell sextet $Ge_{12}^{1-}$ that obeys the $2n^2 + 2n + 1$ ($S = n + ½$) rule with $n = 2$ have much larger and negative NICS and much greater MCI values, a clear indication of their enhanced aromatic character [39].

Finally, let us mention that oxide cluster $Ce_6O_8$ was shown to exhibit *d*-atomic based spherical aromaticity [62]. NBO and AdNDP procedures revealed the presence of a 6c-2e σ bond of spherical shape spread over six cerium atoms that is mostly composed of $5d_{z2}$ atomic orbitals (AOs) of cerium atoms.

### 13.2.2 Aromaticity Inside the Sphere

Not all spherical aromatic molecules follow the Hirsch's rule. A clear example is represented by the icosahedral closo borane $B_6H_6^{2-}$. Neither the total valence electrons of this compound, that is, 26, nor its total valence electrons except those electrons that form the localized B—H bonds, that is, 14, obey the $2(n+1)^2$ rule.

For this system, the delocalized electrons that give rise to aromaticity are the 14 electrons located *inside* the cage, not those on the surface of the cage that are localized forming B—H bonds. When the electrons that contribute to aromaticity are not delocalized on the surface but inside the cluster, the Hirsch's and the $2n^2 + 2n + 1$ ($S = n + ½$) rules are not obeyed and other rules of aromaticity apply. In the following sections, we discuss the two most important cases.

#### 13.2.2.1 *Closo* Boranes

*Closo* borohydride clusters or *closo* boranes are stable and aromatic boron clusters with a general formula of $[B_nH_n]^{2-}$ ($5 \leq n \leq 14$) arranged in a polyhedral structure [4]. As is discussed in Section 2.10, the rules that apply in these clusters are Wade's $2n + 2$ electron rule, where $n$ is the number of vertexes or boron atoms of the polyhedron and Mingos' $4n + 2$ rule [63, 64]. Wade's rule refers to the skeletal electrons, that is, all valence electrons except those of the B—H bonds, whereas Mingos' rule also incorporates the exo electrons of the B—H bonds and, consequently, the $4n + 2$ in Mingos' rule counts the total number of valence electrons. Let us consider the smallest borane $B_6H_6^{2-}$. For this species, the total number of valence electrons is 26 ($n = 6$ in Mingos' rule) and if we remove the 12 electrons of the B—H bonds we get 14 electrons ($n = 6$ in Wade's rule). Both Wade's and Mingos' rules are obeyed and, in fact, these two rules are equivalent. The 14 electrons of the cage are located in the molecular orbitals depicted in Figure 13.5. The boron atom in B–H units is assumed to be *sp* hybridized. By combining the *sp* outward orbitals with *1s* orbitals of H, the exo electron pairs of the B—H bonds are formed. The rest of the *n sp* inward orbitals form a radial in-phase orbital that is the most stable skeletal bonding molecular orbital and $n - 1$ unoccupied orbitals higher in energy. The *2n p* unhybridized orbitals yield *n* bonding and *n* antibonding tangential orbitals. $[B_nH_n]^{2-}$ clusters with $n + 1$ electron pairs ($2n + 2$ electrons) have closed-shell structures that result from filling the lowest-lying *sp* radial orbital and the *n* bonding tangential orbitals. The process is shown in Figure 13.5 for the molecular orbitals of $B_6H_6^{2-}$. Interestingly, by adding four electrons to $B_6H_6^{2-}$ one forms $B_6H_6^{6-}$ that is isoelectronic to benzene and it is planar and aromatic [65].

The high symmetry and the closed-shell electronic structure of $B_6H_6^{2-}$ borane cluster are symptomatic of aromaticity. Its aromatic character is further substantiated by the large and negative NICS value in the center of the cage and by the equalization of bond lengths. Not only of $B_6H_6^{2-}$ but also for all $B_nH_n^{2-}$ borane clusters ($n$ = 5–14), NICS values are large and negative (see Figure 13.6). This aromaticity is in line with the high thermal stability of some of these clusters. For instance, $Li_2[B_{12}H_{12}]$ [67] and $Na_2[B_{12}H_{12}]$ [68] do not decompose below 600°C. The aromatic character of *closo* borane clusters is nowadays out of discussion. NICS results [66, 69–71], bond length alternation values [70], resonance

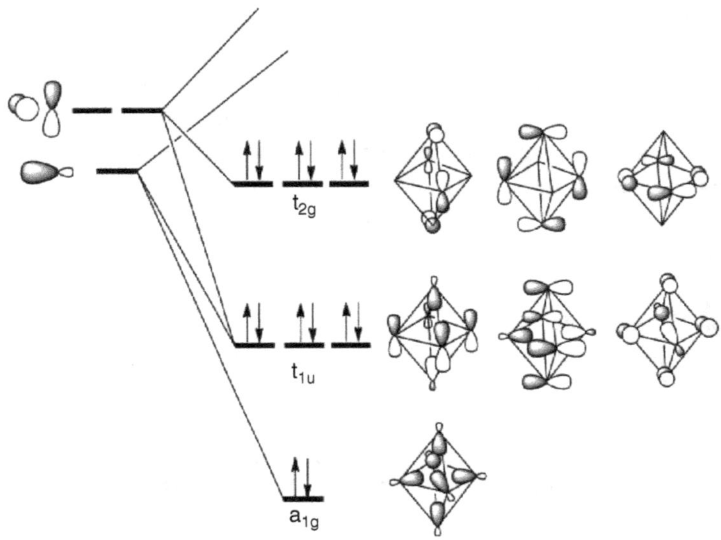


**Figure 13.5** Schematic representation of the skeletal bonding molecular orbitals of $B_6H_6^{2-}$ with the seven skeletal electron pairs.

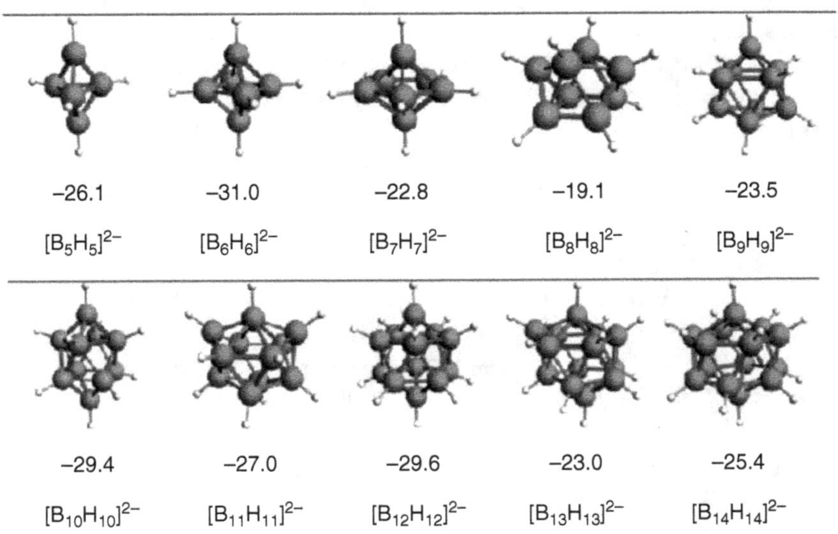


**Figure 13.6** The molecular structure and the NICS values (ppm) calculated at the center of the cage for the $B_nH_n^{2-}$ borane clusters ($n$ = 5–14). Source: Values from Ref. [66].

energy calculations [72], and ring currents [73] support the aromatic character of these clusters.

Although Wade-Mingos' rule was initially developed for borane clusters, it has been shown that it is applicable to other main group clusters as Zintl phases and ions [74, 75]. Finally, let us mention that a direct connection between the Hückel's $4n + 2$ π-electron rule for two-dimensional aromaticity and the Wade-Mingos' $2n + 2$ rule for three-dimensional aromaticity has been established [71].

#### 13.2.2.2 Jellium Cluster Model

The distribution observed in mass spectrum experiments of alkaline metal clusters produced in a supersonic expansion with a noble gas carrier is not uniform. Certain metal clusters, the ones that are more stable, are found in more abundance than others. For example, the most abundant $Na_N$ clusters are found for the some magic numbers like $N = 2, 8, 18, 20, 34, 40, 58, 68, 90, 92\ldots$ [76]. These type of distribution is also observed for other metals such as Li, K, Be, Cu, Ag, Au, or lithium-, potassium-, cesium-, and copper-doped aluminum clusters [77–84]. The high stability of certain clusters is usually explained with the jellium cluster model, also known as free electron shell model [76]. This model is often used in solid-state physics and chemistry as a simple model of delocalized electrons in a metal. It is a quantum mechanical model of interacting electrons in a solid where the atomic nuclei and the electron density are assumed to be uniformly distributed in space. Solution of the Schrödinger equation for this model system leads to the energy levels $1S^21P^61D^{10}2S^21F^{14}2P^61G^{18}2D^{10}1H^{22}3S^2\ldots$ that show that the aforementioned magic numbers correspond to closed-shell structures in the jellium cluster model. For instance, the electronic configuration of the aromatic $Mg_{17}$ cluster (see Figure 13.7) with 34 valence electrons can be written as $1S^21P^61D^{10}2S^21F^{14}$ leading to a closed-shell electronic structure and aromaticity that makes it relatively stable [85]. At variance with fullerenes, the

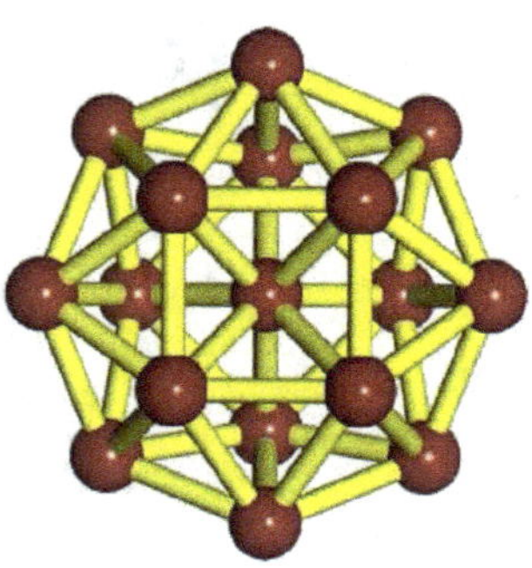

$Mg_{17}$

**Figure 13.7** The molecular structure of $Mg_{17}$. Source: Reprinted with permission from Ref. [85]. Copyright 2016 American Chemical Society.

electrons are not distributed on the surface of the sphere but through the whole sphere. Another example is the $T_d$ cluster $Au_{10}^{2+}$ [86].

Finally, let us mention that in real systems, the energy ordering of the shell orbitals can be changed according to the specific potential determined by the shape and charge of the cluster analyzed.

## 13.3 Octahedral Aromaticity

Experimentally, only two octahedral aromatic compounds have been observed, namely, $B_6H_6^{2-}$ and $Al_6^{2-}$, the latter in the form of $LiAl_6^{2-}$ [87]. $Al_6^{2-}$ has a singlet ground state with the $(1a_{1g})^2(1t_{1u})^6(1e_g)^2(2a_{1g})^2(1t_{2g})^6$ electronic configuration (see Figure 13.8). Theoretically, the $H_6^{2-}$ cluster was described as octahedral σ-aromatic [89], although since it has nine imaginary frequencies, it is not a stable species. The metalla-aromatic $Be_6$ in the quintet state [90] and singlet $Au_6^{2-}$ species [91] are also two octahedral aromatic species described in computational work. All these species have electronic structures that are either closed-shell or

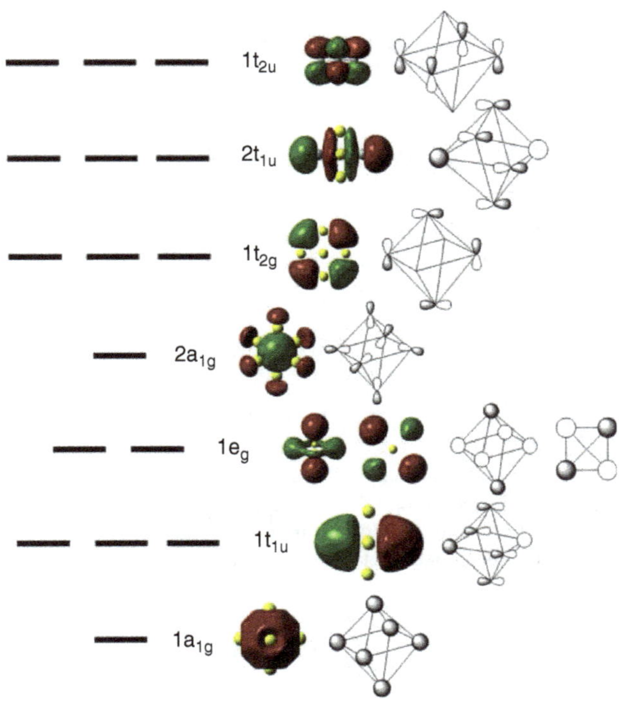


**Figure 13.8** The molecular orbitals of a typical octahedral cluster. Source: Reprinted from Ref. [88]. Copyright 2016 Royal Society of Chemistry.

**Table 13.2** Molecular structures of some octahedral clusters of the type $^{2S+1}A_{1g}\ X_6{}^q$ that are minima on the potential surface. Values of X–X bond distance in Å, MCI and PDI in electrons, and NICS in ppm.

| System | q | Spin | Electronic state | d(X–X) | $MCI_6$ [a] | $MCI_4$ [a] | $MCI_3$ [a] | $NICS(0)_6$ [b] | $NICS(0)_3$ [b] | PDI |
|---|---|---|---|---|---|---|---|---|---|---|
| $Li_6$ | 1 | S = 3/2 | $^4A_{1g}$ | 3.007 | 0.068 | 0.038 | 0.078 | −19.17 | −14.07 | 0.260 |
| | 1 | S = 5/2 | $^6A_{1g}$ | 3.112 | 0.069 | 0.035 | 0.077 | −18.63 | −13.72 | 0.258 |
| $Be_6$ | 2 | S = 3 | $^7A_{1g}$ | 2.129 | 0.061 | 0.055 | 0.137 | −58.66 | −45.88 | 0.584 |
| | −1 | S = 3/2 | $^4A_{1g}$ | 2.050 | 0.121 | 0.066 | 0.175 | −35.72 | −17.16 | 0.656 |
| | 0 | S = 2 | $^5A_{1g}$ | 2.040 | 0.080 | 0.054 | 0.158 | −31.54 | −15.38 | 0.603 |
| | 2 | S = 0 | $^1A_{1g}$ | 2.135 | 0.140 | 0.068 | 0.158 | −50.45 | −37.13 | 0.521 |
| | 3 | S = 1/2 | $^2A_{1g}$ | 2.267 | 0.097 | 0.054 | 0.136 | −43.21 | −33.00 | 0.461 |
| $B_6$ | 0 | S = 1 | $^3A_{1g}$ | 1.675 | 0.085 | 0.060 | 0.174 | −72.97 | −53.40 | 0.769 |
| | 2 | S = 0 | $^1A_{1g}$ | 1.628 | 0.109 | 0.063 | 0.196 | −47.64 | −24.90 | 0.800 |
| | 3 | S = 1/2 | $^2A_{1g}$ | 1.677 | 0.072 | 0.039 | 0.175 | −36.67 | −19.02 | 0.742 |
| $Na_6$ | −2 | S = 0 | $^1A_{1g}$ | 3.605 | 0.022 | 0.035 | 0.078 | −17.79 | −16.99 | 0.330 |
| | 1 | S = 3/2 | $^4A_{1g}$ | 3.602 | 0.061 | 0.038 | 0.064 | −20.12 | −15.64 | 0.237 |
| $Mg_6$ | −2 | S = 3 | $^7A_{1g}$ | 3.096 | 0.057 | 0.063 | 0.105 | −52.10 | −44.09 | 0.511 |
| $Al_6$ | −2 | S = 0 | $^1A_{1g}$ | 2.710 | 0.081 | 0.087 | 0.125 | −80.73 | −68.12 | 0.678 |
| | 2 | S = 2 | $^5A_{1g}$ | 2.948 | 0.046 | 0.049 | 0.078 | −24.86 | −18.88 | 0.491 |
| $Si_6$ | 1 | S = 3/2 | $^4A_{1g}$ | 2.461 | 0.055 | 0.054 | 0.146 | −43.57 | −48.33 | 0.726 |

a) $MCI_6$, $MCI_4$, and $MCI_3$ are the multicenter indices computed for the six atoms of the octahedron, for the four equatorial atoms and for the three atoms of an octahedron face, respectively.
b) $NICS(0)_6$ and $NICS(0)_3$ are the NICS in the center of the octahedron and in the center of an octahedron face, respectively.

Source: Adapted from Ref. [88].

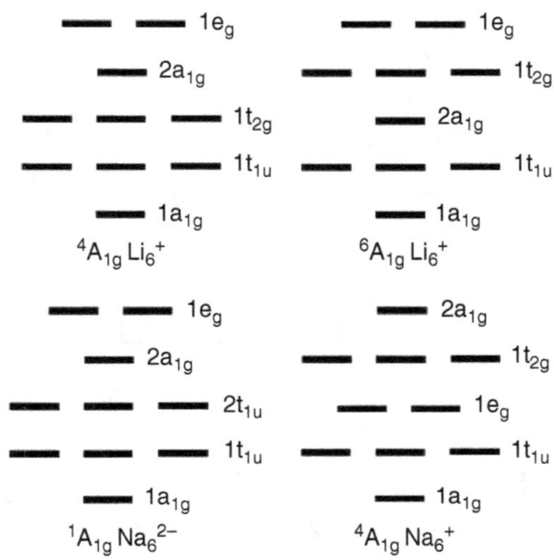


**Figure 13.9** The ordering of the molecular orbitals in $^{4}A_{1g}$ and $^{6}A_{1g}$ $Li_6^+$, and $^{1}A_{1g}$ $Na_6^{2-}$ and $^{4}A_{1g}$ $Na_6^+$ clusters.

half-filled shell with the same spin electrons. These two types of structures offer good prospects of aromaticity [91].

So, to find new octahedral aromatic species, one can start from the typical valence molecular orbitals of an octahedral compound shown in Figure 13.8 and fill in with electrons until closed-shell or half-filled shell electronic structures are obtained. With this recipe, 15 new octahedral aromatic molecular clusters (plus the already known $Al_6^{2-}$) of the type $^{2S+1}A_{1g}$ $X_6^q$ with X = Li–C and Be–Si, q = −2 to +4, and spin S were found [88]. The molecular structure, electronic state, and some aromaticity indicators of these clusters are collected in Table 13.2. On the other hand, some clusters, like $Mg_6^{-2}$ and $B_6^{+4}$ in the $^{1}A_{1g}$ state or $Na_6^{2-}$ and $Mg_6^{2-}$ both in the $^{7}A_{1g}$ electronic state, were found to be nonaromatic or antiaromatic despite having closed-shell or half-filled shell electronic structures. Therefore, this condition is not sufficient for aromaticity, although in most cases it works. Despite, the existence of octahedral aromaticity is confirmed from these studies, it was not possible to propose a rule for octahedral aromaticity [88]. The reason is that the ordering of molecular orbitals does not remain the same for all octahedral clusters as can be seen in Figure 13.9.

## 13.4 Cubic Aromaticity

In 2015, Cui et al. [92] reported the synthesis and characterization of a polyzinc compound $[Zn^{I}_8(HL)_4(L)_8]^{12-}$ (L = tetrazole dianion) composed of eight Zn(I) ions in a cubic arrangement with short Zn—Zn bond lengths of 2.2713 Å. The molecular orbitals of the $[Zn_8]^{+8}$ unit are represented in Figure 13.10. They resemble

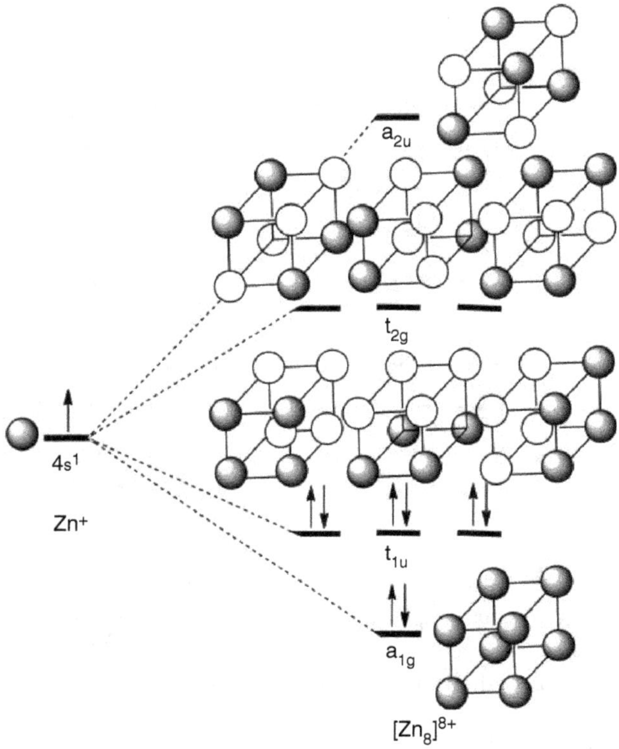


**Figure 13.10** The molecular orbitals of the $[Zn_8]^{+8}$ cluster.

those of $B_6H_6^{2-}$ (Figure 13.5). The eight Zn(I) $4s^1$ electrons in the $[Zn_8]^{+8}$ cluster are distributed in four bonding molecular orbitals $(a_{1g})^2(t_{1u})^6$ leading to an extensive delocalization over the cube that brings significant stabilization. Calculated NICS indices at the cubic-, face-, and bond-center display negative values comparable to those in benzene, thus indicating aromaticity. This situation is referred by the authors to as cubic aromaticity and the counting rule is $6n+2$. However, as can be seen in Figure 13.10 this rule only leads to closed-shell electronic configurations for $n = 0$, 1, and 2, and, therefore, it is limited to these $n$ values.

It is worth noting that, for $n = 0$ and 1, the $6n+2$ counting rule leads to the same result as the $2(n+1)^2$ Hirsch's rule. In fact, the $H_8$ cluster of $O_h$ symmetry (d(H–H) = 1.129 Å at MP2(fc)/6-311++G(d,2p) level of theory), which has eight delocalized σ-electrons in equivalent orbitals to those of Figure 13.10, was found to have a large and negative magnetic susceptibility [89, 93] and it was considered a case of a spherical (not cubic) aromatic molecule [89]. Similarly, the distorted cubic $Si_8^{2-}$ $T_d$ dianion (the $O_h$ $Si_8^{2-}$ cluster is a transition state and their cube's squares are transformed into bent rhombuses in the optimization process to form $T_d$ $Si_8^{2-}$ species) [94], as well as the related silicon clusters Be@$Si_8$, B@$Si_8^+$, and C@$Si_8^{2+}$, with 34 valence electrons and energy levels $(1a_{1g})^2(1t_{1u})^6(1t_{2g})^6(1a_{2u})^2(1e_g)^4(2a_{1g})^2(1t_{2u})^6(2t_{1u})^6$, are also found aromatic. The molecular orbitals $1t_{1u}$, $1t_{2g}$, $1a_{2u}$, $1_{eg}$, and $1t_{2u}$, are responsible for the 12 Si–Si σ bonds of the cube, whereas the $(2a_{1g})^2(2t_{1u})^6$ correspond to delocalized electrons that bring aromatic stabilization. However, despite these species have cubic geometry they were considered spherical and not cubic aromatic species. So, in general, cubic species with $n = 0$ and 1 can also be considered a particular case of spherical aromaticity.

## 13.5 Tetrahedral Aromaticity

The most basic system that exhibits tetrahedral aromaticity is $H_4^{2+}$ ($T_d$). This system has two σ-electrons delocalized in an $a_1$ orbital (see Figure 13.11), that is, the bonding occurs through a four-center two-electron (4c-2e) interaction. However, this $H_4^{2+}$ ($T_d$) cluster is a stationary higher order saddle point in the potential energy surface, and, therefore, is not a stable species [89]. Some authors consider $H_4^{2+}$ ($T_d$) an example of spherical aromaticity [93]. A similar 4c-2e bonding pattern is found in $Li_4^{2+}$ and $Na_4^{2+}$ in $T_d$ symmetry [95, 96]. The main difference is that, in $Li_4^{2+}$ and $Na_4^{2+}$ ($T_d$), $np$ atomic orbitals play a relevant role. At variance with $H_4^{2+}$ ($T_d$), $Li_4^{2+}$ and $Na_4^{2+}$ ($T_d$) are minima and kinetically stable species. However, they are thermodynamically unstable with respect to dissociation into $2M_2^+$ or $M^+ + M_3^+$ (M = Li, Na). The estimation of the resonance energy in $Li_4^{2+}$ ($T_d$) of 137.2 kcal/mol by Alexandrova et al. [96]

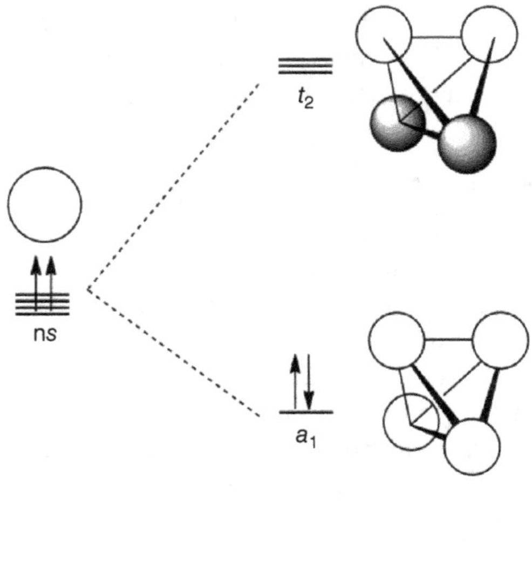


**Figure 13.11** The lowest valence molecular orbitals of $H_4^{2+}$.

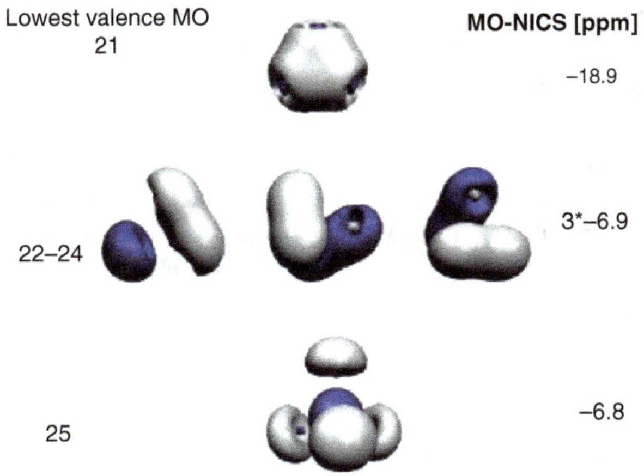


**Figure 13.12** The lowest valence molecular orbitals of $P_4$ with the NICS orbital components in ppm. Source: Reprinted from Ref. [97]. Copyright 2006 American Chemical Society.

provides support to the aromatic character of this dication. Experimentally, these type of 4c-2e interaction was first reported in tetrahedral homoaromatic 1,3-dehydro-5,7-adamantanediyl dication [52].

$P_4$ [74, 97] is a common allotrope of phosphorous that has been found to exhibit tetrahedral aromaticity (NICS(0) = −52.9 ppm). The lowest valence molecular orbitals of $P_4$ with their corresponding NICS orbital components are shown in Figure 13.12. Not only $P_4$, but also valence isoelectronic $N_4$ (NICS(0) = 69.6 ppm), $As_4$ (NICS(0) = −53.3 ppm), $Sb_4$ (NICS(0) = −38.8 ppm), or $Bi_4$ (NICS(0) = −36.3 ppm) species have the same tetrahedral aromatic character [74]. Similarly, Zintl anions [98] like $Si_4^{4-}$ (NICS(0) = −41.8 ppm), $Ge_4^{4-}$ (NICS(0) = −39.9 ppm), $Sn_4^{4-}$ (NICS(0) = −32.3 ppm), and $Pb_4^{4-}$ (NICS(0) = −29.1 ppm) with the same electronic structure are described as aromatic species [74].

Finally, tetrahedral cages of copper and silver tetra capped by lithium cations, $M_4Li_4$ (M = Cu, Ag) were found to be minima [97]. The NICS values on the center of the tetrahedral cage of −50.24 ppm ($Cu_4Li_4$ $T_d$) and −33.8 ppm ($Ag_4Li_4$ $T_d$) are indicative of tetrahedral aromaticity.

## 13.6 Cylindrical Aromaticity

In some large cylindrical molecules such as nanotubes [99–101], cycloparaphenylenes [102, 103], or certain boron and boron nitride clusters [104–106], among others, the aromaticity can be described by the hollow cylinder model (HCM) [107, 108]. In this model, the Schrödinger equation is solved for a particle moving in a hollow cylinder box defined by the parameters given in Figure 13.13. The molecular orbitals (MOs) obtained can be separated into and radial (located in the different layers of the cylinder and perpendicular to the cylinder axis – they

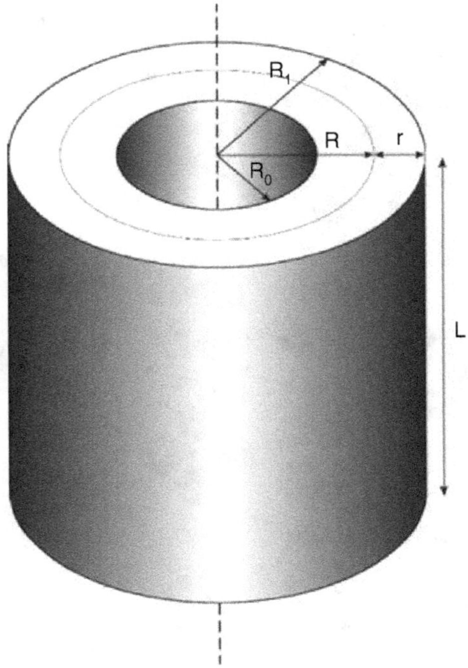


**Figure 13.13** Parameters considered in the hollow cylinder model. Source: Reprinted from Ref. [107]. Copyright 2014 American Chemical Society.

can be considered π-MOs) and tangential (located in the outer part of the cylinder and parallel to the cylinder axis – they can be considered σ-MOs). In the HCM, both radial and tangential orbitals are found to be doubly degenerated except some of them that are not degenerated. Then to properly fill the electron shells to generate a closed shell electronic structure, depending on the number of nondegenerated orbitals occupied, the rule of radial and tangential aromaticity is $4N + 2M$, N and M being the number of degenerated and nondegenerated MOs.

The wave functions of the radial and tangential orbitals in the HCM are characterized by three quantum numbers: $k$, $l$, and $n$. For the radial orbitals, $n$ is the principal quantum number of the valence $ns$ and $np$ atomic orbitals involved in the bonding (e.g., for cylinders formed by B or C, $n = 2$). If p is the number of rings in the tube, then $k = 1, , \text{p}-1, \text{p}$. Finally, if q is the number of atoms in each ring of the tube, then $l = 0, \pm 1, \ldots \pm\text{q}-1, \pm\text{q}$. For the tangential orbitals, $n = 1$, $k$ must be greater than the quantum principal number of the valence $ns$ and $np$ atomic orbitals involved in the bonding (e.g., for cylinders formed by B or C, $k = 3, 4, 5, \ldots$), and, as for the radial orbitals, $l = 0, \pm 1, \ldots \pm\text{q}-1, \pm\text{q}$.

In the particular case of $B_{20}$ [107, 109] (see Figure 13.14), we have 20 electrons in delocalized MOs, 10 in radial and 10 in tangential MOs, and the classical $4n + 2$ Hückel's counting rule is thus recovered in both MOs components with $n = 2$. The shape of the five occupied radial and tangential HOMOs and of some unoccupied LUMOs of $B_{20}$ are displayed in Figure 13.15. Ring currents maps of $B_{20}$ shows

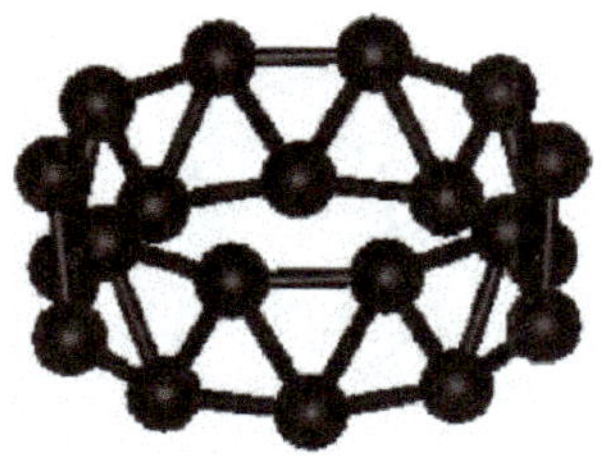

**Figure 13.14** The molecular structure of $B_{20}$. Source: Reprinted from Ref. [109]. Copyright 2006 American Institute of Physics.

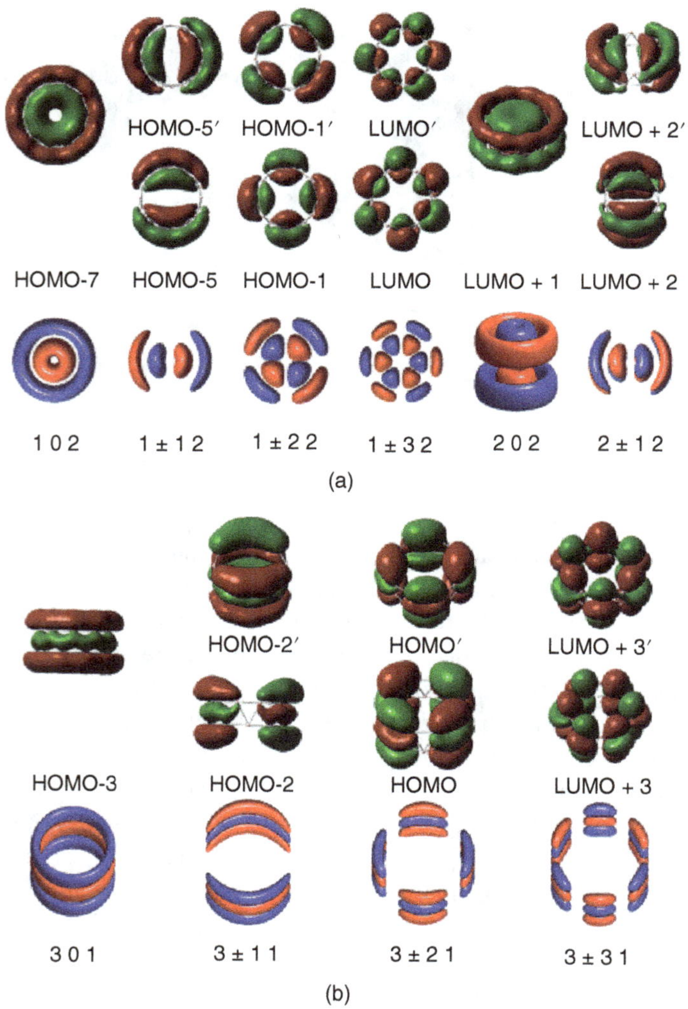


**Figure 13.15** Shape of the delocalized molecular orbitals of $B_{20}$ obtained with the TPSSh/6-311+G(d) method and the corresponding wave functions obtained by solving the hollow cylinder model for (a) radial and (b) tangential orbitals. The quantum numbers *k*, *l*, and *n*, are also given. Source: Reprinted from Ref. [107]. Copyright 2014 American Chemical Society.

a clear diatropic ring current, giving evidence to the aromatic character of this species [107]. Not only $B_{20}$, but also $B_{24}$, $B_{27}^+$ and $B_{28}$ are found to have cylindrical aromatic character [107, 108].

## References


**1** Lipscomb, W.N., Pitochelli, A.R., and Hawthorne, M.F. (1959). *J. Am. Chem. Soc.* 81: 5833–5834.
**2** Pitochelli, A.R. and Hawthorne, F.M. (1960). *J. Am. Chem. Soc.* 82: 3228–3229.
**3** Knoth, W.H., Miller, H.C., England, D.C. et al. (1962). *J. Am. Chem. Soc.* 84: 1056–1057.
**4** Lipscomb, W.N. (1963). *Boron Hydrides*. New York: W.A. Benjamin.
**5** King, R.B. (2001). *Chem. Rev.* 101: 1119–1152.
**6** Núñez, R., Tarrés, M., Ferrer-Ugalde, A. et al. (2016). *Chem. Rev.* 116: 14307–14378.
**7** Garcia-Borràs, M., Osuna, S., Luis, J.M. et al. (2014). *Chem. Soc. Rev.* 43: 5089–5105.
**8** Chen, Z.F. and King, R. (2005). *Chem. Rev.* 105: 3613–3642.
**9** Bühl, M. and Hirsch, A. (2001). *Chem. Rev.* 101: 1153–1183.
**10** Kroto, H.W., Heath, J.R., O'Brien, S.C. et al. (1985). *Nature* 318: 162–163.
**11** Klein, D.J., Schmalz, T.G., Hite, G.E., and Seitz, W.A. (1986). *J. Am. Chem. Soc.* 108: 1301–1302.
**12** Liu, S.Z., Lu, Y.J., Kappes, M.M., and Ibers, J.A. (1991). *Science* 254: 408–410.
**13** Haddon, R.C. (1993). *Science* 261: 1545–1550.
**14** Fowler, P. (1991). *Nature* 350: 20–21.
**15** Hirsch, A. (1993). *Angew. Chem. Int. Ed. Engl.* 32: 1138–1141.
**16** Hirsch, A. (1994). *The Chemistry of Fullerenes*. Stuttgart: Thieme.
**17** Hirsch, A. (1999). *Top. Curr. Chem.* 199: 1–65.
**18** Martin, N. (2006). *Chem. Commun.* 2093–2104.
**19** Thilgen, C., Herrmann, A., and Diederich, F. (1997). *Angew. Chem. Int. Ed. Engl.* 36: 2269–2280.
**20** Cyrański, M.K., Howard, S.T., and Chodkiewicz, M.L. (2004). *Chem. Commun.* 2458–2459.
**21** Wodrich, M.D., Wannere, C.S., Mo, Y. et al. (2007). *Chem. Eur. J.* 13: 7731–7744.
**22** Bühl, M. (1998). *Chem. Eur. J.* 4: 734–739.
**23** Chen, Z., Wu, J.I., Corminboeuf, C. et al. (2012). *Phys. Chem. Chem. Phys.* 14: 14886–14891.

24 Fallah-Bagher-Shaidaei, H., Wannere, C.S., Corminboeuf, C. et al. (2006). *Org. Lett.* 8: 863–866.

25 Saunders, M., Cross, R.J., Jiménez-Vázquez, H.A. et al. (1996). *Science* 271: 1693–1697.

26 Bühl, M., Patchkovskii, S., and Thiel, W. (1997). *Chem. Phys. Lett.* 275: 14–18.

27 Saunders, M., Jiménez-Vázquez, H.A., Cross, R.J. et al. (1994). *Nature* 367: 256–258.

28 Chen, Z., Cioslowski, J., Rao, N. et al. (2001). *Theor. Chem. Acc.* 106: 364–368.

29 Pasquarello, A., Schlüter, M., and Haddon, R.C. (1992). *Science* 257: 1660–1661.

30 Pasquarello, A., Schlüter, M., and Haddon, R.C. (1993). *Phys. Rev. A: At. Mol. Opt. Phys.* 47: 1783–1789.

31 Zanasi, R. and Fowler, P.W. (1995). *Chem. Phys. Lett.* 238: 270–280.

32 Haddon, R.C. (1995). *Nature* 378: 249–255.

33 Johansson, M.P., Jusélius, J., and Sundholm, D. (2005). *Angew. Chem. Int. Ed.* 44: 1843–1846.

34 Ruoff, R.S., Beach, D., Cuomo, J. et al. (1991). *J. Phys. Chem.* 95: 3457–3459.

35 Haddon, R.C., Schneemeyer, L.F., Waszczak, J.V. et al. (1991). *Nature* 350: 46–47.

36 Poater, J., Fradera, X., Duran, M., and Solà, M. (2003). *Chem. Eur. J.* 9: 1113–1122.

37 Van Lier, G., Fowler, P., De Proft, F., and Geerlings, P. (2002). *J. Phys. Chem. A* 106: 5128–5135.

38 Krygowski, T.M. and Ciesielski, A. (1995). *J. Chem. Inf. Comput. Sci.* 35: 1001–1003.

39 Poater, J. and Solà, M. (2011). *Chem. Commun.* 47: 11647–11649.

40 Matito, E., Salvador, P., Duran, M., and Solà, M. (2006). *J. Phys. Chem. A* 110: 5108–5113.

41 Boltalina, O.V., Sidorov, L.N., Borshchevsky, A.Y. et al. (1993). *Rapid Commun. Mass Spectrom.* 7: 1009–1011.

42 Murata, M., Ochi, Y., Tanabe, F. et al. (2008). *Angew. Chem. Int. Ed.* 47: 2039–2041.

43 Sternfeld, T., Thilgen, C., Hoffman, R.E. et al. (2002). *J. Am. Chem. Soc.* 124: 5734–5738.

44 Sternfeld, T., Hoffman, R.E., Saunders, M. et al. (2002). *J. Am. Chem. Soc.* 124: 8786–8787.

45 Shabtai, E., Weitz, A., Haddon, R.C. et al. (1998). *J. Am. Chem. Soc.* 120: 6389–6393.

46 Campanera, J.M., Bo, C., and Poblet, J.M. (2005). *Angew. Chem. Int. Ed.* 44: 7230–7233.

47 Kroto, H.W. (1987). *Nature* 329: 529–531.

**48** Garcia-Borràs, M., Osuna, S., Swart, M. et al. (2013). *Angew. Chem. Int. Ed.* 52: 9275–9278.

**49** Rabilloud, F. (2010). *J. Phys. Chem. A* 114: 7241–7247.

**50** Hirsch, A., Chen, Z., and Jiao, H. (2000). *Angew. Chem. Int. Ed.* 39: 3915–3917.

**51** Chen, Z., Jiao, H., Hirsch, A., and Thiel, W. (2001). *J. Mol. Model.* 7: 161–163.

**52** Bremer, M., Schleyer, P.v.R., Schötz, K. et al. (1987). *Angew. Chem. Int. Ed. Engl.* 26: 761–763.

**53** Chen, Z., Jiao, H., Hirsch, A., and Schleyer, P.v.R. (2002). *Angew. Chem. Int. Ed.* 41: 4309–4312.

**54** Chen, Z.F., Hirsch, A., Nagase, S. et al. (2003). *J. Am. Chem. Soc.* 125: 15507–15511.

**55** Williams, R.V. (2001). *Chem. Rev.* 101: 1185–1204.

**56** Kiely, A.F., Haddon, R.C., Meier, M.S. et al. (1999). *J. Am. Chem. Soc.* 121: 7971–7972.

**57** Prato, M., Lucchini, V., Maggini, M. et al. (1993). *J. Am. Chem. Soc.* 115: 8479–8480.

**58** Smith, A.B., Strongin, R.M., Brard, L. et al. (1993). *J. Am. Chem. Soc.* 115: 5829–5830.

**59** Suzuki, T., Li, Q.C., Khemani, K.C., and Wudl, F. (1992). *J. Am. Chem. Soc.* 114: 7301–7302.

**60** Castro, A.C., Osorio, E., Jiménez-Halla, J.O.C. et al. (2010). *J. Chem. Theory Comput.* 6: 2701–2705.

**61** Shao, N., Bulusu, S., and Zeng, X.C. (2008). *J. Chem. Phys.* 128: 154326.

**62** Yu, X., Oganov, A.R., Popov, I.A., and Boldyrev, A.I. (2016). *J. Comput. Chem.* 37: 103–109.

**63** Wade, K. (1971). *J. Chem. Soc. D* 792–793.

**64** Mingos, D.M.P. (1972). *Nat. Phys. Sci.* 236: 99–102.

**65** Alexandrova, A.N., Birch, K.A., and Boldyrev, A.I. (2003). *J. Am. Chem. Soc.* 125: 10786–10787.

**66** Poater, J., Solà, M., Viñas, C., and Teixidor, F. (2016). *Chem. Eur. J.* 22: 7437–7443.

**67** Pitt, M.P., Paskevicius, M., Brown, D.H. et al. (2013). *J. Am. Chem. Soc.* 135: 6930–6941.

**68** Udovic, T.J., Matsuo, M., Unemoto, A. et al. (2014). *Chem. Commun.* 50: 3750–3752.

**69** Schleyer, P.v.R., Subramanian, G., and Dransfeld, A. (1996). *J. Am. Chem. Soc.* 118: 9988–9989.

**70** Schleyer, P.v.R., Najafian, K., and Mebel, A.M. (1998). *Inorg. Chem.* 37: 6765–6772.

71 Poater, J., Solà, M., Viñas, C., and Teixidor, F. (2014). *Angew. Chem. Int. Ed.* 53: 12191–12195.
72 Aihara, J. (1978). *J. Am. Chem. Soc.* 100: 3339–3342.
73 Poater, J., Viñas, C., Bennour, I. et al. (2020). *J. Am. Chem. Soc.* 142: 9396–9407.
74 Hirsch, A., Chen, Z., and Jiao, H. (2001). *Angew. Chem. Int. Ed.* 40: 2834–2838.
75 Liu, C., Popov, I.A., Chen, Z. et al. (2018). *Chem. Eur. J.* 24: 14583–14597.
76 Knight, W.D., Clemenger, K., de Heer, W.A. et al. (1984). *Phys. Rev. Lett.* 52: 2141–2143.
77 Katakuse, I., Ichihara, T., Fujita, Y. et al. (1985). *Int. J. Mass Spectrom. Ion Processes* 67: 229–236.
78 Thomas, O.C., Zheng, W., and Bowen, K.H. Jr., (2001). *J. Chem. Phys.* 114: 5514–5519.
79 Thomas, O.C., Zheng, W.-J., Lippa, T.P. et al. (2001). *J. Chem. Phys.* 114: 9895–9900.
80 Zheng, W.-J., Thomas, O.C., Lippa, T.P. et al. (2006). *J. Chem. Phys.* 124: 144304.
81 Hoshino, K., Watanabe, K., Konishi, Y. et al. (1994). *Chem. Phys. Lett.* 231: 499–503.
82 Cerowski, V., Rao, B.K., Khanna, S.N. et al. (2005). *J. Chem. Phys.* 123: 074329.
83 Höltzl, T., Veszprémi, T., Lievens, P., and Nguyen, M.T. (2010). *Aromaticity and Metal Clusters* (ed. P.K. Chattaraj), 271–296. Boca Ratón: CRC Press.
84 Cohen, M.L., Chou, M.Y., Knight, W.D., and De Heer, W.A. (1987). *J. Phys. Chem.* 91: 3141–3149.
85 Xia, X., Kuang, X., Lu, C. et al. (2016). *J. Phys. Chem. A* 120: 7947–7954.
86 Petrar, P.M., Sárosi, M.B., and King, R.B. (2012). *J. Phys. Chem. Lett.* 3: 3335–3337.
87 Kuznetsov, A.E., Boldyrev, A.I., Zhai, H.-J. et al. (2002). *J. Am. Chem. Soc.* 124: 11791–11801.
88 El Bakouri, O., Duran, M., Poater, J. et al. (2016). *Phys. Chem. Chem. Phys.* 18: 11700–11706.
89 Jiao, H., Schleyer, P.v.R., and Glukhovtsev, M.N. (1996). *J. Phys. Chem.* 100: 12299–12304.
90 Srinivas, S. and Jellinek, J. (2004). *J. Chem. Phys.* 121: 7243–7252.
91 Chen, Z., Neukermans, S., Wang, X. et al. (2006). *J. Am. Chem. Soc.* 128: 12829–12834.
92 Cui, P., Hu, H.-S., Zhao, B. et al. (2015). *Nat. Commun.* 6: 6331.
93 Jursic, B.S. (2000). *J. Mol. Struct. THEOCHEM* 498: 159–165.
94 Ngan, V.T. and Nguyen, M.T. (2010). *J. Phys. Chem. A* 114: 7609–7615.

**95** Glukhovtsev, M.N., Schleyer, P.v.R., and Stein, A. (1993). *J. Phys. Chem.* 97: 5541–5546.

**96** Alexandrova, A.N. and Boldyrev, A.I. (2003). *J. Phys. Chem. A* 107: 554–560.

**97** Corminboeuf, C., Wannere, C.S., Roy, D. et al. (2006). *Inorg. Chem.* 45: 214–219.

**98** Corbett, J.D. (2000). *Angew. Chem. Int. Ed.* 39: 670–690.

**99** Gao, C., Guo, Z., Liu, J.-H., and Huang, X.-J. (2012). *Nanoscale* 4: 1948–1963.

**100** Su, D.S., Perathoner, S., and Centi, G. (2013). *Chem. Rev.* 113: 5782–5816.

**101** Li, Z., Liu, Z., Sun, H., and Gao, C. (2015). *Chem. Rev.* 115: 7046–7117.

**102** Lewis, S.E. (2015). *Chem. Soc. Rev.* 44: 2221–2304.

**103** Hirst, E.S. and Jasti, R. (2012). *J. Organomet. Chem.* 77: 10473–10478.

**104** Yang, X., Ding, Y., and Ni, J. (2008). *Phys. Rev. B: Condens. Matter* 77: 041402.

**105** Wang, J., Lee, C.H., and Yap, Y.K. (2010). *Nanoscale* 2: 2028–2034.

**106** Golberg, D., Bando, Y., Huang, Y. et al. (2010). *ACS Nano* 4: 2979–2993.

**107** Pham, H.T., Duong, L.V., and Nguyen, M.T. (2014). *J. Phys. Chem. C* 118: 24181–24187.

**108** Duong, L.V., Pham, H.T., Tam, N.M., and Nguyen, M.T. (2014). *Phys. Chem. Chem. Phys.* 16: 19470–19478.

**109** An, W., Bulusu, S., Gao, Y., and Zeng, X.C. (2006). *J. Chem. Phys.* 124: 154310.

# 14

# Excited State Aromaticity

*"He who loves practice without theory is like the sailor who boards ship without a rudder and compass and never knows where he may cast"*

Leonardo da Vinci

## 14.1 Introduction

Aromaticity plays a major role to rationalize the stability, molecular structure, unusual chemical reactivity, particular spectroscopic features, and distinctive magnetic properties related to strong induced currents of ground state stable molecules with cyclic electron delocalization like [*N*]annulenes or *closo* borane clusters [1–5]. In 1938, Evans and Warhurst [6] already became aware of the similarity between the π-electrons of benzene and the six delocalized electrons in the cyclic transition state (TS) of Diels-Alder (DA) reactions. Today, it is generally accepted that most thermally allowed pericyclic reactions take place preferentially through concerted aromatic TSs [7–10]. Therefore, it is widely accepted nowadays that aromaticity influences the properties of not only stable molecules, but also of some TSs. What is probably less known is that the properties and reactivity of some molecules in certain excited states are strongly influenced by aromaticity [11]. Studies of aromaticity in the excited states are relatively scarce. Experimentally, this is the result of the inherent difficulty to study the properties of molecules in their excited states. From a theoretical point of view, the reason for the scarcity of studies of excited state aromaticity is twofold: (i) the correct treatment of excited states requires in many cases the use of expensive high-level multiconfigurational methods, and (ii) it is not clear whether the typical reference compound used by many indicators of aromaticity, that is, the ground state of benzene, can be used as a reference in excited states. Despite the difficulties encountered in the study of excited state aromaticity, there is now enough evidence showing that, in many cases, the concept aromaticity is required to understand excited state chemistry.

*Aromaticity and Antiaromaticity: Concepts and Applications*, First Edition.
Miquel Solà, Alexander I. Boldyrev, Michał K. Cyrański, Tadeusz M. Krygowski, and Gabriel Merino.

As is discussed in Chapter 2, [$N$]annulenes, that is, monocyclic conjugated hydrocarbons such as benzene, in their electronic ground state ($S_0$) obey the Hückel's rule of aromaticity [12–15]. This rule states that [$N$]annulenes of $D_{nh}$ symmetry with $4n + 2$ π-electrons are aromatic, whereas [$N$]annulenes with $4n$ π-electrons are antiaromatic. $4n + 2$ aromatic [$N$]annulenes are stabilized with respect to their open chain polyene analogues due to cyclic electron delocalization. The situation is reversed with [$N$]annulenes having $4n$ π-electrons. In 1973, Breslow [16] proposed the term *antiaromatic* for these [$N$]annulenes because cyclic delocalization of π-electrons, in this case, leads to a strong destabilization. For the lowest-lying triplet excited state ($T_1$) of [$N$]annulenes, the rule for aromaticity and antiaromaticity, called Baird's rule [17], is opposite to that for their closed-shell singlet ground states. Thus, [$N$]annulenes with $4n$ π-electrons are aromatic in the $T_1$ state, while those with $4n+2$ π-electrons exhibit antiaromatic character. There are many evidences showing that excited state aromaticity is not only found in $T_1$ states of [$N$]annulenes with $4n$ π-electrons, but also in their lowest-lying singlet excited states ($S_1$) corresponding to the occupation of the same orbitals as in the $T_1$ state but with unparalleled spins [18–22]. Soncini and Fowler [23] generalized the Hückel's and Baird's rules suggesting that [$N$]annulenes of $4n+2$ π-electrons at the lowest-lying electronic states with even spin (singlet, quintet, ...) and those of $4n$ π-electrons at the lowest-lying states with odd spin (triplet, septet, ...) are aromatic.

The existence of excited state aromaticity helps to rationalize the molecular properties and the chemical reactivity of some molecules in their $T_1$ and $S_1$ excited states. In the next sections of this chapter, we review some theoretical and experimental studies that verify the existence of excited state aromaticity, and we discuss some selected examples of the influence of aromaticity in excited state properties and reactions. To complement the information given in this chapter, we refer the reader to Ref. [11].

## 14.2 Theoretical and Experimental Studies of Excited State Aromaticity

The first mention to the aromaticity of excited states can be traced back to the work by Baird in 1972 [17]. Since then, several studies have provided further support to the concept of (anti)aromaticity in the $T_1$ and $S_1$ excited states. In the following subsections, we review first the most relevant theoretical and computational works and then the experimental studies that sustain the presence of excited state aromaticity.

### 14.2.1 Theoretical and Computational Studies

In his landmark study of 1972, Baird used perturbational molecular orbital theory to suggest that annulenes that are antiaromatic in their singlet ground state are aromatic in their lowest-lying triplet state and vice versa for annulenes that are aromatic in the ground state [17]. In the same paper, Baird performed neglect of diatomic differential overlap (NDDO) calculations to confirm the validity of his rule. Baird's rule was further substantiated theoretically from the calculation of resonance energies with the Hückel's method by Aihara who found a sign reversal for the resonance energies of small annulenes when going from the ground state to the lowest-lying triplet excited state [8]. A few years later, Fratev et al. using a limited configuration interaction treatment showed that the equilibrium structure of the $T_1$ state of cyclobutadiene presents bond length equalization and $D_{4h}$ symmetry, which is an indication of aromaticity [18]. The $4n$ $\pi$-electron Baird's rule for triplet state aromaticity was further confirmed through nucleus-independent chemical shifts (NICSs), magnetic susceptibility, and aromatic stabilization energy (ASE) calculations by Schleyer et al. [24, 25]. In the work by Gogonea and coworkers, the $T_1$ state of $C_4H_4$, $C_5H_5^+$, $C_7H_7^-$, $C_8H_8$, and $C_9H_9^+$ was found aromatic, the optimized geometry being of $D_{nh}$ symmetry with C–C bond lengths close to those of benzene [25]. The study of ring currents in $4n$ $\pi$-electron monocycles [26] provided further support to the aromaticity of their lowest-lying triplet states. Zilberg and Hass [27], using valence bond theory, concluded that [*N*]annulenes with $4n$ $\pi$-electrons in their $T_1$ state can be considered $4n$-$2$ $\pi$-electron Hückel's aromatic systems with two nonbonding same-spin $\pi$-electrons. The validity of the Baird's rule was additionally substantiated by Krygowski and Cyrański [28] through the calculation of the harmonic oscillator model of aromaticity (HOMA) descriptor for the following series of $4n$ $\pi$-electron annulenes: $C_4H_4$, $C_5H_5^+$, $C_6H_6^{2+}$, $C_7H_7^-$, $C_8H_8$, and $C_9H_9^+$. They found that the triplet state of $4n$ $\pi$-electron annulenes exhibit an aromatic character that intensifies as the size of the ring increases. In 2008, a theoretical work [29] based on the analysis of the bifurcation in the $\pi$-contribution to the electron localization function ($ELF_\pi$) for the lowest-lying triplet state of $4n$ $\pi$-electron monocycles also confirmed the validity of Baird's rule, showing that for these species the maximal difference between the bifurcation values is small and the ring-closure bifurcation value is relatively high. But excited state aromaticity is not only ascribed to triplet state aromaticity for $4n$ $\pi$-electron monocycles. For instance, the lowest-lying singlet excited state ($S_1$) of square cyclobutadiene and cyclooctatetraene was reported to be aromatic according to NICS descriptors of Karadakov [20, 21] computed at the complete active space self-consistent field (CASSCF) level. Feixas et al. [22] analyzed the electron delocalization and (anti)aromaticity of a series of low-lying excited states of cyclobutadiene, benzene, and cyclooctatetraene with different multiplicities at the CASSCF level by means of electron delocalization indices [30]. The results obtained proved that benzene is antiaromatic in the $T_1$ state, whereas cyclobutadiene and cyclooctatetraene are aromatic in their vertical $S_2$ ($S_2$ is the second lowest-lying singlet excited stated but it has the same electronic configuration as $T_1$, except that the unpaired electrons have different spins) and $T_1$ excited states in line with the predictions from Baird's rule for triplet state aromaticity. The electronic indices obtained at the CASSCF level of the lowest-lying quintet state of benzene and septet state of cyclooctatetraene indicated that these states are not aromatic, and, therefore, did not support the Soncini and Fowler generalization of Baird's rule. Finally, let us mention that Zhu et al. [31, 32] confirmed the triplet aromaticity of *4n* π-electron monocycles using two variants of the isomerization stabilization energies (ISEs) method. These methods were based on two different homodesmotic reactions that compare the methyl derivative with the exocyclic methylene isomers ($ISE_I$), and the indene with the isoindene isomers ($ISE_{II}$, see Figure 14.1). In 2011, Dominikowska and Palusiak [33] analyzed the applicability of Clar's aromatic π-sextet theory to a series of doubly charged polycyclic aromatic hydrocarbons (PAHs) in the lowest-lying singlet and triplet states. They found that the Clar model can be applied to discuss the aromaticity of doubly-charged PAHs in the lowest-lying singlet state. However, it fails when applied to the triplet states. Interestingly, Zhu and coworkers [34] showed that the "hyperconjugatively" antiaromatic 5,5-difluorocyclopentadiene compound [35] becomes "hyperconjugatively" aromatic in the lowest-lying triplet state.

A summary of some indicators of aromaticity used to assess the aromaticity of the ground and the lowest-lying triplet excited states can be found in Table 14.1. Results in Table 14.1 show that the lowest-lying triplet state of butadiene and cyclooctatetraene are aromatic and that of benzene is antiaromatic. As a whole, the aromaticity of the lowest-lying triplet and singlet excited states of [*N*]annulenes

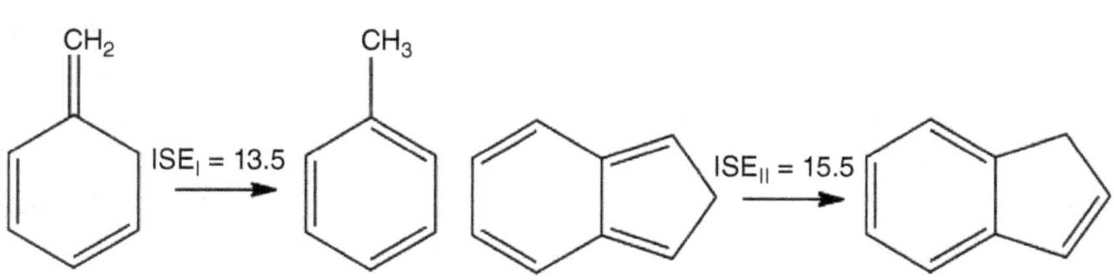


**Figure 14.1** The ISE (in kcal/mol) of benzene in the $T_1$ antiaromatic state determined with two different methods: (a) methyl benzene–5-methylene-1,3-cyclohexadiene ($ISE_I$) and (b) isoindene–indene ($ISE_{II}$).

**Table 14.1** Hückel's resonance energies (HRE in units of β), isomerization stabilization energies (ISE in kcal/mol), nucleus-independent chemical shifts (NICS in ppm), *para*-delocalization index (PDI in electrons), aromatic fluctuation index (FLU), multicenter indices ($I_{ring}$ and MCI in electrons), bifurcation value for ring closure of the π basin (RCBV[$ELF_\pi$]) and span in the bifurcation values (ΔBV[$ELF_\pi$]).

| | Cyclobutadiene | | Benzene | | Cyclooctatetraene | |
|---|---|---|---|---|---|---|
| **Aromaticity index** | $S_0$ | $T_1$ | $S_0$ | $T_1$ | $S_0$ | $T_1$ |
| HRE[a)] | −1.226 | 0.305 | 0.273 | −0.692 | −0.595 | 0.186 |
| $ISE_I$ | — | −16.4[b)] | −33.2[c)] | 13.5[b)] | — | −15.6[b)] |
| $ISE_{II}$[d)] | 35.2 | −16.2 | −21.8 | 15.5 | 10.2 | −17.0 |
| NICS(0) | 36.4[e)] | −3.7[e)] | −8.2[e)] | 39.6[e)] | 40.7[f)] | −8.9[f)] |
| $NICS(1)_{zz}$ | 88.1[e)] | −16.5[e)] | −27.8[e)] | 90.6[e)] | 98.2[f)] | −25.6[f)] |
| PDI[g)] | — | — | 0.050 | 0.015 | — | — |
| FLU[g)] | 0.036 | 0.009 | 0.000 | 0.020 | 0.024 | 0.003 |
| $I_{ring}$[g)] | 0.006 | 0.033 | 0.031 | 0.003 | 0.001 | 0.003 |
| MCI[g)] | 0.009 | 0.036 | 0.044 | 0.002 | 0.001 | 0.005 |
| RCBV($ELF_\pi$)[h)] | 0.082 | 0.648 | 0.907 | 0.144 | 0.359 | 0.845 |
| ΔBV($ELF_\pi$)[h)] | 0.908 | 0.000 | 0.000 | 0.745 | 0.637 | 0.000 |

a) Reference [8];
b) Reference [31];
c) Reference [36];
d) Reference [32];
e) Reference [20];
f) Reference [21];
g) Reference [22], PDI and FLU obtained using the Fulton expression for delocalization indices;
h) Reference [29].

with 4*n* π-electrons has been confirmed by theoretical studies based on the geometric, energetic, magnetic, and electronic behavior of these species.

### 14.2.2 Experimental Studies

As to the experimental proofs, the identification by Breslow et al. [37–39] of the planar triplet ground states of $C_5H_5^+$ and $C_5Cl_5^+$ was a first indication of the validity of Baird's triplet-state aromaticity. This study was more recently complemented by a pulsed-field-ionization zero-kinetic-energy photoelectron spectroscopic study of the triplet ground state of $C_5H_5^+$ that confirmed the $D_{5h}$ symmetry of this cation in this state [40]. The relative stability of photochemically obtained cyclobutadiene at cryogenic temperatures that slowly decomposes into two acetylene molecules [41, 42] was taken by Fratev et al. [18] as an indication of the aromaticity of its $T_1$ state according to Baird's rule. Similarly, Paquette et al. [43] reported in 1974 the synthesis and spectroscopic characterization of all isomeric dimethylcyclooctatetraenes. The authors irradiated the 1,2- and 1.4-dimethylcyclooctatetraenes in acetone solutions for more than 100 hours without almost decomposition of the starting material. The photostability of these species was attributed to the aromaticity of their $T_1$ states. In 2015, Osuka and coworkers [44] found that two-related bis-rhodium hexaphyrins (**R26H** and **R28H**, Figure 14.2) containing 26 and 28 $\pi$-electrons peripheries exhibit spectroscopic properties consistent with Baird's rule. **R28H** is the doubly reduced and diprotonated analogue of **R26H**. **R26H** and **R28H** are Hückel's aromatic and antiaromatic, respectively. The ground state absorption spectrum of **R26H** show two intense B-like bands and several Q-like bands as normally found in aromatic porphyrinoids (see Figure 10.10). Conversely, **R28H** presents relatively broad and weak absorption spectra in the visible region that is normally associated to antiaromatic porphyrinoids. The authors were able to record the spectra of the $T_1$ states of these two bis-rhodium hexaphyrins. They found that the spectrum of **R26H** in the $T_1$ state becomes broader and its intensity weaker, that is, similar to that of antiaromatic porphyrinoids in the ground state. Similarly, the absorption spectrum of **R28H** is sharpened and more intense resembling that of aromatic porphyrinoids in the ground state. In addition, anisotropy of the induced current density (ACID), NICS, and HOMA calculations for **R26H** and **R28H** agree with the change in aromatic character suffered by these bis-rhodium hexaphyrins when going from the ground to the lowest-lying triplet excited state in line with the expectations based on Baird's rule. The same group of Osuka and coworkers [45] demonstrated aromaticity reversal in the singlet excited state of internally 1,3-phenylene-strapped [26]- and [28]hexaphyrins, in agreement with the fact that

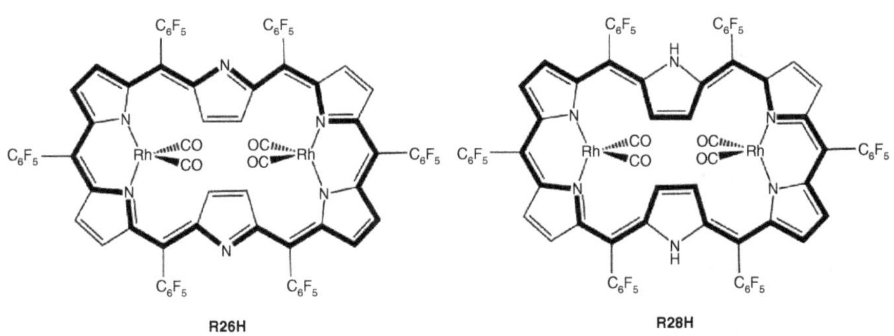


**Figure 14.2** The bis-rhodium hexaphyrins (**R26H** and **R28H**) containing 26 and 28 $\pi$-electrons peripheries marked in bold. Source: Adapted from Ref. [44].

the lowest-lying singlet excited state of $4n + 2$ and $4n$ π-electrons are antiaromatic and aromatic, respectively (Figure 14.2).

Further evidence in favor of Baird's rule came from the properties and the reactivity of excited states that are discussed in the next sections.

## 14.3 Influence of Aromaticity in the Excited State Properties

Excited state aromaticity has an important effect on the excited state properties of some molecules. An obvious example is the molecular structure that, as shown in Section 14.2, is highly symmetric for most Baird aromatic species in the lowest-lying singlet and triplet excited states. Another well-known example is the ring current reversal when going form aromatic $S_0$ (diatropic) to antiaromatic $T_1$ (paratropic) or the other way round [26]. In the following subsections, we describe some other selected less obvious examples.

### 14.3.1 Molecular Dipole Moments

Triplet state aromaticity is useful to rationalize dipole moments of fulvenes, fulvalenes, and azulenes [46]. As can be seen in Figure 14.3, in the ground

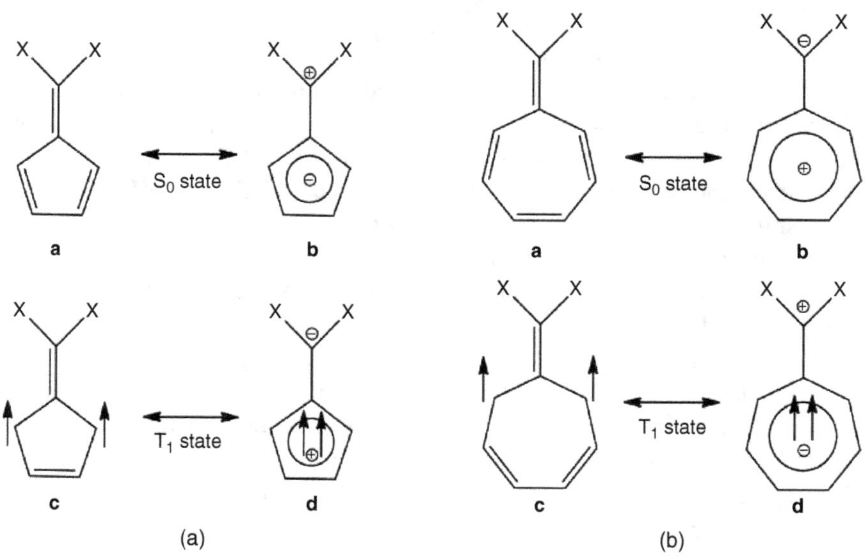


**Figure 14.3** The two most important resonant structures for the ground state ($S_0$) and the lowest-lying triplet state ($T_1$) of (a) penta- and (b) heptafulvenes. Resonances structures **b** and **d** are particularly important in the $S_0$ and $T_1$ states, respectively, due to aromaticity. Source: Adapted from Ref. [11].

state of penta- and heptafulvenes, the zwitterionic resonance structure **b** plays a nonnegligible role in the description of the electronic structure due to the Hückel's aromaticity generated in the 5- and 7-MRs. Consequently, both penta- and heptafulvenes have relatively large dipole moments that point in opposite directions. When these molecules are excited to the lowest-lying triplet ($T_1$) or singlet ($S_1$) states, the zwitterionic resonance structures **d** are stabilized by Baird aromaticity and there is a reversal in the direction of the dipole moments. At the CASSCF level [46], the dipole moment of the ground state of pentafulvene is 0.3 D (pointing in the direction of the exocyclic carbon atom) and that of the $T_1$ state is −0.5 D. On the other hand, for heptafulvene, the dipole moments are −0.2 and 0.3 D for the $S_0$ and $T_1$ states, respectively. This change in the direction of the dipole moments of penta- and heptafulvenes when going from $S_0$ to $T_1$ is a manifestation of the reversal of the electron-counting rule for aromaticity. Interestingly, resonance structures **b** of pentafulvene and **d** of heptafulvene are stabilized with electron-donating X substituents, while electron-withdrawing X substituents favor resonance structures **d** of pentafulvene and **b** of heptafulvene. Therefore, the dipole moment of these species in the $S_0$ and $T_1$ or $S_1$ states can be tuned through the choice of substituents with suitable electronic character [47].

### 14.3.2 Singlet-Triplet Energy Gaps

Not only the dipole moment of the penta- and heptafulvenes, but also the singlet-triplet energy gap ($\Delta E_{ST}$) can be tuned by introducing appropriate substituents in the exocyclic carbon atom. For pentafulvenes, electron-donating groups in the exocyclic carbon atom stabilize the ground state by increasing the contribution of the resonance structure **b** and destabilize the triplet state by diminishing the weight of resonance structure **d**, thus increasing the singlet-triplet energy gap compared to parent pentafulvene. The effect of electron-withdrawing groups is the opposite. For heptafulvenes, we have a similar situation, but now electron-donating substituents in the exocyclic carbon atom reduce the singlet-triplet energy gap. Ottosson et al. proved this change in $\Delta E_{ST}$ experimentally for a limited set of substituted fulvenes and computationally for larger sets of penta- and heptafulvenes, benzofulvenes, and dibenzofulvenes [48, 49]. Since the $T_1$ and $S_1$ states have a similar aromatic behavior in these compounds, the energy of the $S_0 \rightarrow S_1$ transition can also be tuned through substitution.

In 2015, Tovar, Casado, and coworkers [50] presented a quinoidal 1,6-methano [10]annulenes compound (**TMTQ**, Figure 14.4) which exhibits a small singlet-triplet energy gap of only 4.9 kcal/mol, the singlet being more stable than the triplet. This was tentatively attributed to the stabilization of the $T_1$ state through Baird-aromaticity of the central 1,6-methano[10]annulene (M10A) fragment.

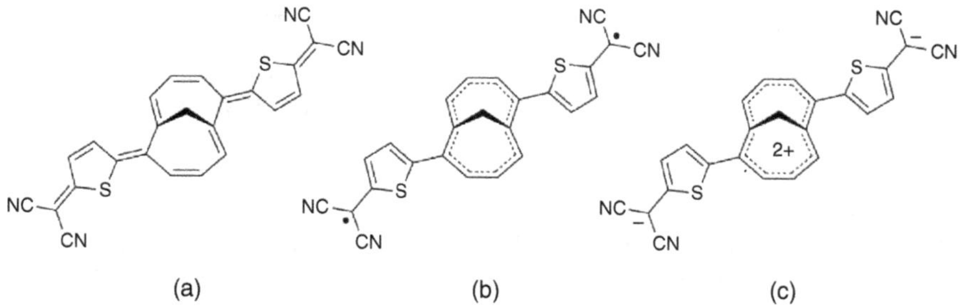


**Figure 14.4** The covalent (a), diradical (b), and ionic (c) resonance structures of TMTQ. Source: Adapted from Ref. [51].

According to these authors, the resonance structure **c** in Figure 14.4 with an 8π-electron 1,6-methano[10]annulenyl dication moiety ($M10A^{2+}$) is essential for the accurate description of the electronic structure of **TMTQ** in the $T_1$ state ($^3$**TMTQ**). However, this state could also be influenced by structure **b** that has a closed-shell 10π-electron Hückel-aromatic ring and two terminal dicyanomethyl radicals. In two subsequent works, Ottosson, Solà et al. [51, 52], using different functionals and multiconfigurational calculations, found that the Hückel-aromatic structure **b** dominates the triplet state description. It was found from charge and spin density distributions and separation of the aromatic fluctuation index (FLU) aromaticity index into α and β electron contributions that the Baird-aromatic character contributes at most 12 percent. Therefore, the small singlet-triplet energy gap in this particular system cannot be attributed to Baird-aromaticity but to Hückel-aromaticity of the M10A ring. Interestingly, by substituting the thiophene rings by phenyl rings or by adding substituents to the thiophene groups, we can tune the closed-shell $4n+2$ Hückel and open-shell $4n$ Baird aromatic character of the M10A ring in the $^3$**TMTQ** species. Consequently, TMTQ and related systems in their triplet states have to be considered Hückel-Baird hybrids [51].

### 14.3.3 Photoacidity

The acid-base behavior of (anti)aromatic compounds can be substantially different in the ground and lowest-lying triplet states. Consider the example of 9H-fluorene and 5H-suberene in Figure 14.5 discussed by Wan and coworkers [53, 54]. 9H-Fluorene in the ground state forms $4n + 2$ π-electron annulenyl anion when deprotonated. Accordingly, 9H-fluorene is relatively acidic in the ground state. However, upon excitation to the $S_1$ state, this compound becomes nonacidic. The reverse is true for 5H-suberene. In the ground state, 5H-suberene yields a $4n$ π-electron annulenyl anion by deprotonation and, accordingly, it is nonacidic.

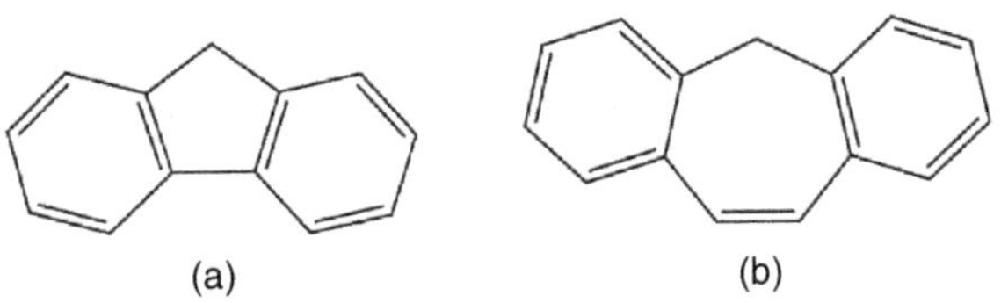

Figure 14.5 The molecular structure of (a) 9H-fluorene and (b) 5H-suberene.

On the other hand, irradiation of 5H-suberene excites this compound to the $S_1$ state where loss a proton is favored by the formation of a Baird aromatic $4n$ $\pi$-electron annulenyl anion. Therefore, 5H-suberene is acidic under irradiation. In general, $4n$ $\pi$-electron annulenyl anions in the $T_1$ state have lower proton affinities than $4n+2$ $\pi$-electron annulenyl anions [55]. And $4n+2$ $\pi$-electron annulenyl anions have higher proton affinities in their $T_1$ than in their $S_0$ states. Thus, the (anti)aromaticity of the $T_1$ and $S_0$ states has an important effect on the acid-base properties of aromatic compounds.

## 14.4 Influence of Aromaticity in the Excited State Reactivity

There are a number of photochemical reactions that can only be understood by resorting to the concept of excited state (anti)aromaticity. This fact has not been widely recognized until the last years [11]. The first evidence that the reactivity of the excited states of certain molecules was affected by (anti)aromaticity was given in 1985 by Wan and Krogh [56] in their study of the photolysis of 9H-fluoren-9-ol in aqueous methanol solution (see Figure 14.6). The driving force for the efficient formation of 9-methoxy-9H-fluorene was considered to be the formation of an aromatic $4\pi$ cationic system in the excited-state. This cationic species in the excited state could be easily formed because of aromatic stabilization, while its formation is destabilized by antiaromaticity in the ground state and this justify the absence of reaction in the dark. A very simple example concerns the hydrogenation of

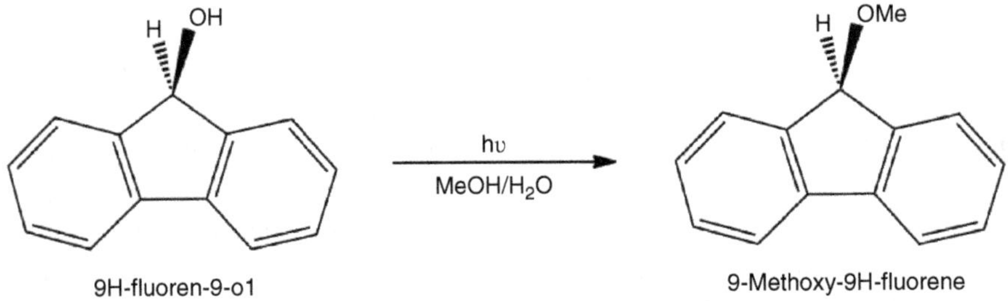


Figure 14.6 The photolysis of 9H-fluoren-9-ol.

annulenes. It has been recently shown that the hydrogenation of benzene is endergonic in the ground state and exergonic in the first triplet state [57]. In contrast, cyclooctatetraene is unreactive in the first triplet state. This is in agreement with the Baird's rule stating that benzene is antiaromatic and destabilized in the $T_1$ and $S_1$ states, whereas cyclooctatetraene is aromatic and stabilized in these states. Like this photochemical hydrogenation, there are many photochemical reactions that are influenced by excited state (anti)aromaticity. In the following subsections, we discuss some selected examples of photochemical processes driven by the deantiaromatization of $4n + 2$ π-electron species or the aromatization due to formation of $4n$ π-electron rings in excited states. A more exhaustive discussion can be found in two reviews by Ottosson et al. [11, 58].

### 14.4.1 Photoisomerizations

Already in benzene, it is possible to observe photoisomerizations that are attributed to the strongly antiaromatic character of its $S_1$ state as previously discussed. In particular, the main products of the photoisomerization upon excitation of benzene to the $S_1$ state are benzvalene and fulvene (see Figure 14.7), although the quantum yields are low [8, 59, 60].

One of the most interesting examples of photoisomerizations is the Z/E photoisomerization of olefins with $4n\pi$- or $(4n + 2)\pi$-electron annulenyl substituents. This photoisomerization shows an interesting a zigzag variation (from allowed to forbidden) with $n$ that can be explained by taking into account the aromaticity of the annulenyl groups in the $T_1$ state [61–66]. Experimentally, the photoisomerization, when allowed, takes place preferably in the $S_1$ state. As discussed before, the aromaticity behavior of the $S_1$ and $T_1$ states in annulenes is similar, and, on this basis, Ottosson et al. [63–66] performed calculations on the $T_1$ state to rationalize the Z/E photoisomerization of annulenyl substituted olefins. As can be seen in Figure 14.8, the shape of the potential energy surface (PES) of the triplet state determines the isomerization quantum yields. After photoexcitation to the $T_1$

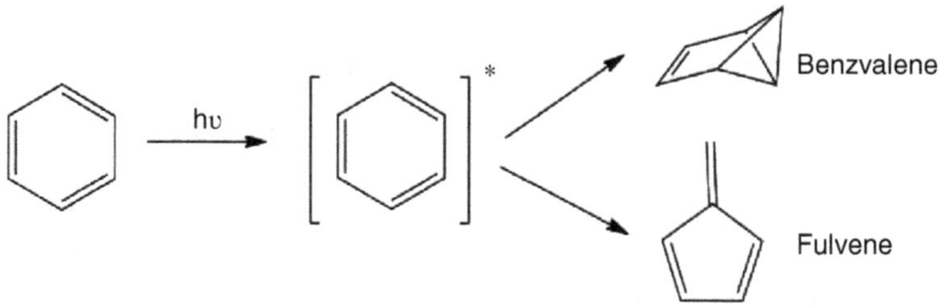


**Figure 14.7** The photorearrangements observed in benzene upon excitation to its $S_1$ state.

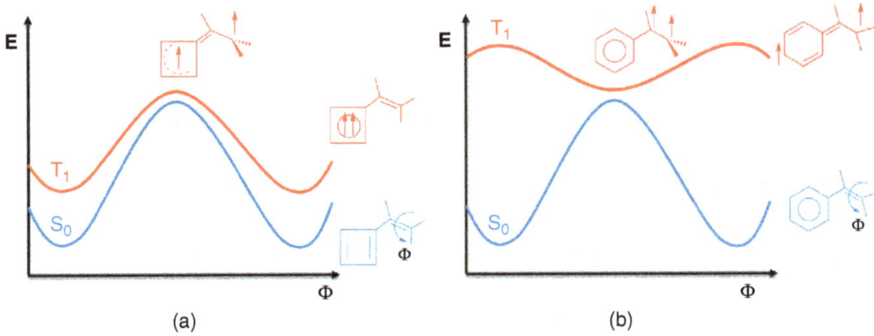


**Figure 14.8** The Z/E photoisomerization of (a) vinylcyclobutadiene and (b) styrene. Source: Adapted from Ref. [11].

state, the biradical can be mainly localized in the double bond or in the annulenyl ring. Let us consider first vinylcyclobutadiene. In its $T_1$ state, the biradical is mainly located in the 4π-electron aromatic 4-MR. Consequently, the double bond character of the vinyl group is quite preserved in the $T_1$ state. This fact, together with the aromatic character of the ring that stabilizes vinylcyclobutadiene in the $T_1$ state, makes photoisomerization difficult for this system. Now, let us discuss the situation in styrene. In the $T_1$ state, the biradical species cannot generate an aromatic ring, and, therefore, the unpaired electrons are localized mostly in the olefinic bond. The most stable situation in this case corresponds to the resonance structure with the biradical in the olefinic bond and the ring with a closed-shell structure with Hückel aromatic character. In this species, the double bond character of this bond is highly diminished and the C=C bond twist is facile, thus leading to photoisomerization.

The effect previously described for cyclobutadienyl and phenyl substituents can be generalized to any annulenyl substituent in the following way: a twist away from planar structures in olefins with $T_1$ antiaromatic annulenyl substituents is energetically favorable, but that in olefins with $T_1$ aromatic annulenyl groups is unfavorable.

### 14.4.2 Excited State Intramolecular Proton Transfer

The molecules that contain both an acid and a basic group in close proximity may rearrange via an intramolecular proton transfer upon photoexcitation, a process usually referred to as excited-state intramolecular proton transfer (ESIPT) [67]. ESIPT processes have attracted considerable attention due to their implication in biology, their role in many laser dyes, in materials able to store information at the molecular level, in high-energy radiation detectors, and in fluorescent probes [68].

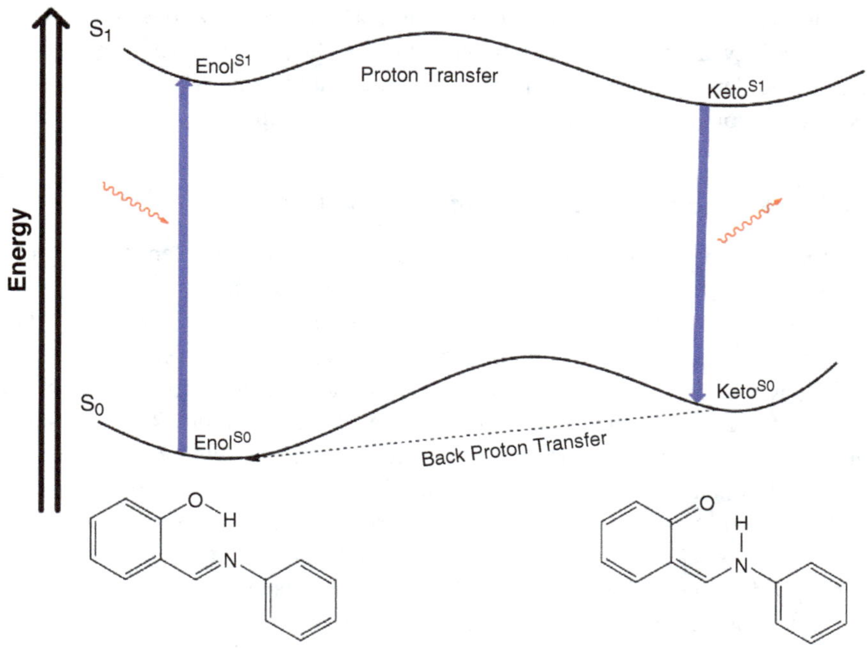


**Figure 14.9** Excited state intramolecular proton transfer in salicylideneaniline.

One of the molecules that exhibits ESIPT is salicylaldimine [69] and related compounds such as the salicylidenaniline (SA) [70]. Figure 14.9 depicts the ESIPT process in SA. In the ground state, the most stable isomer is the enol form. The tautomerization in this state is not possible because it is an endergonic process that has to surpass a relatively high-energy barrier. After the molecule is excited to the first singlet electronic state, there is an important change in the relative stability of the two isomers. In the $S_1$ state, the most stable isomer is the keto form and the proton transfer from the enol to the keto form takes place through a small energy barrier in few femtoseconds. Radiative deactivation leads to the keto form in the $S_0$ state that evolves to the enol form. Nonradiative deactivation can also takes place through E–Z isomerization in the keto form of the $S_1$ state. Roncha-Rinza and coworkers [70] showed that, not surprisingly, the aromaticity of the six-membered ring (6-MR) in the initial phenol group in the $S_0$ state is clearly reduced when going from the enol to the keto form. Moreover, after photoexcitation the phenol 6-MR in the enol tautomer becomes much less aromatic. The aromaticity of this ring keeps approximately constant when going from the enol to the keto form in the $S_1$ state as indicated by the $\theta$ aromaticity index of Matta and the $I_{ring}$, $I_{NG}$, MCI, and $I_{NB}$ multicenter indices of aromaticity. The results indicate that the differences in proton transfer reactivity in the $S_0$ and $S_1$ states of SA are due to a considerable loss of aromaticity after photoexcitation that facilitates the proton transfer reaction in the excited state. Therefore, changes in the aromaticity of a molecule after electron excitation explains the ESIPT process in SA and related species.

### 14.4.3 Photochemical Formation of *Ortho*-Xylylenes

Photoexcitation of benzene derivatives can generate *o*-xylylenes that can react further to different products. The initial formation of *o*-xylylene derivatives is driven by the release of antiaromatic destabilization of the lowest-lying singlet excited state of the benzene ring. Three selected reactions are shown in Figure 14.10. In the first example (Figure 14.10a), photoexcitation of 1,2-dihydronaphthalene leads to a C—C bond break that results in a transient *o*-xylylene species. The excited state antiaromatic character of 1,2-dihydronaphthalene explains the formation of *o*-xylylene species. After E-Z isomerization of this *o*-xylylene intermediate, the formation of benzocyclo[3.1.0] hexane occurs [71]. The first part of the second reaction (Figure 14.10b) is similar to the previous one. When 1-methyl-1,2-dihydronaphthalene is irradiated, a C—C bond cleavage occurs leading to a transient *o*-xylylene species. This intermediate undergoes a [1, 7] sigmatropic H-shift to form the final stable product [72]. In the last example (Figure 14.10c), 1-phenyl-1-*o*-tolylethylene on irradiation undergoes an intramolecular [1,5]-hydrogen shift generating a short-live *o*-xylylene intermediate. Subsequently, this intermediate is trapped in a thermal DA reaction with maleic anhydride [72].

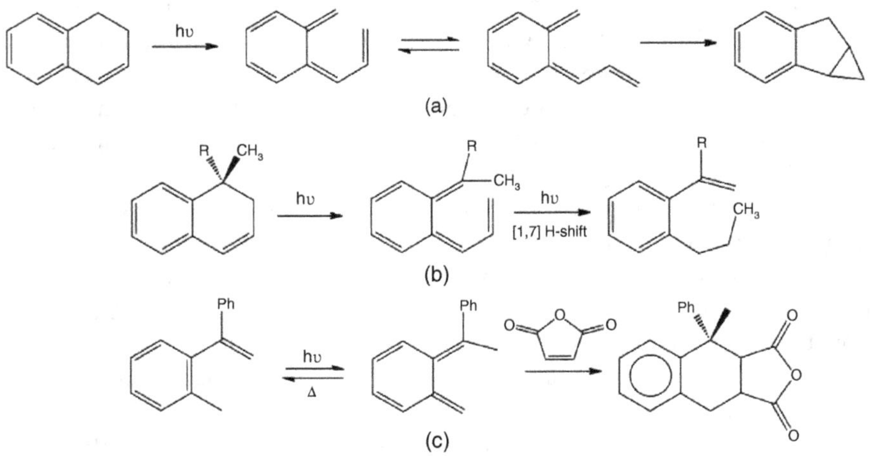


**Figure 14.10** Photoreactivity of benzene derivatives that upon irradiation produce *o*-xylylene intermediates.

### 14.4.4 Photochemical Pericyclic Reactions

The classical photochemical pericyclic reactions are cycloadditions and electrocyclizations. Rationalization of these photochemical reactions comes from the Dewar-Zimmerman's rules [7, 73] that state that photochemical processes are allowed if the TS involves a cyclic orbital array of Hückel topology (with an even number of phase orbital inversions) with $4n$ $\pi$-electrons and it is forbidden if comprises $4n+2$ $\pi$-electrons. For benzene, there are three types of alkene photocycloadditions [74–77], that is, [2+2], [3+2], and [4+2] photocycloadditions (see Figure 14.11). In all cases, benzene has to be irradiated and promoted to the $S_1$ excited state [78]. In this antiaromatic state, benzene is highly reactive at variance with its ground state. This is why Papadakis and Ottosson [58] consider that benzene could be viewed as a "Dr. Jekyll and Mr. Hyde" molecule with the reactive antiaromatic excited state representing its "Mr. Hyde" character. These photocycloadditions are driven by the relief of antiaromatic character of the benzene ring in the $S_1$ state. Concerted [3 + 2] and [4 + 2] photocycloadditions of olefins are forbidden to occur from the first excited-state. The fact that [3 + 2] and [4 + 2] cycloadducts are nevertheless observed to be formed can be explained because the reaction occurs in a nonconcerted fashion.

Cyclooctatetraene (COT) can be generated with reactions that proceed entirely through the ground state, but also with photochemical processes starting

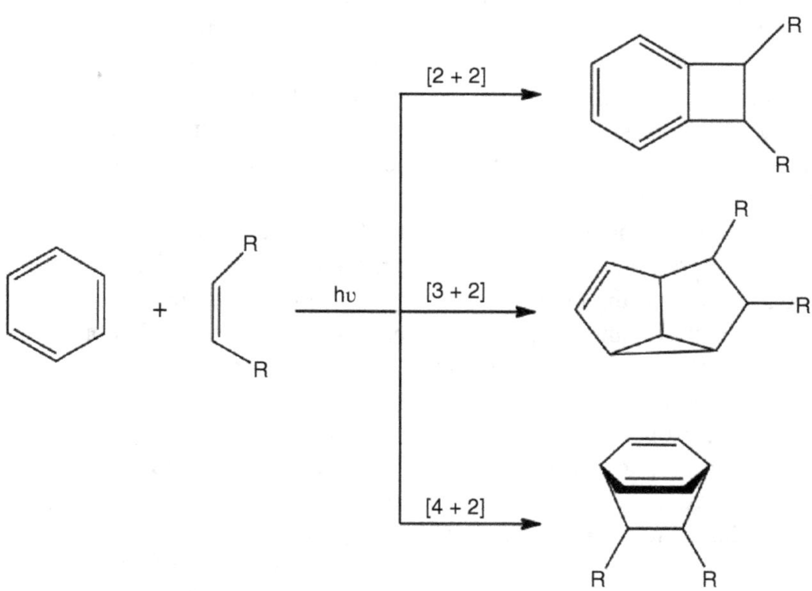


**Figure 14.11** Photocycloadditions involving benzene and alkenes.

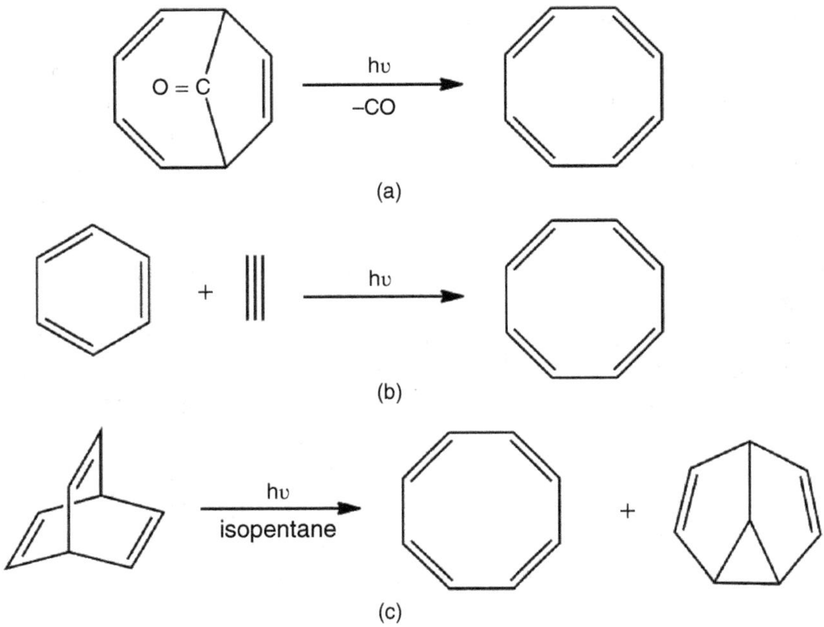


**Figure 14.12** Some photocycloadditions leading to cyclooctatetraene.

from suitable precursors. Three of these photochemical syntheses are depicted in Figure 14.12. In these cases, gain of aromaticity in the excited state is a driving force for the COT formation. The first one uses bicyclo[4.2.1]nona-2,4,7-trien-9-one as a precursor and represents the most efficient route to COT. It consists in a cheletropic photoextrusion of CO that produces COT in 82 percent yield upon irradiation in a tetrahydrofuran solution [79]. The second one corresponds to the [2 + 2] cycloaddition of acetylene to benzene to form a bicyclo[4.2.0]octa-2,4,7-triene followed by a photochemically allowed disrotatory cyclobutene ring opening to generate the COT [80]. It occurs upon irradiation of an acetylene-saturated benzene solution, but the process is not very efficient and the quantum yield is very low. In the third route, the precursor is barrelene that upon irradiation forms COT in low yields, the main product being semibullvalene formed through a di-π-methane rearrangement [81].

Another interesting example of a reaction that is facilitated by the relief of excited state antiaromaticity of the benzene ring is the photoreaction that forms benzofulvenes from benzannelated enynes [82] shown in Figure 14.13. Photoexcitation of aromatic enynes is followed by alleviation of the destabilizing antiaromatic character of the benzene ring to form a twisted alkene diradical in an analogous process to that observed in the Z/E photoisomerization of annulenyl substituted olefins. The diradical formed attacks the triple bond and gives a cyclic

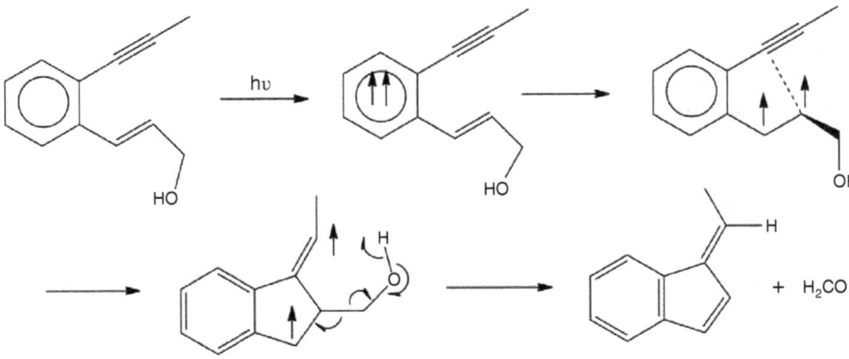


**Figure 14.13** $C_1$–$C_5$ cycloaromatization reaction of enynes to form benzofulvenes.

1,4-diradical intermediate. The 1,4-diradical product of the C1–C5 cyclization undergoes an internal H atom transfer coupled with the fragmentation of an exocyclic C—C bond to form the corresponding benzofulvene and to release formaldehyde.

## References


1 Schleyer, P.v.R. (2001). *Chem. Rev.* 101: 1115–1118.
2 Schleyer, P.v.R. (2005). *Chem. Rev.* 105: 3433–3435.
3 Schleyer, P.v.R. and Jiao, H. (1996). *Pure Appl. Chem.* 68: 209–218.
4 Krygowski, T.M., Cyrański, M.K., Czarnocki, Z. et al. (2000). *Tetrahedron* 56: 1783–1796.
5 Stanger, A. (2009). *Chem. Commun.* 1939–1947.
6 Evans, M.G. and Warhust, E. (1938). *Trans. Faraday Soc.* 34: 614–624.
7 Dewar, M.J.S. (1971). *Angew. Chem. Int. Ed. Engl.* 10: 761–776.
8 Aihara, J. (1978). *Bull. Chem. Soc. Jpn.* 51: 1788–1792.
9 Houk, K.N., Gonzalez, J., and Li, Y. (1995). *Acc. Chem. Res.* 28: 81–90.
10 Schleyer, P.v.R., Wu, J.I., Cossio, F.P., and Fernandez, I. (2014). *Chem. Soc. Rev.* 43: 4909–4921.
11 Rosenberg, M., Dahlstrand, C., Kilså, K., and Ottosson, H. (2014). *Chem. Rev.* 114: 5379–5425.
12 Hückel, E. (1931). *Z. Phys.* 70: 104–186.
13 Hückel, E. (1931). *Z. Phys.* 72: 310–337.
14 Hückel, E. (1932). *Z. Phys.* 76: 628–648.
15 Hückel, E. (1937). *Z. Elektrochem.* 43: 752–788, 827–849.
16 Breslow, R. (1973). *Acc. Chem. Res.* 6: 393–398.
17 Baird, N.C. (1972). *J. Am. Chem. Soc.* 94: 4941–4948.

**18** Fratev, F., Monev, V., and Janoschek, R. (1982). *Tetrahedron* 38: 2929–2932.

**19** Zilberg, S. and Haas, Y. (1998). *J. Phys. Chem. A* 102: 10843–10850.

**20** Karadakov, P.B. (2008). *J. Phys. Chem. A* 112: 7303–7309.

**21** Karadakov, P.B. (2008). *J. Phys. Chem. A* 112: 12707–12713.

**22** Feixas, F., Vandenbussche, J., Bultinck, P. et al. (2011). *Phys. Chem. Chem. Phys.* 13: 20690–20703.

**23** Soncini, A. and Fowler, P.W. (2008). *Chem. Phys. Lett.* 450: 431–436.

**24** Jiao, H.J., Schleyer, P.v.R., Mo, Y. et al. (1997). *J. Am. Chem. Soc.* 119: 7075–7083.

**25** Gogonea, V., Schleyer, P.v.R., and Schreiner, P.R. (1998). *Angew. Chem. Int. Ed.* 37: 1945–1948.

**26** Fowler, P.W., Steiner, E., and Jenneskens, L.W. (2003). *Chem. Phys. Lett.* 371: 719–723.

**27** Zilberg, S. and Haas, Y. (1998). *J. Phys. Chem. A* 102: 10851–10859.

**28** Krygowski, T.M. and Cyrański, M.C. (1999). *Tetrahedron* 55: 11143–11148.

**29** Villaume, S., Fogarty, H.A., and Ottosson, H. (2008). *ChemPhysChem* 9: 257–264.

**30** Feixas, F., Matito, E., Poater, J., and Solà, M. (2015). *Chem. Soc. Rev.* 44: 6434–6451.

**31** Zhu, J., An, K., and Schleyer, P.v.R. (2013). *Org. Lett.* 15: 2442–2445.

**32** An, K. and Zhu, J. (2014). *Eur. J. Org. Chem.* 2014: 2764–2769.

**33** Dominikowska, J. and Palusiak, M. (2011). *Phys. Chem. Chem. Phys.* 13: 11976–11984.

**34** Sun, H., An, K., and Zhu, J. (2015). *Chem. Asian J.* 11: 234–240.

**35** Fernández, I., Wu, J.I., and Schleyer, P.v.R. (2013). *Org. Lett.* 15: 2990–2993.

**36** Schleyer, P.v.R. and Pühlhofer, F. (2002). *Org. Lett.* 4: 2873–2876.

**37** Breslow, R., Chang, H.W., Hill, R., and Wasserman, E. (1967). *J. Am. Chem. Soc.* 89: 1112–1119.

**38** Saunders, M., Berger, R., Jaffe, A. et al. (1973). *J. Am. Chem. Soc.* 95: 3017–3018.

**39** Breslow, R. (1982). *Pure Appl. Chem.* 54: 927–938.

**40** Worner, H.J. and Merkt, F. (2006). *Angew. Chem. Int. Ed.* 45: 293–296.

**41** Chapman, O.L., McIntosh, C.L., and Pacansky, J. (1973). *J. Am. Chem. Soc.* 95: 614–617.

**42** Bally, T. and Masamune, S. (1980). *Tetrahedron* 36: 343–370.

**43** Paquette, L.A., Ley, S.V., Meisinger, R.H. et al. (1974). *J. Am. Chem. Soc.* 96: 5806–5815.

**44** Sung, Y.M., Yoon, M.-C., Lim, J.M. et al. (2015). *Nat. Chem.* 7: 418–422.

**45** Sung, Y.M., Oh, J., Kim, W. et al. (2015). *J. Am. Chem. Soc.* 137: 11856–11859.

**46** Möllerstedt, H., Piqueras, M.C., Crespo, R., and Ottosson, H. (2004). *J. Am. Chem. Soc.* 126: 13938–13939.

**47** Hafner, K. (1964). *Angew. Chem. Int. Ed. Engl.* 3: 165–173.

**48** Rosenberg, M., Ottosson, H., and Kilsa, K. (2011). *Phys. Chem. Chem. Phys.* 13: 12912–12919.

**49** Ottosson, H., Kilså, K., Chajara, K. et al. (2007). *Chem. Eur. J.* 13: 6998–7005.

**50** Streifel, B.C., Zafra, J.L., Espejo, G.L. et al. (2015). *Angew. Chem. Int. Ed.* 54: 5888–5893.

**51** Jorner, K., Feixas, F., Ayub, R. et al. (2016). *Chem. Eur. J.* 22: 2793–2800.

**52** Escayola, S., Tonnelé, C., Matito, E. et al. (2021). *Angew. Chem. Int. Ed.* 60: 10255–10265.

**53** Wan, P., Krogh, E., and Chak, B. (1988). *J. Am. Chem. Soc.* 110: 4073–4074.

**54** Wan, P. and Shukla, D. (1993). *Chem. Rev.* 93: 571–584.

**55** Rosenberg, M., Ottosson, H., and Kilså, K. (2010). *J. Organomet. Chem.* 75: 2189–2196.

**56** Wan, P. and Krogh, E. (1985). *J. Chem. Soc., Chem. Commun.* 1207–1208.

**57** Papadakis, R., Li, H., Bergman, J. et al. (2016). *Nat. Commun.* 7: 12962.

**58** Papadakis, R. and Ottosson, H. (2015). *Chem. Soc. Rev.* 44: 6472–6493.

**59** Ward, H.R. and Wishnok, J.S. (1968). *J. Am. Chem. Soc.* 90: 5353–5357.

**60** Kaplan, L. and Wilzbach, K.E. (1968). *J. Am. Chem. Soc.* 90: 3291–3292.

**61** Waldeck, D.H. (1991). *Chem. Rev.* 91: 415–436.

**62** Arai, T. and Tokumaru, K. (1993). *Chem. Rev.* 93: 23–39.

**63** Brink, M., Möllerstedt, H., and Ottosson, C.-H. (2001). *J. Phys. Chem. A* 105: 4071–4083.

**64** Villaume, S. and Ottosson, H. (2009). *J. Phys. Chem. A* 113: 12304–12310.

**65** Kato, H., Brink, M., Möllerstedt, H. et al. (2005). *J. Org. Chem.* 70: 9495–9504.

**66** Zhu, J., Fogarty, H.A., Möllerstedt, H. et al. (2013). *Chem. Eur. J.* 19: 10698–10707.

**67** Zhao, J., Ji, S., Chen, Y. et al. (2012). *Phys. Chem. Chem. Phys.* 14: 8803–8817.

**68** Padalkar, V.S. and Seki, S. (2016). *Chem. Soc. Rev.* 45: 169–202.

**69** Forés, M., Duran, M., and Solà, M. (2000). *Chem. Phys.* 260: 53–64.

**70** Gutiérrez-Arzaluz, L., Cortés-Guzmán, F., Rocha-Rinza, T., and Peón, J. (2015). *Phys. Chem. Chem. Phys.* 17: 31608–31612.

**71** Cookson, R.C., Costa, S.M.D.B., and Hudec, J. (1969). *J. Chem Soc. D* 1272–1272.

**72** Pratt, A.C. (1974). *J. Chem. Soc., Chem. Commun.* 183–184.

**73** Zimmermann, H. (1971). *Acc. Chem. Res.* 4: 272–280.

**74** Angus, H.J.F. and Bryce-Smith, D. (1960). *J. Chem. Soc. (Resumed)* 4791–4795.

**75** Bryce-Smith, D., Gilbert, A., and Orger, B.H. (1966). *Chem. Commun. (London)* 512–514.

**76** Wilzbach, K.E. and Kaplan, L. (1966). *J. Am. Chem. Soc.* 88: 2066–2067.

**77** Wilzbach, K.E. and Kaplan, L. (1971). *J. Am. Chem. Soc.* 93: 2073–2074.

**78** Streit, U. and Bochet, C.G. (2011). *Beilstein J. Org. Chem.* 7: 525–542.

**79** Antkowiak, T.A., Sanders, D.C., Trimitsis, G.B. et al. (1972). *J. Am. Chem. Soc.* 94: 5366–5373.

**80** Bryce-Smith, D., Gilbert, A., and Grzonka, J. (1970). *J. Chem. Soc. D* 498–499.

**81** Zimmerman, H.E. and Grunewald, G.L. (1966). *J. Am. Chem. Soc.* 88: 183–184.

**82** Mohamed, R.K., Mondal, S., Jorner, K. et al. (2015). *J. Am. Chem. Soc.* 137: 15441–15450.

# Index



*Aromaticity and Antiaromaticity: Concepts and Applications*, First Edition.
Miquel Solà, Alexander I. Boldyrev, Michał K. Cyrański, Tadeusz M. Krygowski, and Gabriel Merino.

## e

## j



## k



## l



## m



## n